Micronutrient Deficiencies in Global Crop Production

Brian J. Alloway
Editor

Micronutrient Deficiencies in Global Crop Production

Springer

Prof. Dr. Brian J. Alloway
University of Reading
Department of Soil Sciences
Whiteknights
Reading RG6 6DW
United Kingdom

Cover photos

Top Photos from left to right:

1. Three leaves of a zinc-deficient citrus tree (in Turkey) showing chlorosis. *Photograph supplied and reproduced by kind permission of Dr. V.M. Shorrocks*
2. Zinc-deficient citrus tree (in Turkey) showing chlorotic leaves. *Photograph supplied and reproduced by kind permission of Dr. V.M. Shorrocks*
3. Zinc-deficient maize plant (in England) showing characteristic chlorosis ("white bud") and stunting. *Photograph by Prof. B.J. Alloway*

Bottom photos from left to right:

1. Ears of copper-deficient wheat plants in France showing empty grain places at the tip and base ("rat-tail" symptom). *Photograph by Prof. B.J. Alloway*
2. Zinc-deficient wheat plant (in Australia) showing chlorosis and stunting. *Photograph supplied and reproduced by kind permission of Dr. V.M. Shorrocks*
3. Ears of copper-sufficient wheat plants, showing all grain places filled (on the left to the photo), and ears of copper-deficient wheat plants, showing empty grain places and some melanism (on the right). These wheat ears are from a field experiment on copper-deficient site in France where the copper-sufficient ears were taken from plants in the copper-treated plots. *Photograph by Prof. B.J. Alloway*

ISBN 978-1-4020-6859-1 e-ISBN 978-1-4020-6860-7

Library of Congress Control Number: 2007941497

Printed on acid-free paper.

9 8 7 6 5 4 3 2 1

springer.com

Preface

There are eight micronutrient elements which are essential for the healthy growth of higher plants. If one or more of these elements is deficient, crops will fail to achieve their optimum yields and the quality of their food products is likely to be impaired. In order to provide adequate food for the world's rapidly increasing population, micronutrient deficiencies in agricultural and horticultural crops should be identified and treated wherever they are found. Micronutrient deficiencies occur all over the world but variations in soil conditions, climate, crop genotypes and management result in marked variations in the incidence of these problems. In addition to yields, the contents of micronutrients in crop products, such as staple grains, are also of great importance to the health of human and livestock consumers.

This book covers the occurrence of deficiencies of the plant micronutrients, their causes, effects and treatment in different countries, regions and continents around the world. Eight chapters deal, respectively, with micronutrient problems in Australia (Chapter 3), India (Chapter 4), China (Chapter 5), the Near East (Chapter 6), Africa (Chapter 8), Europe (Chapter 9), South America (Chapter 10) and the United States of America (Chapter 11). Of the remaining four chapters, Chapter 1 provides an introduction to the plant micronutrients and their deficiencies, and the soils associated with these problems. Chapter 2 follows on with a wide-ranging overview of the causes of deficiencies, the agronomy of micronutrient fertiliser use, the links between deficiencies and human nutrition and the biofortification of staple foods with elements essential for humans. This latter theme is further developed in Chapter 12, which concentrates on the links between the trace elements essential for humans in crops and human health problems, such as iron and zinc deficiencies, and explores ways in which the micronutrient content of food crops can be increased and their bioavailability to humans improved. Chapter 7 differs from the other chapters in that it provides an in-depth case study of an investigation into zinc deficiency in wheat in Central Anatolia, Turkey. This resulted in the widespread adoption of zinc fertilisation of wheat crops in the region, leading to greatly increased yields and consequent improvements in human health.

Apart from covering the occurrence, causes and treatment of deficiencies in the respective countries, each chapter deals in greater detail with one or more related topics, adding depth to the treatment of this broad subject. Examples include soil testing and plant analysis, field experiments, innovative treatments, micronutrients in the subsoil, nutrient interactions, changes in cropping systems, micronutrient budgets and hidden deficiencies.

In addition to local names for soil types, the equivalent soil group in either the FAO–UNESCO/World Reference Base for Soil Resources, or the USDA Soil Taxonomy classification is also given. The equivalent soil groups in both of these classifications are listed in Appendix II. The botanic names of the crops are given in the chapters, but are also listed in Appendix 1. Average world and national yields of maize, rice and wheat are given in Appendix 3, to put the yield data given in the chapters into a broader perspective.

The book is primarily based on chapters developed from papers presented at the Special Symposium on "Micronutrient Deficiencies in Global Crop Production" held in May 2005 at the 8th International Conference on the Biogeochemistry of Trace Elements (ICOBTE) in Adelaide, South Australia. This symposium was organised by the editor with sponsorship from the International Copper Association Ltd (ICA), the International Council on Mining and Metals (ICMM) and the International Zinc Association (IZA). However, due to the limited number of papers which could be presented, additional authors were subsequently sought for chapters on Africa and the Near East. Their contribution has ensured that the crops, types of management, climate and soil conditions representative of the most of the world's agricultural land have been covered.

The editor wishes to acknowledge the following, who have assisted him in the task of bringing the book to completion. Firstly, thanks to Murray Cook, formerly of IZA (now of the Galvanisers Association), for help with organising the sponsorship for the ICOBTE symposium, and to the ICA, ICMM and IZA for providing funds to enable most of the team of authors to be brought together at the ICOBTE Symposium. Secondly, thanks are extended to Professor Iain Thornton (Imperial College, London), Professor Michael McLaughlin (University of Adelaide, Australia), Professor Ronald McLaren (Lincoln University, New Zealand) and Professor Volker Römheld (University of Hohenheim, Stuttgart) for help with reviewing chapters. Finally, and most important of all, the editor would like to thank the authors of all the chapters for the high quality of their contributions.

Contents

Contributors

Brian J. Alloway,
Department of Soil Science, School of Human and Environmental Sciences,
University of Reading, PO Box 233, Reading RG6 6DW, United Kingdom,
b.j.alloway@reading.ac.uk

Patrick H. Brown,
Department of Plant Sciences, University of California-Davis, MS2,
One Shields Avenue, Davis, CA 95616, USA, phbrown@ucdavis.edu

Ismail Cakmak,
Faculty of Engineering and Natural Sciences, Sabanci University,
34956 Istanbul, Turkey, cakmak@sabanciuniv.edu

Anthony C. Edwards,
Nether Backhill, Ardallie, Aberdeenshire AB42 5BQ, Scotland,
t.edwards@macaulay.ac.uk

Nand K. Fageria,
National Rice & Bean Research Center of EMBRAPA, Caixa Postal 179,
Santo Antônio de Goiás, GO, CEP 75375-000, Brazil, Fageria@cnpaf.embrapa.br

Xiaoyun Fan,
Department of Plant Nutrition, China Agricultural University,
Yuanmingyuan West Road No. 2, Beijing, 100094, P.R. China

Xiaoping Gao,
Department of Plant Nutrition, China Agricultural University,
Yuanmingyuan West Road No. 2, Beijing, 100094, P.R. China

Robin D. Graham,
Department of Crop and Atmospheric Sciences, School of Agriculture and Wine,
University of Adelaide, 5005 South Australia, robin.graham@adelaide.edu.au

Robert E. Holloway,
Soil and Land Systems, Department of Earth and Environmental Sciences, University of Adelaide, Waite Road, Urrbrae, South Australia 5064, bholloway@arris.co.au

Michiel C. Laker,
Department of Plant Production and Soil Science, University of Pretoria, Pretoria 0002, South Africa

Abdul Rashid,
National Agricultural Research Center (NARC), Park Road, Islamabad-45500, Pakistan, abdul.reashid@comsats.net.pk

John Ryan,
International Center for Agricultural Research in the Dry Areas (ICARDA), PO Box 5466, Aleppo, Syria, j.ryan@cgiar.org

Rongli Shi,
Department of Plant Nutrition, China Agricultural University, Yuanmingyuan West Road No. 2, Beijing, 100094, P.R. China

Alex H. Sinclair,
SAC Environmental Group, Ferguson Building, Craibstone Estate, Aberdeen, AB21 9YA, Scotland, alex.sinclair@sac.co.uk

Maha V. Singh,
Indian Institute of Soil Science, Nabibagh, Berasia Road, Bhopal-462038, India drmvsingh@gmail.com

Samuel P. Stacey,
Soil and Land Systems, Department of Earth and Environmental Sciences, University of Adelaide, Waite Road, Urrbrae, South Australia 5064, samuel.stacey@adelaide.edu.au

Luís F. Stone,
National Rice & Bean Research Center of EMBRAPA, Caixa Postal 179, Santo Antônio de Goiás, GO, CEP 75375-000, Brazil, stone@cnpaf.embrapa.br

Johan H. van der Waals,
Department of Plant Production and Soil Science,
University of Pretoria and
Terra Soil Science, PO Box 32201, Glenstantia, Pretoria 0010, South Africa, jvdw@cleanstream.co.za

Ross M. Welch,
USDA-Agricultural Research Service, US Plant, Soil and Nutrition Laboratory, Tower Road, Ithaca, New York 1485302901, USA, rmw1@cornell.edu

Fusuo S. Zhang,
Department of Plant Nutrition, China Agricultural University, Yuanmingyuan West Road No. 2, Beijing, 100094, P.R. China, zhangfs@cau.edu.con

Chunquin Zou,
Department of Plant Nutrition, China Agricultural University, Yuanmingyuan West Road No. 2, Beijing, 100094, P.R. China, zcq@cau.edu.con

List of Abbreviations

a^{-1}	per year
AAPFCO	American Association of Plant Food Coordinators
AB-DTPA	ammonium bicarbonate-DTPA extraction
ADAS	ADAS UK Ltd – research, consultancy and knowledge transfer in the cropped environment (formerly Agricultural Development and Advisory Service of Ministry of Agriculture, Fisheries and Food, UK)
AE	agronomic efficiency (i.e., kg of seed produced per kg of micronutrient (e.g., Zn) applied)
APP	ammonium polyphosphate (fluid fertiliser: 10% N and 34% P_2O_5)
ARC-ISCW	Agricultural Research Council, Institute for Soil, Climate and Water (South Africa)
ASC	Australian Soil Classification System
CGIAR	Consultative Group on International Agricultural Research
CIAT	International Centre for Tropical Agriculture (Centro Internacional de Agricultura Tropical) (in Cali, Columbia)
CIMMYT	International Maize and Wheat Improvement Centre (Centro Internacional de Mejoramiento de Maiz y Trigo)
DAP	diammonium phosphate (fertiliser – 18% N and 47% P_2O_5)
DM	dry matter, or dry weight (DW) (in plant tissue analysis)
DMT1	divalent metal transporter 1 – the major transferrin-independent iron uptake system of intestinal cells
DTPA	diethylene triamine pentaacetic acid (chelating agent – used in some soil tests and sometimes also as a fertiliser ligand)
DRIS	Diagnostic and recommendation integrated system
EDTA	ethylene diamine tetraacetic acid (chelating agent used in some soil tests and in some fertilisers)
EU	European Union (27 constituent countries, since January 2007)
FAO	Food and Agriculture Organization of the United Nations
FYM	farm yard manure (cattle urine and faeces mixed with straw)
GAEC	Good agricultural and environmental condition (farmland) (Scottish Exec', 2004)
ha	hectare (equivalent to 2.471 acres)
HWE	hot water extraction soil test for boron (see also HWS)

HWS	hot water soluble (soil test extractant for boron), i.e., HWS B
HYV	High yielding varieties programme ("Green Revolution")
ICARDA	International Centre for Agricultural Research in Dry Areas
ICP-MS	Inductively Coupled Plasma Mass Spectrometry (multi-element analytical method, generally more accurate with a lower detection limit than ICP-OES)
ICP-OES	Inductively Coupled Plasma-Optical Emission Spectrometry (multi-element analytical method)
IFPRI	International Food Policy Research Institute (Washington, DC, USA)
IITA	International Institute for Tropical Agriculture (in Nigeria)
IRRI	International Rice Research Institute (the Philippines)
MAP	monoammonium phosphate (fertiliser 11% N and 49% P_2O_5)
mg kg^{-1}	milligrams per kilogram (equivalent to micrograms per gram $\mu g g^{-1}$, or parts per million – ppm)
mg L^{-1}	milligrams per litre (equivalent to ppm in liquids)
MIR	Mid infra-red (spectroscopy) technique, which enables soil minerals and organic matter species to be identified
NATO-SFS	North Atlantic Treaty Organization, Science for Stability Program
ng	nanogram (10^{-9}g)
NP	combined nitrogen and phosphorus fertilisers
NPK	nitrogen, phosphorus and potassium compound fertilisers
NSW	New South Wales (State of, in Australia)
NUE	nutrient-use-efficiency
NVZ	nitrate vulnerable zone
OC	organic carbon
OSP	ordinary superphosphate (fertiliser, 21% P_2O_5)
Qld	Queensland (Australia)
QTL	quantitative trait loci (for traits which are influenced by multiple genes)
RDA	recommended daily allowance (of nutrients for humans)
RDI	recommended daily intake (of micronutrients by humans)
ROS	reactive oxygen species
SA	South Australia (State of, in Australia)
SAC	Scottish Agricultural Colleges – agricultural advice, consultancy and research in Scotland
SOD	superoxide dismutase
TEA	triethanolamine (chelating agent) used with DTPA in a soil test at pH 7.3
TFI	The Fertilizer Institute (USA)
t ha^{-1}	tonnes per hectare (equivalent to 1.1 tons per acre)
TSP	triple superphosphate (fertiliser 47% P_2O_5)
UNESCO	United Nations Educational, Social and Cultural Organization
USDA	United States Department of Agriculture
WA	Western Australia (State of, in Australia)
WHO	World Health Organization
YEB	youngest emerged leaf blade (for plant tissue analysis)
YOB	youngest open blade (for plant tissue analysis)
yr^{-1}	per year
Z	Zadoks Scale (see glossary for more details)

List of Figures

List of Tables

Chapter 1
Micronutrients and Crop Production: An Introduction

Brian J. Alloway

Abstract Eight trace elements are essential for higher plants: boron (B), chlorine (Cl), copper (Cu), iron (Fe), manganese (Mn), molybdenum (Mo), nickel (Ni) and zinc (Zn). Whenever the supply of one or more of these elements is inadequate, yields will be reduced and the quality of crop products impaired, but crop species and cultivars vary considerably in their susceptibility to deficiencies. Zinc deficiency is the most ubiquitous micronutrient problem throughout the world affecting many crops including the staples maize, rice and wheat. Boron deficiency is the second most widespread micronutrient problem and dicotyledon species tend to be more sensitive to B deficiency than graminaceous crops. Iron deficiency is important in some regions, especially those with a Mediterranean climate and calcareous soils. Copper deficiency is important in some parts of the world, such as Europe and Australia where cereals are most affected. Likewise, Mn and Mo deficiencies vary in importance around the world. Acute micronutrient deficiencies in plants are accompanied by distinct symptoms, but hidden deficiencies without obvious symptoms are generally more widespread.

1.1 Introduction

Micronutrients are those trace elements which are essential for the normal healthy growth and reproduction of plants and animals. The trace elements essential for plants are: boron (B), chlorine (Cl), copper (Cu), iron (Fe), manganese (Mn), molybdenum (Mo), nickel (Ni) and zinc (Zn). Although, cobalt (Co) is known to be essential for the bacterial fixation of atmospheric nitrogen (N) in leguminous plants, it is not considered to be essential for all higher plants. Nevertheless, it has been shown to have beneficial effects on crops in other plant families, such as the *Graminae* (e.g., wheat, *Triticum* spp.) (Asher, 1991) and is referred to as a "beneficial" element. Other beneficial elements, which have not yet been proved to be "essential", include silicon (Si), sodium (Na), selenium (Se), vanadium (Va) and aluminium (Al) (Barker and Pilbeam, 2007).

The trace elements recognised as being essential for animals are: Co, Cu, chromium (Cr), fluorine (F), iodine (I), Fe, Mn, Mo, Se and Zn. However, an additional

B.J. Alloway (ed.), *Micronutrient Deficiencies in Global Crop Production.*

seven elements are also regarded as being essential for humans (see Graham, Chap. 2 and Welch, Chap. 12).

For a trace element to be essential for either plants or animals (i.e., a micronutrient), it needs to satisfy three criteria: (1) the organism cannot grow and reproduce normally without the element, (2) its action must be specific and unable to be replaced by any other element, and (3) its action must be direct (Arnon and Stout, 1939). However, Epstein (1965) advocated that an element can also be regarded as essential if it is a component of a molecule known to be an essential metabolite, even if it cannot be demonstrated that it fulfils all of the criteria proposed by Arnon and Stout.

In geochemistry, the term "trace element" is given to elements which normally occur in trace amounts (usually <1,000 mg kg^{-1}) in rocks and soils. However, the biological use of the term "trace element" applies to elements occurring at relatively low concentrations (usually <100 mg kg^{-1}) in the dry matter of living organisms. The macro elements carbon (C), hydrogen (H), oxygen (O) and N, which form the main organic compounds in plants and animals, are present in the highest concentrations (at percentage levels). Potassium (K) tends to be present in similar concentrations to N (1.4–5.6%) but phosphorus (P), calcium (Ca), magnesium (Mg), Na and Cl are present in intermediate concentrations (0.1–2.5%) (Wild and Jones, 1988; Marschner, 1995).

It is very important that the micronutrient element requirements of crops are met as well as their macronutrient needs if they are to yield satisfactorily and bear products (e.g., grains and fruits) of acceptable quality. The dose response curves for all micronutrients show that, just as yields can be affected by deficiencies, they can also be reduced by toxicity due to excessive concentrations of the same elements. It is therefore important that soils and/or crops are monitored to ensure that the available micronutrient concentrations in soils are in the optimum range, being neither too low, nor too high. Typical dose–response graphs for micronutrients and non-essential elements are shown in Fig. 1.1.

It is only during the last 70 years that most micronutrient deficiency problems have been widely recognised and treated in the field. This has been largely due to the increased intensification of arable farming in many parts of the world and also to the cultivation of virgin and/or reclaimed land. Intensification involves the increased use of N, P, K and other fertilisers, growing new and higher yielding crop cultivars, liming to create more optimal soil pH conditions, increased use of agrichemicals to control pests and diseases and, in more arid areas, increased use of irrigation. Prior to this intensification, much lower crop yields were usually accepted as the norm in many parts of the world and the crop cultivars grown were generally well adapted to local soil and climatic conditions.

With the adoption of more intensive methods of crop husbandry, it was frequently found that crops began to exhibit various symptoms of stress which had not previously been known on the same area of land. Many of these micronutrient deficiencies were brought about by the increased demands of more rapidly growing crops for available forms of micronutrients. In some cases, several of the elements were rendered less available due to changes in soil conditions, such as increased pH through liming. Perhaps the most important and ubiquitous cause has been the

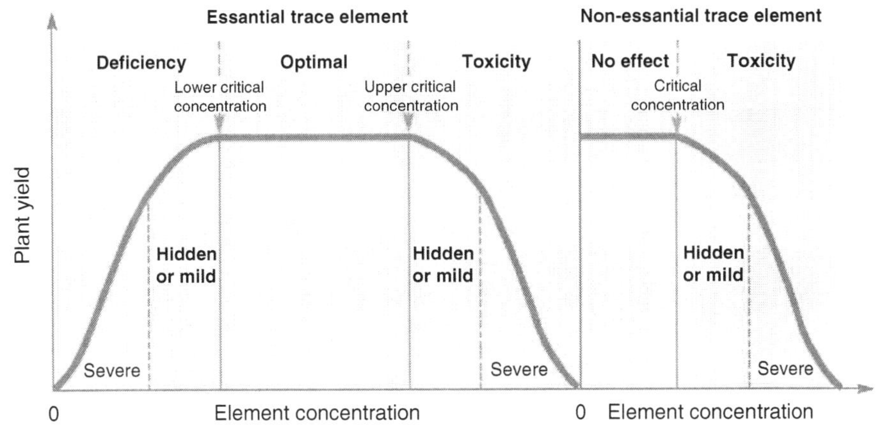

Fig. 1.1 Typical dose–response curves for essential and non-essential trace elements in crops (Alloway, 2004)

introduction of new species and/or cultivars of crops which have a greater requirement for certain micronutrients. The functions of the plant micronutrients are briefly summarised in Table 1.1 (see also Chap. 10).

Although the amounts of micronutrients required by plants are relatively low, individual species and varieties can vary considerably in their requirements for specific elements. Differences in the efficiency with which crop varieties are able to utilise low supplies of B, Cu, Mn, Fe and Zn have resulted in them being labelled as being either "efficient" or "inefficient" for a specific micronutrient. Under field conditions, where the available supply of a micronutrient may be marginal or low, efficient varieties will be better able to grow and yield more satisfactorily than inefficient ones.

When the supply of a micronutrient to plants is deficient, in addition to crop yields and quality being affected, there may also be visible symptoms of physiological stress, especially in cases of severe deficiency. Although plant species differ in the nature of the symptoms of micronutrient deficiencies which they display, there are several generalizations which can be made. In most cases, severe deficiencies will cause stunted growth, discoloration and, in some cases, necrotic spots on the leaves. The discolouration will usually commence as chlorosis when, instead of the normal green colour of chlorophyll, either all or part of the leaves turn yellow, even white, but leaves can also turn brown. Deficiency symptoms can also include smaller or twisted leaves, and loss of turgor. Leaf symptoms are usually seen only on old leaves in the case of Mo, on new leaves with Fe, Mn and Cu, on both young and old leaves with Zn, and on terminal buds with B deficiency. Green veins are seen on new leaves with Fe and Mn deficiency and yellow veins with Cu deficiency (Reddy and Reddi, 1997).

The stunted appearance of plants caused by some deficiencies, such as Zn, is due to reduced internodal expansion and this can give rise to a "rosette" appearance

Table 1.1 Brief summary of the essential functions of micronutrients in plants (Adapted from Srivastava and Gupta, 1996; Heckman, 2007; Xu et al., 2000.)

Element	Functions
Boron	Metabolism and transport of carbohydrates, regulation of, meristematic tissue, cell wall synthesis, lignification growth regulator metabolism, phenol metabolism, integrity of membranes, root elongation, DNA synthesis, pollen formation and pollination
Chlorine	Involved (as Cl^-) in the light reaction in photosynthesis, charge compensation and osmoregulation of the whole plant and individual cells, such as stomatal guard cells, and the activity of certain enzymes
Cobalt	Only proved to be essential in symbiotic N fixation (in roots of legumes) but has other beneficial roles in some other plant families
Copper	Constituent of several enzymes, with roles in photosynthesis, respiration, protein and carbohydrate metabolism, lignification and pollen formation
Iron	Constituent of cytochromes and metalloenzymes. Roles in photosynthesis, symbiotic N fixation, N metabolism, and redox reactions
Manganese	Photolysis of water in chloroplasts, regulation of enzyme activities, protection against oxidative damage of membranes
Molybdenum	N fixation, constituent of enzymes including nitrate reductase and sulphite oxidase
Nickel	Constituent of urease enzyme, role in N assimilation, protection of nitrate reductase against inactivation
Zinc	Constituent of several enzymes with roles in carbohydrate and protein synthesis; maintaining the integrity of membranes, regulating auxin synthesis and in pollen formation

where the whorls of leaves are situated more or less on top of each other. A summary of the main types of symptoms associated with each of the plant micronutrients is given in Table 1.2.

Less severe deficiencies may not manifest themselves until later stages in the development of the plant. In the case of mild to marginal Cu deficiency in cereals, leaf growth can appear normal and the only obvious symptoms appear when the ear, or spike, develops (anthesis). These symptoms can include late development of the ears and abnormal looking ears due to empty grain positions, giving a "rat tailed" appearance in the case of Cu-deficient wheat plants (Fig. 1.3 in Colour Section).

In many cases, visible symptoms provide a convenient and low-cost means of identifying micronutrient deficiency problems, especially in areas where recurrent deficiency problems are found. However, these symptoms are usually only clearly expressed in cases of acute deficiency. These acute deficiencies will often be found for the first time when either new land is put into arable use or when new crop species or cultivars are grown for the first time. For example, Cu deficiency was called "reclamation disease" when it was observed in cereal crops grown on newly reclaimed peaty soils in the Netherlands and in Florida, USA. In the case of Zn, severe deficiencies were found in the Central Anatolia region of Turkey when new,

Table 1.2 Some common symptoms of micronutrient deficiency in widely grown crops (Adapted from Kabata-Pendias, 2001 with additions from Brown, 2007a and Heckman, 2007. See also Chap. 10, Table 10.9)

Element	Symptoms	Sensitive crops
Boron	Chlorosis and browning of young leaves; death of growing points; distorted blossom development; lesions in pith and roots ("heart rot") and multiplication of cell division. Empty grain positions in wheat ears	Legumes, *Brassicae*, beets, celery, grapes, fruit trees (apples and pears) and wheat
Chlorine	Wilting of leaves, especially at margins, shriveling and necrosis of leaves, frond fracture and stem cracking in coconut. Sub-apical swelling in roots	Oil palm, kiwi fruit, sugar, beet, wheat, barley, subterranean clover
Copper	Wilting, melanism, white twisted tips, reduction in panicle formation, disturbance of lignification and of development and fertility of pollen	Cereals, sunflower, spinach, onions, carrots and alfalfa
Iron	Interveinal chlorosis of young leaves	Fruit trees (citrus), grapes, peanut, soya bean, sorghum and calcifuge species
Manganese	Chlorotic spots and necrosis of young leaves and reduced turgor. Necrotic spots on cotyledons of peas	Cereals, legumes and fruit trees (apples, cherries and citrus)
Molybdenum	Chlorosis of leaf margins, "Whiptail" of leaves and distorted curding of cauliflower; "fired" margin and deformation of leaves due to NO_3 excess and destruction of embryonic tissues	*Brassiceae* and legumes
Nickel	Leaf tip necrosis (legumes), chlorosis and patchy necrosis (*Gramineae*)	Pecan, wheat, potato, bean, soya bean
Zinc	Interveinal chlorosis (mainly in monocotyledons), stunted growth, "little leaf". Rossette of trees and violet–red points on leaves	Cereals (especially maize and rice), grasses, flax/linseed and fruit trees (citrus)

high yielding varieties of wheat were grown with heavier applications of NPK fertilisers (see Cakmak, Chap. 7).

However, where the available supplies of micronutrients in the soil have been gradually depleted by repeated cropping, especially with higher yielding varieties, the degree of deficiency may be less severe and the manifestation of symptoms less distinct. Another problem with relying on visible symptoms to diagnose deficiencies is that they can often be confused with symptoms of deficiency of certain other micro or macronutrients, or with symptoms of disease, drought or heat stress and damage by herbicides. In most cases, it is advisable to carry out either soil or plant analysis to confirm the deficiency diagnosis.

In the case of marginal deficiencies, it is possible for the yields of many crops, especially cereals, to be significantly reduced (sometimes by 20% or more), and the quality of crop products to be impaired, without the manifestation of distinct visible symptoms. These are usually referred to as hidden deficiencies or "hidden hunger", latent, and/or subclinical deficiencies. In many parts of the world, this type of deficiency is likely to be more widespread and have a greater economic impact than more severe deficiencies. This is due to the fact, that without obvious symptoms, farmers are often not sure of the causes of the disappointing yields and quality in their crops. Poor yields are sometimes ascribed to inadequate supplies of N and P, with the result that more of these macronutrients may be applied to successive crops. This will often exacerbate the micronutrient deficiency and also lead to increased leaching of N and P into ground waters, possibly causing both ecological and water resource problems.

It is important to realize that the absence of symptoms of a deficiency of a particular micronutrient does not necessarily imply that the supply of this micronutrient is adequate. This is discussed in relation to possible deficiencies of micronutrients in the tropical cropping zone of Australia by Holloway et al. (Chap. 3). Another possibility is that more than one micronutrient may be deficient at a particular site (multi-micronutrient deficiencies). In correcting a diagnosed deficiency of one element, there is a risk that the available concentration of another micronutrient may be reduced in some way, thereby inducing a deficiency of this element instead. This has been found with Cu and Mn, Cu and Zn, and Mn and Fe and other combinations of micronutrients (Alloway, 1976).

As shown in Table 1.3, crop species vary considerably in their susceptibility to deficiencies of different micronutrients. However, intra-specific variations (between varieties/cultivars) can sometimes be even greater than differences in susceptibility between species. Nevertheless, all crops will be affected by a severe deficiency of any micronutrients. The main difference between genotypes is in the critical concentrations at which the supply of a particular micronutrient becomes inadequate. These will be significantly lower for the more tolerant genotypes. This is illustrated by Zn deficiency in wheat. Although, wheat is shown to be much less sensitive to Zn deficiency than other species, such as maize (as shown in Table 1.3), there is still a point at which the supply of available Zn becomes low enough to bring about the onset of physiological stress in wheat due to Zn deficiency. The widespread occurrence of Zn deficiency in wheat in Central Anatolia, Turkey, reported by Cakmak in Chap. 7 is a good example of this.

Cultivars which are able to grow normally in soils with marginally low available concentrations of a micronutrient are classed as being "efficient" for that particular micronutrient. Those cultivars which are unable to tolerate such low levels of this micronutrient are classed as "inefficient". The relative level of efficiency is usually expressed as an "efficiency index" and an example of this index for Zn is shown below (Graham and Rengel, 1993):

$$\text{Zn efficiency} = \frac{\text{Yield (without Zn)}}{\text{Yield (with Zn)}} \times 100$$

Table 1.3 The relative susceptibility of selected crop species to deficiencies of micronutrients (Martens and Chesterman, 1991; Loué, 1986; Prasad and Power, 1997; Follet et al., 1981 and others. The inclusion of two classes in a box reflects differences between sources of information. See also Chap. 6, Table 6.3)

Crop	B	Cu	Fe	Mn	Mo	Zn
Alfalfa	High	High	Medium	Medium/low	Medium	Low
Apple	High	Medium	–	High	Low	High
Barley	Low	Medium /high	High/ medium	Medium	Low	Medium
Bean	Low	Low	High	High	Medium	High
Cabbage	Medium	Medium	Medium	Medium	Medium	–
Carrot	Medium	High	–	Medium	Low	Low
Citrus	Low	High	High	High	Medium	High
Corn (maize)	Low/ medium	Medium	Medium	Low	Low	High
Cotton	High	Medium	Medium	Medium/low	–	High
Grass	Low	Low	High	Medium/low	Low	Low
Linseed/ flax	Medium	–	High	Low	–	High
Oat	Low	High	Medium	High	Low/ medium	Low
Pea	Low	Low/ medium	Medium	High	Medium	Low
Potato	Low	Low	–	High	Low	Medium
OS Rape/canola	High	Low	–	–	–	–
Rice	Low	Low	Medium/low	Med	Low	Medium/ high
Rye	Low	Low	Low	Low	Low	Low
Sorghum	Low	Medium	High	High/medium	Low	High/ medium
Soya bean	Low	Low	High	High	Medium	Medium
S. Beet	High	Medium	High	Medium/high	Medium	Medium
Spinach	Medium	High	High	High	High	Medium
Vine (grapes)	High	Medium	High	High	Low	Low
Tomato	High/ medium	Medium	High	Medium	Medium	Medium
Wheat	Low	High	Medium/ low	High	Low	Low

For a given genotype, nutrient efficiency is reflected by the ability to produce a high yield in a soil that is limiting in one of more nutrient elements (Graham, 1984)

Genotypic variations in efficiency have been reported for B, Cu, Fe, Mn and Zn in crop plants. It is claimed by Rerkasem and Jamjod (2004) that wheat shows a wider range of genotypically related B efficiency than any other crop for any other macro or micronutrient element.

1.1.1 Explanations for Variations in Micronutrient Efficiency

There are several possible explanations for the differences in micronutrient efficiency between cultivars and these are discussed by Graham and Rengel (1993) and Marschner (1995). These are briefly summarized below and comprise differences in:

- The volume and length of roots
- Presence, or not, of proteoid roots
- Root-induced changes in rhizosphere pH
- Increased absorption through vesicular mycorrhizae, if present
- Release of root exudates to facilitate uptake (e.g., phytosiderophores), triggered by low Fe or Zn, organic acids, such as malic acid (Gao et al., 2007)
- Efficiency of utilization of the micronutrients once absorbed into plants
- Recycling of elements within the tissues of the growing plant
- Tolerance of inhibiting factors, e.g., bicarbonate ions $[(HCO_3)^-]$ inhibiting Zn uptake in rice

1.2 Individual Micronutrient Deficiencies

1.2.1 Boron

Over a period of nearly 70 years, B deficiency has been recognised in at least 80 countries on 132 plant species (Shorrocks, 1997). Boron is generally required in greater amounts by dicotyledon plant species than by monocotyledons (Srivistava and Gupta, 1996). Boron deficiency is often a problem in grape vines (*Vitis vinifera* L.) and in tree fruits, especially apple (*Malus sylvestris* Mill) and olive (*Olea europea* L.). In field crops, it affects sunflowers (*Helianthus annuus* L.), sugar beet (*Beta vulgaris* L.), black gram (*Vigna mungo* L.), and oilseed rape (canola) (*Brassica napus* L.). Although generally less sensitive to deficiency than dicotyledons, monocotyledon cereals including maize (*Zea mays* L.), sorghum (*Sorghum bicolor* L.), wheat, barley and oats can all be affected by B deficiency. Yield responses in wheat to B applications have been reported in Bangladesh, Brazil, Bulgaria, China, Finland, India, Madagascar, Nepal, Pakistan, South Africa, Sweden, Tanzania, Thailand, USA, Russia, countries of the former Yugoslavia and Zambia (Shorrocks, 1997; Rerkasem and Jamjod, 2004). The contiguous upland cereal growing areas of India, Nepal and Bangladesh comprise the world's largest known area affected by B-deficiency in wheat (Rerkasem and Jamjod, 2004).

Root elongation is usually the first effect of B deficiency in most plant species, but it is rarely seen in wheat. It is the reproductive phase which is the most sensitive to deficiency in wheat and this results in male sterility due to disrupted pollen formation. This causes a major reduction in the numbers of grain set and, thus a reduction in grain yield (Rerkasem and Jamjod, 2004). The lack of any obvious symptoms in the vegetative growth phase mean that many of the cases of male sterility caused by a shortage of B in wheat are hidden deficiencies, at least for a large part of the growth period. This implies that this condition can only be diagnosed by either soil or plant analysis or by yield responses to B fertilisation.

In addition to wheat, other crops known to be relatively sensitive to B deficiency, such as oilseed rape (canola) and sunflower, have been found to have a greater requirement for B during their reproductive phase than during their period of vegetative growth. The beneficial effect of B on grape vines is mainly due to its effect on flowering and fruit set. Rerkasem and Jamjod (2004) consider that wheat is more prone to B deficiency than rice, maize, soya bean or mung bean. In South Asia, where alternating rice and wheat is now the most common cropping system, the wheat in this system is generally more prone to B deficiency. The following rice crop on the same land is rarely affected by this deficiency. This may be partly due to wheat being grown in the coolest months in this subtropical region (Rekasem and Jamjod, 2004). Boron deficiency is a particular problem on alkaline and heavily limed soils and on highly leached sandy soils. The physiology of B in plants and the requirements of crops, especially high value tree crops are discussed by Brown in Chap. 11.

Excess B, causing toxicity is found in various low rainfall areas of the world, especially southern Australia, where it is of geochemical origin (Holloway et al., Chap. 3). Boron toxicity from excessive application of B fertilisers has also been reported in other low rainfall areas, including India and the Near East as discussed by Singh (Chap. 4) and Rashid and Ryan (Chap. 6). In low rainfall areas, there is insufficient percolation of water down the soil profile to leach away accumulations of soluble B salts.

1.2.1.1 Treatment of Boron Deficiency

The fertiliser compounds available for the treatments of B deficiency are given in Table 1.4.

Table 1.4 Boron compounds used for treating boron deficiency in plants (Martens and Westerman, 1991; Borax Ltd.)

Compound	Formula	Boron content (%)
Boric acid	H_3BO_3	17.5
Disodium tetraborate decahydrate (borax)	$Na_2B_4O_7.10H_2O$	11.3
Disodium tetraborate pentahydrate (borax, Granubor II)	$Na_2B_4O_7.5H_2O$	14.8
Anhydrous sodium tetraborate	$Na_2B_4O_7$	21.5
Sodium pentaborate	$Na_2B_{10}O_{16}.10H_2O$	18.3
Solubor	$Na_2B_8O_{13}.4H_2O$	20.9

Boron compounds can be mixed with fertilisers to produce "boronated fertilisers" which tend to be used in areas of high rainfall, such as South America and parts of north-western Europe where there is marked leaching of B and little danger of this highly soluble element accumulating to possibly harmful levels.

Foliar sprays of B compounds are widely used throughout the world on perennial crops such as nuts, vines and fruit orchards because they consistently give better results than soil applications. Rates are usually 10–50% of broadcasting rates: 0.08–0.38 kg B ha^{-1} or 0.4–1.9 kg Solubor ha^{-1} for grapes (Martens and Westerman, 1991). Solubor is normally applied to crops at concentrations of up to 1% (w:v) (Shorrocks, 1997). Foliar sprays have the advantage that they enable an existing deficiency problem to be treated rapidly. They overcome problems of low availability in soil and tend not to make a significant input of B (or any other micronutrient) to the soil.

In India, B fertilisers are broadcast and cultivated into the soil before seeding, or banded (0.5–2 kg B ha^{-1}) (Singh, Chap. 4). In China, B fertilisers are applied either to the seeds, the soil, or to foliage of oilseed rape and cotton (Zou et al., Chap. 5).

1.2.2 Chlorine

Although classed as a micronutrient, Cl is often found in relatively high concentrations in plants (Heckman, 2007). Symptoms of deficiency of Cl are rarely seen in field crops and are most likely to be found in plants grown in solution culture in the greenhouse. Rainfall is usually an adequate source of Cl but in regions remote from the sea, inputs from this source are generally much lower and deficiencies can occur in sensitive cultivars. Amounts of Cl deposited from rain range from 18 to 36 kg ha^{-1} in continental areas to greater than 100 kg ha^{-1} in coastal areas (Heckman, 2007). In the Great Plains of the USA, where chloride salts, such as KCl are rarely used as fertilisers and where the input from rainfall is low, crops have been found to respond to Cl fertilisation. In general, soils do not adsorb chloride ions in significant amounts. therefore regular inputs of this element in rainfall, from weathering minerals, groundwater or fertilisers are necessary to maintain an adequate supply for crops. Symptoms of Cl deficiency are wilting, marginal chlorosis, followed by bronzing and stunting. Chlorine-deficient sugar beet shows interveinal chlorosis and stunting of secondary roots (Srivastava and Gupta, 1996). Kiwi fruit, sugar beet and coconut have relatively high Cl requirements (Marschner, 1995; Xu et al., 2000).

Highly leached, permeable, sandy texture soils such as Arenosols, Ferralsols, and Acrisols, with no weathering minerals releasing Cl, in areas remote from the sea, are the most likely to give rise to Cl deficiency in sensitive crops. In a recent review, Xu et al. (2000) reported that the concentration range below which deficiencies can occur varies from 0.1 to 6 mg g^{-1} (Dry Matter – DM) or between 0.03 and 0.17 mmol L^{-1} of the plant tissue water content for different species. In wheat at heading, the critical concentration lies between 1.5 and 4 mg g^{-1} (DM) with no response above this value. In pot experiments, positive responses to Cl have been found in potato, peanuts, tomato and sugar beet. Yield increases have been found in maize and in wheat. There is also a beneficial effect of Cl in reducing foliar and root diseases of wheat (Xu et al., 2000).

1.2.2.1 Treatment of Chlorine Deficiency

The fertiliser compounds available for treating chlorine deficiency are shown in Table 1.5.

Table 1.5 Chlorine compounds used for treating chlorine deficiency in plants (Heckman, 2007)

Compound	Formula	Cl content (%)
Potassium chloride	KCl	47
Sodium chloride	NaCl	60
Ammonium chloride	NH_4Cl	66
Calcium chloride	$CaCl_2$	64
Magnesium chloride	$MgCl_2$	74

1.2.3 Copper

Copper deficiency can be a major problem in cereals, especially wheat (*Triticum aestivum* L.), barley (*Hordeum vulgare* L.), oats (*Avena sativa* L.), and a wide range of other crops, including alfalfa (*Medicago sativa* L.), oilseed rape (canola) (*Brassica napus* L.), onions (*Allium cepa* L.), spinach (*Spinacea oleracea* L.) and lettuce (*Lactuca sativa* L.). In cereals, vegetative growth can be affected by Cu deficiency and symptoms shown, but in cases of marginal or hidden deficiency, male sterility resulting in grain yield losses of up to 20% or more, without the appearance of obvious symptoms is commonly found. This hidden deficiency, like the male sterility caused by B deficiency in wheat in South Asia, is of major economic importance because it is relatively difficult to detect before yields are affected.

In addition to reduced yields, Cu deficiency can also adversely affect the quality of crop products. These include shrivelled grains and reduced viability of seeds in cereals. In citrus, abnormal shaped fruits with a rough exterior, low juice content and poor flavour and in apples, small fruits of poor quality are found due to Cu deficiency. In sugar beet, juice purity is reduced due to elevated concentrations of nitrogenous compounds. In vegetables, small size, chlorotic leaves, apparent wilting and discolouration of edible portions tend to render them less marketable (Alloway and Tills, 1984).

An example of acute Cu deficiency, with characteristic symptoms, in a mature wheat crop on a Rendzina soil in north-western France is shown in Figs. 1.2 and 1.3 (in Colour Section).

In the foreground of Fig. 1.2, the Cu-deficient wheat plants are shorter, darker in colour and have a lower density of ears (spikes) per unit area than the blocks of

Fig. 1.2 View of a field trial with copper on wheat growing on a Rendzina soil in France showing the copper-treated area (taller and pale coloured) in the distance. The copper-deficient crop in the foreground shows a lower density of ear-bearing tillers and a darker colour due to melanism (From B.J. Alloway) (*See Colour Plates*)

Cu-treated crop in the distance. Figure 1.3 (Colour Section) shows ears of both the Cu-treated and the Cu-deficient wheat from the two areas shown in Fig. 1.2. The ears from the Cu-treated plants are larger and full of grains. In contrast, the ears from the deficient plants are smaller, with empty grain positions at each end due to the failure to set grain due to pollen sterility. They also show some dark pigmentation (melanism) which is characteristic of Cu deficiency in wheat on Rendzinas (organic-rich calcareous soils) (Shorrocks and Alloway, 1985).

Significant losses in grain yield of up to 20% without prior symptoms are recognized as an important feature of Cu deficiency in cereals (Graham and Nambiar, 1981). Pollen sterility and impaired carbohydrate metabolism (reduced starch formation) are two of the major causes of the yield reductions associated with hidden Cu deficiency in cereals on soils with marginally low concentrations of available Cu (Jewell et al., 1988).

The soils most often associated with Cu deficiency include sandy and highly leached soils (Arenosols, Ferralsols, Acrisols) with low total Cu contents, calcareous (Calcisols, especially Rendzinas) and other high pH soils, such as saline soils (Solonchaks) and soils with high contents of organic matter (Histosols, Podzols), where the Cu is relatively unavailable. Other factors which can cause or exacerbate Cu deficiency include high N and P applications leading to dilution of Cu in tissues, and relatively high concentrations of other micronutrients, including Zn, Fe and Mn (e.g., from the treatment of other deficiencies).

Fig. 1.3 Ears of wheat from the field experiment shown in Fig. 1.2. Normal ears from copper-sufficient plants on the left and partially filled ears showing some melanism from copper-deficient plants on the right (From B.J. Alloway) (*See Colour Plates*)

Copper deficiency occurs in many parts of Europe, due to the widespread occurrence of sandy, calcareous, eluviated and organic-rich soils in this region, There are also favourable growing conditions for cereals and a high level of intensive crop management in many areas. It is estimated that around 30% of arable soils in Scotland are Cu-deficient, with 25% deficient in Germany and Denmark and 20% in Finland (Sinclair and Edwards, Chap. 9). On the basis of soil analysis data, up to 40% of soils in Ireland and Poland are potentially Cu-deficient (Alloway, 2005). Copper deficiency is found in all states in Australia on calcareous and acid soils. The largest areas of Cu deficiency (millions of hectares) are in Western Australia, South Australia and western Victoria and generally coincide with those affected by Zn and Mn deficiency (Holloway et al., Chap. 3). Copper deficiency occurs on the tropical red soils (Ferralsols) in Brazil and is often associated with intensive cropping and associated pH increases due to liming (Fageria and Stone, Chap. 10). Copper deficiency is generally not considered to be a major problem in many parts of Asia and the Near East (Singh, Chap. 4; Zou et al., Chap. 5; Rashid and Ryan, Chap. 6).

1.2.3.1 Treatment of Copper Deficiency

The fertiliser compounds available for the treatment of Cu deficiency are shown in Table 1.6.

In Europe, cereal land is normally treated with between 2 t and 15 t Cu ha^{-1} as $CuSO_4$ applied to the soil at infrequent intervals of between 5 and 15 years. Copper oxychloride is often applied to soils at rates of 10 kg ha^{-1}. Foliar treatments include $3Cu(OH)_2.CuCl_2$ at 0.5–2.2 kg ha^{-1} or 100 g ha^{-1} of Cu EDTA applied to each crop (Sinclair and Edwards, Chap. 9). In Brazil, 1–2 kg Cu ha^{-1} either broadcast, or banded is used, or a 0.1–0.2% Cu solution of copper sulphate in 400 l water/ha as a foliar spray (Fageria and Stone, Chap. 10). In Australia, Cu is applied mixed with phosphatic fertilisers but there is an increasing trend to using liquid fertilisers for both macronutrients and micronutrients (Holloway et al., Chap. 3).

Table 1.6 Copper compounds used for treating copper deficiency in plants (Gilkes, 1981; Martens and Westerman, 1991)

Compound	Formula	Copper content (%)
Copper sulphate	$CuSO_4.5H_2O$	25
Bordeaux mixture	$CuSO_4.3Cu(OH)_2 + 3CaSO_4$	12–13
Basic copper sulphates	$CuSO_4.3Cu(OH)_2$[a]	13–53
Copper oxychloride	$3Cu(OH)_2.CuCl_2.4H_2O$	52
Cuprous oxide	Cu_2O	89
Cupric oxide	CuO	75
Copper EDTA chelate	Na_2EDTA	14
Copper HEDTA chelate	NaCuHEDTA	9
Copper lignosulphonate	–	5–8
Copper polyflavonoid	–	5–7

[a] General formula of basic copper sulphates.

1.2.4 Iron

Iron deficiency is mainly a problem on calcareous and other alkaline soils with pH values of greater than 6 or 7, in which Fe has a low availability. The availability of Fe can also be reduced by relatively high concentrations of P, NO_3-N, high organic matter contents, and root infections (Fageria and Stone, Chap. 10). In general, C_4 plants have a higher requirement for Fe than C_3 species (Marschner, 1995). Fruit trees, grape vines (*Vitis vinifera* and *Vitis labrusca*), cereals, beans (*Vicia* and *Phaseolus* spp.), potato (*Solanum tuberosum* L.), soya (*Glycine max* L.) and sorghum (*Sorghum bicolor* L.) all tend to be susceptible to Fe deficiency in high pH soils. Iron is most available in acid or waterlogged (gleyed) soils and toxicity can occur on these soils. This is a particular problem in flooded (paddy) rice soils where rice yields can be severely reduced by Fe toxicity.

On Mediterranean-type soils in the Middle East, Fe deficiency is the second most important micronutrient deficiency problem, after that of Zn, due to the high pH calcareous soils. Legumes, citrus tree species and deciduous fruits are more susceptible to Fe deficiency than cereals (Rashid and Ryan, Chap. 6).

In China, peanuts are highly susceptible to Fe deficiency when grown in monoculture on calcareous and alkaline soils. However, intercropping peanuts with maize improves the uptake of Fe by peanuts due to root secretions from the maize (Zou et al., Chap. 5).

In addition to soil properties affecting the availability of Fe, the widely used herbicide Glyphosate is known to induce Fe deficiency (and also other micronutrient deficiencies) in some crops growing on soil with residues of this chemical (Eker et al., 2006).

1.2.4.1 Treatment of Iron Deficiency

The fertiliser compounds available for the treatment of Fe deficiency are shown in Table 1.7.

Table 1.7 Iron compounds used for treating iron deficiency in plants (Martens and Westerman, 1991)

Compound	Formula	Iron content (%)
Ferrous sulphate (heptahydrate)	$FeSO_4.7H_2O$	19
Ferrous sulphate (monohydrate)	$FeSO_4.H_2O$	33
Ferrous ammonium sulphate	$(NH_4)_2SO_4.FeSO_4.6H_2O$	14
Iron HEDTA chelate	NaFeHEDTA	5–9
Iron EDDHA chelate	NaFeEDDHA	6
Iron DTPA chelate	NaFeDTPA	10
Iron lignosulphonate	–	5–8
Iron polyflavonoid	–	9–10
Iron methoxypropane	FeMPP	5

In the USA, the most Fe-deficiency sensitive crops are soya bean and high-value crops, such as citrus species, grape vines and peaches. These crops receive the major proportion of Fe fertilisers which are usually applied as foliar sprays (Brown, Chap. 11). On high pH soils in China, acid ferrous sulphate is used as a fertiliser on soils and for soaking soya bean seeds. In fruit trees, neither soil, nor foliar, applications of Fe compounds are very effective so alternative methods of treatment have been developed (Zou et al., Chap. 5). In Brazil, foliar applications of Fe salts are used (1–2% solution of $FeSO_4.7H_2O$) in 400 l of water ha^{-1}. Iron chelates, such as FeEDDHA and FeEDTA, can also be used, but they are more expensive (Fageria and Stone, Chap. 10).

1.2.5 Manganese

Manganese deficiency problems occur on soils with low total contents of Mn (heavily weathered tropical and sandy soils), on peaty soils, or organic-rich soils with a pH above 6, and on mineral soils with pH values of 6.5 or above, such as calcareous soils, or acid soils which have been heavily limed. In addition to the available Mn status of soils, temperature, soil moisture and light intensity can all affect the incidence and severity of Mn deficiency in crops (Moraghan and Mascagni, 1991). According to Marschner (1995), Mn deficiency symptoms include: interveinal or blotched chlorosis in mature and young leaf blades and interveinal necrosis in young leaf blades. Deficiency problems are worst when crops are developing in wet and cold conditions (i.e., worse in spring than later in the growing season). Manganese deficiency is widespread in northern Europe, with its generally cool and humid climate. In the United Kingdom, Mn is the most commonly encountered micronutrient deficiency in field crops. Although it is most severe on organic soils with a pH above 6 and on sandy soils, it can also occur on a wider range of soils in some years, depending on weather conditions and the crops grown. Cereals, sugar beet, potatoes, oilseed rape and peas are the field crops most commonly affected. Cereal crops on ploughed-up grassland are particularly prone to Mn deficiency (Sinclair and Edwards, Chap. 9)

Manganese is more mobile in imperfectly drained (gleyed) soils and so rice grown in paddy fields tends not to be affected by deficiency. With regard to light, Mn deficiency is more likely to occur under low light conditions. The availability of Mn is strongly influenced by reactions in the rhizosphere. Root exudates, including phytosiderophores (which also mobilise Fe and Zn), have the ability to render Mn^{2+} available for uptake into the roots.

Many crops are highly sensitive to Mn deficiency, including: apple (*Malus domestica* Borkh.), cherry (*Prunus avium* L.), raspberry (*Rubus* spp. L.), pea (*Pisum sativum* L.), bean (*Phaseolus Vulgaris* L.), sugar beet (*Beta vulgaris* L.) potatoes (*Solanum tuberosum* L.), soya bean (*Glycine max* Merr.), oat and wheat (Humphries et al., 2007). However, maize (*Zea mays* L.) and rye (*Secale cereale* L.) are considered to be relatively tolerant (Marschner, 1995).

In recent years it has become apparent that prolonged use of the herbicide Glyphosate, is causing increased incidence of deficiencies of Mn and other micronutrients, such as Fe in some crops, including soya, sunflower and citrus, compared with controls grown under organic farming conditions on adjacent land. Research by Neumann et al. (2006) has shown that foliar-applied glyphosate to target (weed) plants can be released into the rhizosphere after translocation through the plant and inhibit the acquisition of micronutrients such as Mn, Zn and B by non-target plants. Glyphosate is widely used around the world and this potential deficiency problem is likely to be of increasing importance with the development of transgenic glyphosate-resistant crops (Brown, Chap. 11). The problem can be overcome by increased use of foliar applications of Mn and any other micronutrient found to be in short supply. Foliar applications are necessary because of the inhibition of root uptake caused by glyphosate. An interesting variation on the problem of glyphosate and Mn nutrition is that the efficacy of glyphosate used on soya beans in the USA can be reduced when it is tank mixed with certain types of Mn fertilisers for foliar application. However, it was found that Mn-EDTA did not interact with the glyphosate (Bernards et al., 2002).

The incidence of Mn deficiency in India is relatively low, with only 4% of advisory soil samples indicated as having low available Mn contents. However, it is becoming an increasing problem in rice–wheat cropping systems on sandy and alkaline soils (Singh, Chap. 4). In some countries, soil testing is not considered to be reliable for predicting Mn deficiency (Sinclair and Edwards, Chap. 9).

1.2.5.1 Treatment of Manganese Deficiency

Fertiliser compounds used for the treatment of Mn deficiency are shown in Table 1.8.

In most parts of the world, $MnSO_4$ is the most widely used Mn fertiliser due to its low cost, high solubility and easy availability. In Brazil, soil applications of 5–50 kg Mn ha^{-1} are used with larger amounts applied by broadcasting and smaller amounts when banding. Foliar applications of 0.1–0.2% $MnSO_4.H_2O$ in 4001 of water/ha are normally used but sometimes several applications may be required (Fageria and Stone, Chap. 10). In India, foliar sprays have generally been found to be more efficient in correcting Mn deficiency in wheat, berseem clover (*Trifolium*

Table 1.8 Manganese compounds used for treating manganese deficiency in plants (Martens and Westerman, 1991; Walter, 1988)

Compound	Formula	Manganese content (%)
Manganese sulphate	$MnSO_4.n\ H_2O$[a]	20–36.4[a]
Manganese oxide	MnO	41–68
Manganese EDTA chelate	$Na_2MnEDTA$	5–12
Manganese lignosulphonate	–	5
Manganese polyflavonoid	–	5–7
Manganese frits	Fritted glass	10–35

[a] Mn content depends on the degree of hydration of $MnSO_4$.

alexandrinum L), oat, sunflower, green gram (*Phaseolus aureus* Roxb.) and other crops (Singh, Chap. 4). In China, Mn deficiencies are corrected either by foliar application of Mn, or by banding Mn with an acid starter fertiliser. With wheat, it has been found that the most effective fertiliser regime is seed treatment with Mn, followed by foliar spraying (Zou et al., Chap. 5). Manganese deficiency in UK crops is usually treated by foliar applications of $MnSO_4$ (5 kg $MnSO_4$ ha^{-1}) but Mn-EDTA and other formulations are also widely used. Although Mn-EDTA is generally more efficient, on the basis of Mn applied, and more convenient to use, it is much more expensive than $MnSO_4$ (Sinclair and Edwards, Chap. 9). In the USA, Mn-EDTA is being used increasingly on crops treated with glyphosate herbicide because $MnSO_4$ and some of the other forms of Mn have been found to interact with this herbicide in tank mixes, rendering it less effective (Brown, Chap. 11).

1.2.6 Molybdenum

Unlike the other plant micronutrients, Mo is most readily taken up by plants in soils with a pH above 7 and is relatively unavailable in acid soils. Molybdenum deficiencies are most likely to occur on acid and severely leached soils and are mainly a problem in brassicas, legumes, such as peanuts, subterranean clover and soya beans, but other crops, including wheat and sunflowers can also be affected.

In Australia, Mo deficiency is the second most ubiquitous micronutrient deficiency problem, after zinc, affecting large areas of cropland with acid soils (Holloway et al., Chap. 3). In China too, Mo deficiency has become an important factor limiting yields in winter wheat and soya beans. Molybdenum-efficient cultivars of both these species have been identified, and growing these on deficient soils would help to reduce yield loss due to deficiency (Zou et al., Chap. 5). On the predominantly acid soils in Africa, Mo deficiency is a widespread problem, particularly in maize, sunflower, groundnuts, dry beans and peas (van der Waals and Laker, Chap. 8). In India, Mo deficiency is not very common but occurs in rice, wheat, soya bean and other legumes on Fe oxide-rich soils (Singh, Chap. 4). In the USA, deficiencies of Mo occur on acid sandy soils of the Atlantic, gulf and Pacific coasts. Legumes, cruciferous crops, grasses and several vegetable crops have responded to Mo fertilisation (Martens and Westerman, 1991).

1.2.6.1 Treatment of Molybdenum Deficiency

Fertiliser compounds used for the treatment of Mo deficiency are shown in Table 1.9.

Although deficiencies in some acid soils can sometimes be corrected by liming, deficiencies in strongly leached soils, with low total and available Mo contents, need to be fertilised with suitable compounds, usually either ammonium or Na-molybdate.

Table 1.9 Molybdenum compounds used treating molybdenum deficiency in plants (Martens and Westerman, 1991)

Compound	Formula	Molybdenum content (%)
Ammonium molybdate	$(NH_4)_6MoO_{22}.4H_2O$	54
Sodium molybdate	$NaMoO_4.2H_2O$	39
Molybdenum trioxide	MoO_3	66
Molybdic acid	$H_2MoO_4.H_2O$	53
Molybdenum frits	Fritted glass	2–3

In Australia, Mo is added to superphosphate fertilisers in many areas and is also applied with Cu and Zn fertilisers on acid, sandy and gravely soils in the south-west of Western Australia (Holloway et al., Chap. 3). In India, soil application of foliar sprays containing 0.05–1% Na molybdate are applied three times to green gram. Basal applications of Mo are applied to peanuts on calcareous soils and the seeds of soya beans and peanuts are also treated with an Mo formulation (Singh, Chap. 4).

1.2.7 Nickel

The first reported evidence of a response to Ni in field crops (potatoes, wheat and beans) was in 1945, but its essentiality was not conclusively demonstrated until 1987 (Brown, 2007a; Brown et al., 1987). A requirement for the use of nickel fertilisers has been established for perennial crops in the south-east of the USA. It has been found that Pecan trees could benefit from the application of Ni because it is effective in controlling pecan scab disease which affects yields. However, there have been few reports of the occurrence of acute Ni deficiency problems in field crops but it is likely that undiscovered (hidden) deficiencies may be important in some areas (Brown, Chap. 11). In contrast to deficiencies, there is generally more concern about the development of Ni toxicity in crops on soils contaminated with Ni from sources such as atmospheric deposition and the recycling of biosolids (sewage sludge).

Like the other cationic elements, Ni will be less available under alkaline conditions and most available in acid soils. Symptoms of Ni deficiency in graminaceous plants include chlorosis, similar to Fe deficiency and patchy necrosis in the youngest leaves. With Ni deficiency there is also a marked enhancement in senescence and reduced tissue Fe concentrations (Brown, 2007a).

1.2.7.1 Treatment of Nickel Deficiency

There are few fertiliser compounds used for the treatment of Ni deficiency owing to the general lack of information about the occurrence of the problem. However, Ni-sulphate ($NiSO_4.7H_2O$, 21% Ni) and other nickel compounds have been used.

Organic complexes containing Ni have also been used and many compound fertilisers and pesticides contain significant concentrations of Ni which may make them effective sources of the element. Biosolids (sewage sludges) can also contain appreciable concentrations of Ni, but must conform to the statutory limits for the state, or country concerned (e.g., <420 mg Ni kg^{-1} dry matter for the USA) (USEPA, 1993). However, biosolids contain elevated concentrations of whole suites of micronutrient and non-essential elements (e.g., Cd) and also organic pollutants. Therefore their use as a source of micronutrients has to be carefully planned and based on a broad-spectrum analysis of the biosolid product from the local sewage treatment plant.

1.2.8 Zinc

Zinc deficiency is by far the most ubiquitous micronutrient deficiency problem in the world as a whole. All crops can be affected by Zn deficiency, but maize (corn), beans, cotton, linseed and citrus fruits are the most sensitive. Although shown to be relatively tolerant to Zn deficiency, yields of rice and wheat, the world's two major staple food crops, are restricted by Zn deficiency over millions of hectares worldwide. It is estimated that up to 50% of the rice grown under lowland (flooded) conditions (paddy rice) may be affected by Zn deficiency (Scharpenseel et al., 1983). In addition to optimising crop yields to produce adequate amounts of food crops, the Zn contents of the crop products, especially cereal grains and beans are of major importance in human nutrition (Hotz and Brown, 2004). There is increasing interest in fortifying crop products with zinc and other micronutrients often in short supply for humans such as Fe, as discussed by Graham (Chap. 2) and Welch (Chap. 12).

Many of the soil and crop management changes involved in the intensification of arable farming increase the risk of Zn deficiency occurring on soils of marginal to low available Zn status. Apart from growing different plant varieties, one of the most common changes with intensification is the application of larger amounts of N, P, K, and S fertilisers. It is widely recognised that high levels of available P in soils of low Zn status can cause Zn deficiency (Loneragan and Webb, 1993).

The soil conditions which are most commonly found to be associated with Zn deficiency in crops include:

- Low total Zn contents (such as in highly weathered and leached tropical soils with low pH, and sandy soils with low contents of organic matter)
- Neutral or alkaline soil pH (including limed soils)
- High salt concentrations (saline soils)
- High calcium carbonate content (calcareous soils with high pH)
- Peat and muck (organic soils)
- High available P status (closely linked with intensive production)
- Prolonged period of waterlogging or flooding (paddy rice soils)
- High magnesium and/or bicarbonate concentrations (in soils and irrigation water)

- Foliar and soil fertilisation with Cu
  Low inputs of livestock manures
  (Alloway, 2004)

For rice grown in flooded (paddy) soils, the possible causes of Zn deficiency can be summarised as:

- Low total Zn content in soil
- Growing highly susceptible (Zn-inefficient) rice varieties (e.g., IR 26)
- High pH (≥7 under anaerobic conditions)
- High bicarbonate (HCO_3^-) concentrations in soils and irrigation water
- Depressed Zn uptake due to increased availability of other micronutrients and P after flooding
- Formation of organic Zn complexes in soils with a high pH and high organic matter contents (e.g., manures/straw)
- Zinc sulphide precipitation when pH decreases in alkaline soil after flooding
- Excessive liming (elevated pH and adsorption of Zn on $CaCO_3$ and $MgCO_3$)
  Wide Ca:Mg ratios (<1.0) (often due to excess Mg from ultrabasic soil parent material)
  (Dobermann and Fairhurst, 2000)

Soil testing in both India and China has shown that 48% and 51% of soils, respectively, are potentially deficient in available Zn. In India, cereals, millet, oilseed rape, fodder crops, vegetables, fruit trees and plantation crops are most affected (Singh, Chap. 4). In China, maize, rice, lentil, pea, cotton, sorghum, apple and peach trees are amongst the crops most affected (Zou et al., Chap. 5). In Australia, Zn deficiency affects crops on sandy and calcareous soils and there are millions of hectares of Zn-deficient land in Western Australia and South Australia. The problem is increasing due to more intensive methods of production and the change from comparatively Zn-rich superphosphate to more pure forms of P fertiliser (Holloway et al., Chap. 3). In South America, Zn deficiency is widespread mainly due to the low total Zn contents of the strongly weathered and leached tropical red soils (Ferralsols). The low Zn status of many of the soils is further exacerbated where the soils are limed and heavily fertilised with P for growing high yielding cultivars of cereals and other crops (Fageria and Stone, Chap. 10).

1.2.8.1 Treatment of Zinc Deficiency

Fertiliser compounds used for the treatment of Zn deficiency are shown in Table 1.10.

The history of the use of Zn fertilisers goes back to 1934 when $ZnSO_4$ was used to treat "white bud" (leaf chlorosis) in maize in Florida. It is generally recognised that, at least in the short-term, the more highly water-soluble Zn fertiliser compounds, such as $ZnSO_4$ (98% soluble), Zn lignosulphonate (91% soluble) and ZnEDTA (100% soluble) are more effective in correcting deficiencies than less soluble compounds, such as ZnO.

Although some traditional fertilisers, such as single superphosphate, are a valuable source of Zn (<600 mg Zn kg^{-1}), the trend towards high analysis fertilisers has necessitated the use of either Zn-fortified fertilisers, or "straight" Zn compounds

Table 1.10 Zinc compounds used for treating zinc deficiency in plants (Martens and Westerman, 1991)

Compound	Formula	Zinc content (%)
Zinc sulphate (monohydrate)	$ZnSO_4.H_2O$	36–37
Zinc sulphate (heptahydrate)	$ZnSO_4.7H_2O$	22–23
Zinc oxysulphate	$xZnSO_4.xZnO$	40–55
Basic zinc sulphate	$ZnSO_4.4Zn(OH)_2$	55
Zinc oxide	ZnO	50–80
Zinc carbonate	$ZnCO_3$	50–56
Zinc nitrate	$Zn(NO_3)_2.3H_2O$	23
Ammoniated zinc sulphate solution	$Zn(NH_3)_4SO_4$	10
Disodium zinc EDTA	$Na_2ZnEDTA$	8–14
Sodium zinc HEDTA	NaZnHEDTA	9–13
Zinc polyflavonoid	–	5–10
Zinc lignosulphonate	–	5–8

being applied to the seedbed directly (Holloway et al., Chap. 3; Cakmak, Chap. 7). Foliar applications can be of $ZnSO_{4,}$ but this can carry a high risk of scorch unless a neutraliser, such as calcium hydroxide is added, or the spray solution is made up in hard water. Chelated forms of Zn are becoming more commonly used for foliar applications and have the advantage of being compatible with many herbicide and fungicide formulations in spray tank mixes, but they are more expensive than inorganic compounds.

In China, $ZnSO_4$ and ZnO are the most widely used fertilisers but $ZnCl_2$ and chelates are also used. The maize crop receives the largest proportion of Zn fertilisation, followed by rice, but lentil, pea, sorghum, cotton, apple and peach trees are also treated widely (Zou et al., Chap. 5). Soil applications of 9–22 kg Zn ha^{-1} on calcareous soils in South Australia have been found to have a beneficial residual effect for about 10 years. More recent practices have been to spray zinc sulphate onto seedbeds at a rate of 1 kg Zn ha^{-1} and cultivate it into the topsoil or as a foliar spray mixed with a compatible cereal fungicide at a rate of 0.2 kg Zn ha^{-1}. The inclusion of urea in foliar sprays of $ZnSO_4$ increases leaf penetration and addition of a sticker can reduce wash off (Holloway et al., Chap. 3).

In South America, application rates of $ZnSO_4.7H_2O$ are 30–50 kg ha^{-1} as a top dressing on both upland and lowland rice (Fageria and Stone, Chap. 10). In India, application rates Zn fertilisers (usually of $ZnSO_4.7H_2O$) are typically 11 kg Zn ha^{-1} for wheat and rice, 5.5 kg Zn ha^{-1} for maize, soybean and sugarcane and 2.5 kg Zn ha^{-1} for peanuts, soya bean and other selected crops (Singh, Chap. 4).

1.3 Hidden Micronutrient Deficiencies in Crops

Hidden deficiencies of micronutrients (also called subclinical, latent, or "symptomless" deficiencies) are often more widely occurring and economically important than severe forms of deficiencies because they are difficult to detect. Nevertheless, subtle differences in colour, height and time of maturity can often be seen in field experiments, on marginally Zn-deficient soils, where control and micronutrient-treated plots are located side by side. In the case of hidden Cu deficiency, numerous second growth tillers can often be found in cereal crops. The farmer is usually aware that crops on some of his land have relatively poor yields, but there are often other possible explanations for this apart from a marginal micronutrient deficiency stress. These can include drought, soil structural problems, herbicide damage, disease and shortage of macronutrients. The same soil and plant causal factors are responsible the onset of both hidden and acute deficiency but, usually, the magnitude of the factor is smaller and the effect more marginal in hidden deficiencies. In most cases, it is the micronutrient-inefficient genotypes which are affected by hidden deficiencies.

Examples of hidden Cu deficiency, where yield responses were obtained with Cu fertilisers in crops not showing obvious symptoms, include:

1. In 20 field experiments with cereals on sandy soils in North East Scotland, Reith (1968) found distinct deficiency symptoms ("white tip") in only 3 of the 18 trials which gave significant positive responses to Cu treatment. The mean yield response was 20% over all the trials. Some untreated control plants showed numerous second growth tillers in the crops and in the stubble after harvest.
2. Experiments by the Agricultural Development and Advisory Service (ADAS) in southern England on shallow chalk soils (Rendzinas/Rendic Leptosols) showed significant increases in yield of around 4% with foliar applications of copper on controls yielding 6.96 t ha^{-1} without any symptoms being apparent (Sinclair and Withers, 1995).
3. An increase of 22.4% (giving a yield of 3.21 t ha^{-1}) was obtained with foliar-applied copper hydroxide before the end of tillering in spring barley growing on a brown sand soil in eastern England. The yield increase was due to a significant increase in the number of grains per ear. The untreated control had received a non-metallic fungicide application (Alloway et al., 1985).
4. Foliar-applied $CuSO_4$ gave a grain yield response of 15.6% (6.68 t ha^{-1} increased to 7.72 t ha^{-1}) in winter wheat growing on an organic-rich Rendzina. The untreated control had received a non-metallic fungicide application. Soil tests indicated a marginal deficiency, but no symptoms of deficiency were observed in the untreated crop (Tills, 1981).
5. Sugar beet (a crop with medium susceptibility to Cu deficiency), growing on a brown sand soil developed on drift over Chalk, gave a significantly increased root yield of 18% and improved juice purity when 12.5 kg Cu ha^{-1} was applied to the seedbed (as $CuSO_4$). No visible symptoms of Cu deficiency were shown by the untreated sugar beet plants (Tills and Alloway, 1981).
6. "Symptom-less" Cu deficiency was reported by Vetter et al. (1985) in 21 field trials with cereals growing on sandy soils, loams developed on loess and on peat soils in the Hanover and Weser-Ems regions of northern Germany. Mean yield responses of up to 30% were obtained with Cu fertilisers, especially when N fertilizers were used and P levels in the soils were optimal.

In parts of India, Nepal and Bangladesh, B deficiency in wheat mainly affects pollen formation and pollination and is hidden during the early stages of growth (Rerkasem and Jamjod, 2004). The yield response of grapevines, on low-B soils in Hungary, to B fertilisation is also indicative of a hidden deficiency since the applied B has the effect of reducing flower-shedding and thereby increasing the amount of fruit set (Györi and Palkovics, 1983).

In addition to physiological responses to micronutrient treatments bringing about increased yields, indirect effects of micronutrients, such as reducing susceptibility of crops to fungal infections, can also help to give increased yields and product quality. This includes the effects of Zn and of Cl on root diseases in wheat (Grewal et al., 1996; Xu et al., 2000). Apart from these specific effects, healthy plants are generally less likely to be infected than those under physiological stress due to micronutrient deficiencies. It is important to note that foliar applications of many micronutrient compounds can have fungicidal, as well as a nutritional, effects. In field experiments to investigate yield responses to micronutrient treatments, it is essential to apply a non-metallic fungicide to control plots in order to rule out this confusing factor. This is not necessary in experiments where fertiliser has been applied to the seedbed.

Finally, it is important to recognize that the absence of symptoms of nutrient deficiency does not necessarily imply optimum nutrition. Unsatisfactory crop performance could still be due to a hidden micronutrient deficiency and these can progress to acute deficiencies with changes in cultivars grown and crop management. The only reliable diagnosis for a hidden deficiency is by soil/plant analysis, or fertiliser responses.

1.4 Worldwide Occurrence of Micronutrient Deficiencies

Details of the occurrence of micronutrient deficiency problems in crops in many different countries or continents are given in Chaps. 2–11 of this volume. However, the amount of information and data available for different countries varies considerably, as do some of the methods used in analytical surveys of soils and crops. However, an analytical survey of the macro- and micronutrient status of arable soils in 30 countries around the world was conducted for the FAO by Sillanpää (1982). The data from this survey, and a subsequent follow-up study, provide a very useful reference with which data from individual countries can be compared. This FAO study was carried out between 1974 and 1982 and comprised a total of 3,538 soil samples collected from representative sites in all 30 participating countries (shown in Table 1.11, except for Ecuador). After collection, the soil samples were sent to The Institute of Soil Science at the Agricultural Research Centre in Jokoinen, Finland where they were prepared and analysed. A cultivar of spring wheat (*Triticum aestivum* cv. "Apu") was grown on samples of each soil under controlled conditions (in a greenhouse at the Institute) to act as an "indicator" of the plant availability of the nutrients. Unfortunately, there was insufficient soil to grow this wheat on the soil samples from Ecuador.

Table 1.11 Percentage distribution of the soil × crop concentration products for six micronutrients in Zones I and II (0–5% and 5–10%, respectively) for wheat grown on soil samples from 29 countries in the study by Sillanpää (1982)

	Element Zone	B		Cu		Fe		Mn		Mo		Zn	
Country	No. of samples	I	II	I	II	I	II	I	II	I	II	I	II
Belgium	36	0	0	0	0	0	0	3	0	0	0	0	0
Finland	90	3	1	6	13	0	0	1	1	0	1	0	0
Hungary	201	0	1	0	0	0	0	4	4	1	3	0	0
Italy	170	0	0	0	0	1	1	12	9	0	2	3	4
Malta	25	0	0	0	0	60	8	80	16	0	0	0	0
New Zealand	35	0	0	17	11	0	0	0	0	9	29	0	0
Argentina	208	0	0	4	19	0	0	0	0	0	0	0	0
Brazil	58	0	0	0	0	0	2	2	0	45	26	0	0
Mexico	242	0	2	2	3	0	2	6	4	1	1	0	0
Peru	68	1	3	1	1	0	3	1	4	1	3	0	0
India	258	12	8	0	0	3	9	14	11	0	2	11	7
Korea	90	3	13	1	7	2	0	8	6	0	2	11	7
Nepal	35	46	23	0	3	0	0	0	9	20	14	3	3
Pakistan	237	2	1	0	0	1	8	11	15	0	0	8	12
Philippines	194	30	25	0	1	0	0	1	1	3	9	0	0
Sri Lanka	18	0	11	0	0	0	11	0	0	0	6	0	0
Thailand	150	10	13	3	1	4	3	0	0	1	2	1	2
Egypt	198	0	0	0	0	0	1	8	13	0	0	0	2
Iraq	150	3	3	0	0	2	7	2	5	0	1	34	23
Lebanon	16	0	0	0	0	13	0	13	0	0	0	6	6
Syria	38	3	0	0	3	8	13	13	11	0	0	5	11
Turkey	298	1	1	0	0	14	9	2	5	0	2	17	18
Ethiopia	125	4	6	21	6	0	1	0	0	0	4	0	1
Ghana	93	0	0	23	24	1	0	0	0	16	22	0	2
Malawi	97	9	14	7	6	0	4	0	0	7	24	1	1
Nigeria	153	11	14	12	20	5	12	1	3	20	12	0	1
S. Leone	48	0	2	68	21	0	4	0	2	81	10	0	0
Tanzania	163	0	1	14	9	6	7	0	0	6	7	3	2
Zambia	44	2	11	36	16	16	11	0	0	50	9	0	2

The soil and wheat samples were analysed for a range of macro- and micronutrients and six key soil properties were measured (pH, organic matter and $CaCO_3$ contents, electrical conductivity, texture and bulk density). From the analytical results, the products of soil concentration × plant concentration were calculated for each nutrient (from here on called "concentration products"). The lowest 5% of these concentration product values were allocated to Zone I and the next lowest 5% to Zone II. The middle 80% of the concentration products were allocated to Zone III, the top 90–95% of concentration products to Zone IV and the uppermost 95–100% of the concentration product values were put in Zone V. This implies that concentration product values in Zones I and II represent low available concentrations of nutrient elements and those in Zones IV and V high available concentrations. Therefore those soil samples with a high probability of severe deficiencies, would be in Zone I and less severe, possibly hidden deficiencies in Zone II. The soils with the greatest risk of crops accumulating high, possibly even toxic, concentrations of some of the nutrient elements monitored would be found in Zone V.

The percentage distribution of concentration products for the soil and wheat sample analyses for six micronutrients in the two lowest zones (I and II) are shown in Table 1.11. From this data, several generalizations can be made regarding the countries with the higher percentages of soils having potentially deficient concentrations of available micronutrients. Owing to insufficient soil in the samples from Ecuador to grow wheat in pots, no concentration products were produced for this country and so the data in Table 1.11 is only for 29 countries.

1.4.1 Countries in the Sillanpää (1982) Study Shown to Have Significant Numbers of Potentially Micronutrient-deficient Soils

1.4.1.1 Boron

From Table 1.11 it can be seen that B deficiencies are most likely to occur in the Far East/South Asian countries included in the survey, especially India, Nepal, Philippines, and Thailand. Significant percentages of low concentration products for B were also found in soil samples from Korea, Nigeria, Malawi, Korea, Zambia and Sri Lanka with smaller numbers of potentially B-deficient soils in Finland, Peru, Pakistan, Iraq, Syria, Turkey and Ethiopia.

1.4.1.2 Copper

All of the African countries in the study had high, or relatively high, percentages of potentially Cu-deficient soils (in Zones I and II). These soils were predominantly acid and some had relatively large organic matter contents. Countries in other regions with significant percentages of potentially Cu-deficient soils include: Finland,

New Zealand, Argentina and Korea. It is interesting to note that, in follow-up field trials in some of the same African countries, responses to Cu were not as high as expected in the light of these low concentration products (Sillanpää, 1990).

1.4.1.3 Iron

Many of the countries with high percentages of Fe concentration products in Zones I and II have calcareous and/or high pH soils and these are: Malta, Lebanon, Syria, Turkey and Iraq. Potentially deficient Fe values were also found in Zambia, Nigeria, Tanzania, India, Sri Lanka, Pakistan and Thailand. However, the African countries differ from the others in having predominantly acid soils which render Fe more available.

1.4.1.4 Manganese

Some of the countries with relatively high percentages of Mn concentration products in Zones I and II were the same as those shown to have potentially Fe-deficient soils, especially Malta and Syria, and the high soil pH is one of the main causes of both low available Mn and Fe. However, higher percentages of potentially Mn-deficient soils were found in India, Egypt, Pakistan, Italy and Korea than were found with Fe.

1.4.1.5 Molybdenum

Molybdenum deficiency is usually associated with acid soils and the countries with the highest percentages of potentially Mo-deficient soils are Sierra Leone, Brazil, Zambia, New Zealand, Ghana, Nepal and Malawi, which all have high percentages of acid soils.

1.4.1.6 Zinc

Zinc is indicated as being most deficient in Iraq, Turkey, Pakistan, India, Korea, Syria and Italy and, with the exception of Italy, these countries all had generally high pH soils. It is interesting to note that follow-up field trials in a second project for FAO (Sillanpää, 1990) in some of the countries included in this study confirmed that Zn deficiency was the most ubiquitous micronutrient problem of all. As shown in Table 1.12, 49% of the 190 field trials in 15 counties showed a response to applications of Zn and, of these, half were "latent" or hidden deficiencies without obvious symptoms.

The data of Sillanpää (1982) only relate to 29 countries which are probably not fully representative of the continents in which they are located. Nevertheless, given

Table 1.12 Summary of the occurrence of acute and latent micronutrient deficiencies in 190 field trials in 15 different countries[a] (Sillanpää, 1990)

	Micronutrients					
Deficiencies	B	Cu	Fe	Mn	Mo	Zn
% Acute	10	4	–	1	3	25
% Latent	21	10	3	9	12	24
Total %	31	14	3	10	15	49

[a]Countries included Ethiopia, Finland, Iraq, Malawi, Mexico, Nepal, Pakistan, Philippines, Sierra Leone, Sri Lanka, Tanzania, Thailand, Turkey, Zaire, Zambia

the logistical problems of carrying out such a large study, this project and its follow-up field experiments in 15 countries is a vitally important source of information on the soil fertility of a large part of the world and it complements the more intensive national surveys carried out in countries such as China (Chap. 5), India (Chap. 4) and various European countries (Chap. 9).

In 1990, Sillanpää reported the results for a series of 190 field trials conducted in 15 countries which, with the exception of Zaire, were surveyed in his earlier 1982 study of soil and plant concentrations of micronutrients. A summary of his results is shown in Table 1.12.

From Table 1.12, it can be seen that the percentage of field experimental sites where deficiencies of the micronutrients were significant, decreased in the order: Zn > B > Mo > Cu > Mn > Fe. Hidden (or "latent") deficiencies of micronutrients were more widespread than acute forms for all the elements except Zn. This helps to confirm the international importance of hidden micronutrient deficiencies as discussed in Sect. 1.3 above.

All of the 18 field experiments conducted in Iraq showed some degree of Zn deficiency, which confirms the predictions from the concentration products derived from the soil and wheat analysis by Sillanpää (1982). In Turkey, 75% of the 21 field trials showed some degree of Zn deficiency (see Cakmak, Chap. 7, for details Zn deficiency problems in Central Anatolia, Turkey).

Either acute or latent Zn deficiencies were found in all of the 15 field experiments in Thailand. Copper was also deficient at many sites but the response of the maize test crop to Cu was not very marked. Maize is only moderately sensitive to Cu deficiency, and so its significance may possibly have been underestimated. However, maize is highly sensitive to Zn deficiency.

In the Philippines, almost half of the 15 trials showed some degree of Zn deficiency. In the five trials conducted in Tanzania, shortages of Cu and/or B were the most common micronutrient deficiency problems. Sri Lanka (10 trials) had relatively few acute micronutrient deficiency problems, but at most sites there were latent deficiencies of B or Zn and occasional latent deficiencies of Cu, Mn or Mo. A majority of the 18 field trials in Nepal showed indications of B and Zn deficiencies. Rerkasem and Jamjod (2004) have discussed the widespread B deficiency in wheat in Nepal and adjacent countries. In Finland, indications of both deficiencies and excesses of B, Cu and Mn were found. The excesses were probably due to the placement of micronutrient fertilisers near to the seed.

The order of occurrence of deficiencies shown in Table 1.12 is Zn (49%) » B (24%) > Mo (15%) ≈ Cu (14%) > Mn (10%) » Fe (3%). This cannot be extrapolated to the whole world because only a relatively small number of countries were involved in the field experiment study (although they had been selected from the earlier survey involving more countries). The countries involved, with the exception of Finland, were all developing nations in tropical or semi-arid regions and therefore not truly representative of the full range of crop production systems around the world. Nevertheless, these data do provide a very important indication of the relative importance of the different deficiency problems in the countries included in the study and many others like them. Many of the more technologically developed countries around the world have more humid, temperate conditions and markedly different soils to those found in the majority of countries involved in this study. In general, crop production techniques in the more developed parts of the world are more intensive and the combination of different soils, crops and farming practices has resulted in individual micronutrient deficiency problems differing locally in importance from the order given in Table 1.12. For example, Zn deficiency is not so important in Europe as in many other parts of the world, but Cu deficiency occurs to a greater extent in many parts of Europe (Sinclair and Edwards, Chap. 9), Australia (Holloway et al., Chap. 3) and some regions in the USA (Brown, Chap. 11) and Canada than was found in the Sillanpää (1992) study. On the more intensively managed farms in the technologically developed countries, both acute and hidden, deficiencies are more likely to have been recognised or detected and treated, so the micronutrient status of many soils is likely to be higher.

These two related studies by Sillanpää (1982, 1990) are particularly useful because they involved standardised sampling and analytical procedures for all of the countries involved and therefore allow comparisons to be made. There are several other valuable sources of information on the percentages of agricultural soils in different countries found to be potentially deficient in micronutrients and these are reported in the chapters on India (Singh, Chap. 4), China (Zou et al., Chap. 5) and Europe (Sinclair and Edwards, Chap. 9).

1.5 Soil Types Commonly Associated with Micronutrient Deficiencies

The available concentrations of micronutrients in soils are determined by several factors, which include:

- Geochemical composition (total micronutrient contents) of the soil parent material
- Pedogenic soil type
- Inputs of trace elements from anthropogenic sources (e.g., atmospheric deposition, pesticides, manures, fertilizers)
- Adsorptive properties of the soil for retaining elements in available/unavailable forms (pH, redox status, organic matter content, calcium carbonate content and salinity)
- Available concentrations of macronutrients and other micronutrients

Table 1.13 Typical total concentrations of micronutrients in soils of different types (mg kg^{-1} DW) (Kabata-Pendias, 2001)

Element	Podzols (sandy)		Cambisols (silty/loamy)		Rendzinas (calcareous)		Histosols (organic)		Chernozems etc. (humus-rich, semi-arid)	
	Range	*Av.*	*Range*	*Av.*	*Range*	*Av.*	*Range*	*Av.*	*Range*	*Av.*
B	<0.1–134	(22)	<1–128	(40)	1–210	(40)	4–100	(25)	11–92	(45)
Cu	1–70	(13)	4–100	(23)	6.8–70	(23)	1–113	(16)	6.5–140	(24)
Mn	7–2000	(270)	45–9200	(525)	50–7750	(445)	7–2200	(465)	100–3907	(480)
Mo	0.17–3.7	(1.3)	0.1–7.2	(2.8)	0.3–7.35	(1.5)	0.3–3.2	(1.5)	0.4–6.9	(2)
Ni	1–110	(13)	3–110	(26)	2–245	(34)	0.2–119	(12)	6–61	(25)
Zn	3.5–220	(45)	9–362	(60)	10–570	(100)	5–250	(50)	20–770	(65)

Typical ranges and average contents of most micronutrients in different types of soils are given in Table 1.13 (Kabata-Pendias, 2001). These show wide ranges in concentrations of most elements in the different soil types, which are largely due to differences in the geological parent material that the soils are derived from. Total contents do not provide an indication of the available concentrations apart from the fact that when the total concentration is low there is a higher chance of sorptive mechanisms holding the elements in unavailable forms. In Table 1.13, the sandy Podzol soils are shown to contain the lowest average total contents of all elements and the Cambisols, with higher clay and silt contents, contain the highest average contents of Mn and Mo. Rendzinas contain the highest average Ni and Zn contents. However, cationic forms of metals tend to have a relatively low availability in these shallow, organic-rich calcareous soils. Likewise, Cu and Mn will tend to have a low availability in Histosols due to binding by organic matter and also in Chernozems, which also tend to have higher pH values than Histosols.

Sometimes, there appear to be no clear links between the incidence of certain deficiencies and soil type, as in the case of low Cu soils in China (Zou et al., Chap. 5). This is likely to be due to variations in the geochemical composition of the soil parent material.

1.5.1 Effects of Soil Parent Material and Pedological Soil Type on the Availability of Micronutrients

The following generalizations can be made with regard to the links between soil parent materials and the geochemical composition and adsorptive properties of the soils derived from them. The equivalent soil groups in the two main international classification systems, the FAO-UNESCO/World Resources Base (ISSS, 1998 a,b) and the USDA Soil Taxonomy (Soil Survey Staff, 1999), are given in Appendix 2.

1.5.1.1 Sandy Soils

Sandy soils developed on sandstones or sandy drift deposits, tend to be have low concentrations of all micronutrients (B, Cl, Cu, Fe, Mn, Mo, Ni, Zn) and can often show multiple micronutrient deficiencies (as shown in Table 1.13). The main world soil types (FAO–UNESCO/World Reference Base classification) comprising sandy soils are Arenosols, Leptosols, Regosols, Podzols and Ferralsols.

1.5.1.2 Tropical Soils

Soils formed on heavily weathered parent materials, such as tropical forest soils will tend to be highly acid and have low available concentrations of almost all micronutrients. However, the low pH favours the availability of cationic forms of elements, especially Fe, but not anionic forms, such as molybdates. Liming of these acid soils will reduce the availability of cationic micronutrient elements and often lead to deficiencies of B, Cu, Mn, and Zn. The natural vegetation of these soils is usually tropical rain forest, or scrub in less humid areas. When the vegetation is cleared to create arable land, after the initial flush of micronutrients present in the ash of the burnt vegetation, both macro and micronutrients often become deficient. The requirement for N, P, K and S is widely recognized but if any micronutrients are deficient, crop performance will be suboptimal. This problem is widely encountered in Brazil and other countries with rain forest (see Fageria and Stone, Chap. 10). The main soil types involved, include: Ferralsols, Acrisols, Lixisols (previously called Red–Yellow Podzols), Nitisols and Plinthosols. In the USDA Soil Taxonomy classification the equivalent orders are: Ultisols, Oxisols and Latosols, and Lateritic soils in other classifications.

1.5.1.3 Organic Soils

Organic soils, such as peats and mucks and humic variants of mineral soils with relatively high contents of organic matter tend to be deficient in Cu, Mn, and Zn and this can often occur over a wide range of pH. The main world soil types comprising organic soils include: Histosols, Chernozems, Kastanozems, Phaeozems, Podzols, Rendzinas (Rendic Leptosols and Calcisols) and Umbrisols and humic variants of several different other types, including: Cambiols, Fluvisols and Gleysols.

1.5.1.4 Calcareous Soils

Calcareous soils have a relatively high content of free $CaCO_3$ (>15%) and include soil developed on limestones of various types and also soils in which $CaCO_3$ has been deposited in voids through the evaporation of Ca-rich groundwater

(Calcification). These calcareous soils tend to give rise to deficiencies of B, Cu, Fe, Mn, Ni and Zn. Several micronutrients can be deficient simultaneously in crops on these soils (multi-element deficiencies). The world soil types comprising these soils include Calcisols, Leptosols, Rendzinas (also included in Calcisols) and calcic variants of several other soil types including Cambisols. Vertisols have a high pH and $CaCO_3$ content and can give rise to deficiencies of Fe and Zn.

1.5.1.5 Saline Soils

Saline soils are characterized by having high concentrations of soluble salts and are normally found in arid and semiarid regions. Inappropriately managed irrigated soils can also become saline (salinisation) due to the accumulation of salts. Where cropped, saline soils can cause deficiencies of Cu, Fe, Mn and Zn. The world soil types comprising saline soils are mainly the Solonchaks and Solonetz. In the USDA Soil Taxonomy classification they include Natrargids, Natrustalfs and Natrixerolls and other soil types which have become salinised through irrigation.

1.5.1.6 Gleyed Soils

Gleyed soils, are soils which are affected by waterlogging ("gleying") for at least part of the year and can occur naturally in low-lying situations in humid regions where groundwater accumulates. Gleying can also occur as a result of impermeable conditions in the upper part of the soil profile, usually in clay-rich soils and can be caused, or exacerbated by poor soil management, leading to compaction. Gleying is characterised by reducing conditions which cause the dissolution of oxides of Fe and Mn and the release of any trace elements sorbed in or on them. Thus, the availability of Fe and Mn can be high in gleyed soils but the total content is usually comparatively low owing to depletion of the mobile Mn^{2+} and Fe^{2+} ions. Zinc (and possibly other micronutrients) may be relatively unavailable in gleyed soils due to the elevated pH and formation of insoluble sulphides (e.g., ZnS). The main soil groups comprising gleyed soils are the Gleysols, Fluvisols, Planosols and gleyed variants of a wide range of soil types, including Cambisols.

The most widely occurring cropping system involving gleyed soils is lowland (flooded) rice production. In this system, many soils which were not naturally gleyed have been converted to gleys in the construction of paddy fields with standing water for most of the growing period of the rice crop. Some of the soils are both calcareous and gleyed and thus the availability of elements like Zn is affected by both pH and redox conditions. Possibly, up to 50% of paddy rice-growing land is affected by Zn deficiency (Scharpenseel et al., 1983).

1.5.1.7 Clay-rich Soils

Clay-rich soils can often show gleying in their uppermost horizons due to impermeability. Most of the clay minerals making up clay-textured soil, except kaolinite, tend to have relatively high adsorptive capacities for many micronutrients and therefore tend to retain trace element inputs (including micronutrients) from various sources, including livestock manures, agrichemicals and atmospheric deposition. In contrast to the low concentrations of most micronutrients found in sandy soils developed on sandstones and sandy drift materials, some parent materials of clay-rich soils can be relatively rich in several micronutrients. Soils developed on shales and clay formations tend to be relatively enriched in a range of trace elements, both essential and non-essential, including: Cu, Mo, Zn and also As, Se, Cd and other elements (Alloway, 1995). The availability of these elements in clay-textured soils will depend on soil factors, including redox conditions, pH, organic matter content and available P status. Soils developed on basalts and other basic igneous rocks can also be relatively enriched in several micronutrients but the high content of adsorptive iron oxides in the soils may affect the availability of some elements. The main groups of clay-rich soils are: Alisols, Chernozems, Luvisols, Vertisols and argic variants of other groups including Cambisols.

1.5.1.8 Contaminated Soils

Any type of soil can become contaminated with micronutrient elements. However, the higher the cation-adsorptive capacity of the soil, the longer the contaminants are likely to be retained against leaching. Some of the main sources of contaminants are atmospheric deposition of localized and long-distance transported industrial air pollutants, recycling and disposal of biosolids (sewage sludges) and animal manures (e.g., Cu and Zn), metal-containing fungicides (e.g., Cu, Zn and Mn), contaminants in agrichemicals and fertilisers (e.g., Zn in superphosphate) and residues of micronutrient fertilisers. Many of these accumulations can be beneficial sources of micronutrients and counteract the risk of deficiencies occurring as they would on similar but uncontaminated soils. However, there is a danger of elements accumulating to potentially toxic levels and also the problem that non-essential elements in the same micronutrient-rich materials may also be accumulating to undesirable levels (e.g., Cd in sewage sludges and P fertilizers) (Alloway, 1995b).

1.5.2 Effects of Soil Conditions on the Availability of Micronutrients to Crops

1.5.2.1 Soil pH

Soil pH is perhaps the most important soil property affecting the availability of trace elements in soils (Alloway, 1995a). Elements existing as cations (e.g., Cu^{2+}, Zn^{2+}) in the soil solution will be more available at low pH, whereas elements in anionic form (MoO_4^{2-} and $HMoO^{4-}$) will be more available at high pH. In humid environments, most mineral soils, except calcareous types, will tend to be naturally acid due to the leaching of bases and higher contents of organic matter. Gleyed soils will normally have slightly higher pH values than their freely drained counterparts. In arid environments, soils will tend to be neutral to alkaline. Although it can be expected that farmers will often attempt to raise the pH of acid soils. Where soils are acid, the micronutrient most likely to be deficient will be Mo. In contrast to deficiency, the availability of Fe and Mn may be so high that some sensitive crops may be affected by toxicity of these elements on acid soils.

In contrast, where soils are neutral to alkaline, the micronutrients likely to be deficient include: B, Cu, Fe, Mn, Ni and Zn (as found on calcareous soils). Apart from naturally calcareous soils, farmers in many areas will have limed their soils to raise the pH and this can have the effect of reducing the availability of the elements listed above, with Mn and Fe being particularly sensitive to liming.

1.5.2.2 Soil Organic Matter Content

Trace elements, including Cu, Mn and Zn can be bound in relatively unavailable forms in soil humic matter. This implies that crops growing on naturally organic-rich soils are likely to be deficient in these elements (Alloway, 1995a). Copper deficiency in particular is strongly correlated with peaty and organic-rich soils (Alloway, 2005). However, mineral soils with relatively high organic matter contents due to land use and/or manure spreading can also give rise to deficiencies of these elements. However, the application of readily decomposable organic material can have the effect of rendering some micronutrients more available as a result of forming water-soluble organic complexes with the elements (Barrow, 1993). Zinc deficiency problems in paddy soils have been rectified by applying manure to the soils to act in this way.

1.5.2.3 Soil Drainage Status (Redox Conditions)

As stated above for gleyed soils, poorly drained soils tend to have higher available concentrations of Fe and Mn and elements such as Co (essential for animals) that are sorbed in iron and manganese oxides. However, gleyed soils also have higher pH values than their more freely drained equivalents and this can affect available concentrations and, to a certain extent, offset the increased availability resulting from the absence of Fe and Mn oxides due to the reducing conditions. In strongly gleyed soils, such as paddy rice soils, some micronutrient elements, including Zn may be present as insoluble sulphides, which renders them virtually unavailable.

The soil properties and the associated micronutrient deficiency problems are summarized in Table 1.14.

1.6 Plant Factors Associated with Micronutrient Deficiencies

The physiological mechanisms and plant stresses involved in causing micronutrient deficiencies in crops are discussed in many of the chapters. However, Table 1.15 provides a brief list of the key factors involved.

Table 1.14 Soil types and properties commonly associated with micronutrient deficiencies

Soil/soil Properties	Micronutrient deficiency
Sandy texture and strongly leached	B, Cl, Cu, Fe, Mn, Mo, Ni, Zn[a]
High organic matter content (>10% OM)	Cu, Mn, Zn
High soil pH (>7)	B, Cu, Fe, Mn, Ni, Zn
High $CaCO_3$ content (>15%) (calcareous soils)	B, Cu, Fe, Mn, Ni, Zn[a]
Recently limed soils	B, Cu, Fe, Mn, Ni, Zn
High salt content (salt-affected soils)	Cu, Zn, Fe, Mn
Acid soils	Mo, Cu, Zn
Gleys	Zn
Heavy Clay	Cu, Mn, Zn

[a] Multi-element deficiencies can occur on sandy and calcareous soils.

Table 1.15 Plant factors associated with micronutrient deficiencies

- Plant genotype (i.e., micronutrient-efficient/inefficient cultivars)
- Nitrogen supply (effects on growth rate, dilution, elements locked up in proteins in foliage)
- Phosphate supply (effects on growth rate – dilution, e.g., Cu and metabolism, e.g., Zn)
- Moisture stress (uptake reduced in drought conditions)
- Temperature stress (high and low temperatures)
- High/low light intensity
- Rooting conditions (restrictions in rooting zone will reduce the volume of soil explored by roots)
- Mycorrhizal infection (increases the effective volume of roots)
- Secretion of root exudates (e.g., phytosiderophores)
- Pathological disease
- Agrichemicals (e.g., glyphosate-induced deficiencies of Mn, Zn, etc.)
- Antagonistic effects of other micronutrients (e.g., Cu–Zn, Fe–Cu, Fe–Mn, and Cu–Mn)
- Previous crop species – there is some evidence that the mineralisation products of some plant species can render certain micronutrients less available in the soil. An example of this is Cu deficiency in wheat following oilseed rape (canola)

1.7 Future Trends in Micronutrient Requirements for Crops

The global incidence of micronutrient deficiencies is likely to increase due to the intensification of arable farming. In many cases, this will necessitate regular applications of micronutrient fertilisers either to the soil, foliage or seed and, ideally, these should be based on regular soil or plant tissue testing. However, this is not likely to be possible in many developing countries so agricultural extension workers will need to be able to advise farmers how often they will need to repeat applications, depending on type of soil, crops and climatic zone, (as discussed by Singh in Chap. 4).

In countries or regions where there is little systematic information available on the micronutrient status of soils, there is a need for rapid surveys to be conducted in order to find out where deficiency, or even toxicity, problems may occur with changes in cropping (as discussed in Chap. 6).

Advances in plant physiological research are likely to provide a better understanding of the importance of hidden deficiencies and lead to increased use of micronutrient fertilisers to ensure optimum yields and quality, especially in high value crops (Brown, Chap. 11). Horticultural crops will probably continue to receive proportionately more micronutrients than field crops owing to their generally greater susceptibility to deficiencies and the relatively high value of their products (Brown, Chap. 11; Zou et al., Chap. 5).

In most parts of the world, especially developing countries, simple trace element compounds such as sulphate salts (e.g., $ZnSO_4.7H_2O$ and $MnSO_4.nH_2O$), will continue to be the most widely used forms of micronutrient fertilisers because they are relatively inexpensive, highly water soluble and easily applied (Chaps. 4 and 10). However, with the trend to intensification, there will be an increasing demand for micronutrient products which can be integrated into arable farming operations, such as compounds which are compatible with herbicides and fungicides in tank mixes for foliar application. Where genetically modified crops, which are resistant to glyphosate, are grown, there will be a need for additional micronutrients, especially Mn, to compensate for any deficiencies induced by this herbicide (Brown, Chap. 11).

Climate change and increasing economic pressures on water resources will affect crop production in many parts of the world, especially subtropical and semi-arid regions. Heat, drought and high light intensity stresses can all affect crop requirements for Zn and possibly other micronutrients (Moraghan and Mascagni, 1991). In China and other rice-growing countries, there will be a trend to growing aerobic and upland rice instead of paddy rice and this has important implications for Zn and possibly other deficiencies (Zou et al., Chap. 5).

As discussed in Chaps. 2 and 12, in addition to optimising crop yields, there will be increasing emphasis on fortifying food crops with micronutrients which are important in human diets, including Fe, Zn, I and Se. It is estimated that 60–80% of the world's population have diets which are deficient in Fe and about 50% deficient in Zn. However, biofortification with micronutrient fertilisers will have to be carefully planned in order to avoid the build up of excessive concentrations in soils. It may be increasingly necessary to use foliar applications to supply the necessary micronutrients without causing significant increases in the total concentrations in soils and hence avoiding possible toxic effects in plants or soil biota.

In conclusion, micronutrients are vitally important for maintaining and increasing food crop production for a growing world population. The micronutrient composition of these crops is also important in providing adequate intakes of these elements for human and animal nutrition. The following chapters in this volume identify the main micronutrient deficiency problems in different countries and continents around the world, the ways in which they are being ameliorated and the challenges posed by changes in crop production brought about by economic pressures and global climate change.

References

Alloway, B.J. (1976). Field studies on the magnesium, copper and zinc nutrition of maize. Journal of Agricultural Science, Cambridge, 86: 93–101.

Alloway, B.J., Tills, A.R. (1984) Copper deficiency in world crops. Outlook on Agriculture, 13: 32–42.

Alloway, B.J., Jewell, A.W., Murray, B.N. (1985) Pollen development in copper deficient cereals. Project 345, International Copper Research Association, New York.

Alloway, B.J. (1995a) Soil processes and the behaviour of metals. In Alloway, B.J (Ed.), Heavy Metals in Soils (2nd Edn.), Blackie, Chapman & Hall, London, pp. 11–37.

Alloway, B.J. (1995b) Origins of heavy metals in soils. In Alloway, B.J (Ed.), Heavy Metals in Soils (2nd Edn.), Blackie, Chapman & Hall, London, pp. 38–57.

Alloway, B.J. (2004) Zinc in Soils and Crop Nutrition. International Zinc Association, Brussels, 130 pp. Available at www.zinc-crops.org.

Alloway, B.J. (2005) Copper Deficient Soils in Europe. International Copper Association, New York, 129 pp.

Arnon, D.I., Stout, P.R. (1939) The essentiality of certain elements in minute quantity for plants with special reference to copper. Plant Physiology, 14: 371–375.

Asher, C.J. (1991) Beneficial elements, functional nutrients and possible new essential elements. In Mortvedt, J.J., Cox, F., Shuman, L.M., Welch, R.M. (Eds.), Micronutrients in Agriculture (2nd Edn.), Soil Science Society of America Book Series No. 4, Madison, WI, pp. 703–724.

Barker, A.V., Pilbeam, D.J. (2007) Introduction. In Barker, A.V., Pilbeam, D.J. (Eds.), Handbook of Plant Nutrition, CRC, Taylor & Francis, Boca Raton, FL, pp. 3–18.

Barrow, N.J. (1993) Mechanisms of reaction of zinc with soil and soil components. In Robson, A.D. (Ed.), Zinc in Soils and Plants, Kluwer, Dordrecht, The Netherlands, pp. 15–32.

Bernards, M., Thelen, K., Penner, D. (2002) The manganese fertilizer antagonism of glyphosate story for 2002. Available at http://www.msu.edu/varietytrials/corn/Cropping%20Systems/mngly.pdf

Brown, P.H., Welch, R.M., Cary, E.E. (1987) Nickel: A micronutrient essential for higher plants. Plant Physiology, 85: 801–803.

Brown, P.H. (2007a) Nickel. In Barker, A.V., Pilbeam, D.J. (Eds.), Handbook of Plant Nutrition, CRC, Taylor & Francis, Boca Raton, FL, pp. 395–410.

Dobermann, A., Fairhurst, T. (2000) Rice: Nutrient Disorders and Nutrient Management. Potash and Phosphate Institute, Potash Institute of Canada, International Rice Research Institute, Los Baños, Philippines, 192 pp.

Eker, S., Ozturk, L., Yazici, A., Erenoglu, B., Romheld, V., Cakmak, I. (2006) Foliar-applied Glyposate substantiallu reduced uptake and transport of iron and manganese in Sunflower (*Helianthus annuus* L.) plants. Journal of Agricultural and Food Chemistry, 54 (26): 10019–10025.

Epstein, E. (1965) Mineral metabolism. In Bonner, J., Varner, J.E. (Eds.), Plant Biochemistry, Academic, London, pp. 438–486.

FAO-UNESCO (1974) Soil Map of the World, Volume 1: Legend, UNESCO, Paris.

Follet, R.H., Murphy, L.S., Donohue, R.L. (1981) Fertilizers and Soil Amendments. Prentice-Hall, Englewood Cliffs, NJ.

Gao, X.P., Kuyper, T.M., Zou, C.Q., Zhang, F.S., Hoffland, E. (2007) Mycorrhizal responsiveness of aerobic rice genotypes is negatively correlated with their zinc uptake when nonmycorrhizal. Plant Soil, 290: 283–291.

Graham, R.D., Nambiar, E.K.S. (1981) Advances in research on copper deficiency in cereals. Australian Journal of Agricultural Research, 32: 1009–10037.

Graham, R.D. (1984) Breeding for nutritional characteristics in cereals. In Tinker, P.B., Läuchli, A. (Eds.), Advances in Plant Nutrition, Volume 1, Praeger, New York, pp. 57–102.

Graham, R.D., Rengel, Z. (1993) Genotypic variation in zinc uptake and utilization by plants. Chap. 8. In Robson, A.D. (Ed.), Zinc in Soils and Plants, Kluwer, Dordrecht, The Netherlands, pp. 107–118.

Grewal, H.S., Graham, R.D., Rengel, Z. (1996) Genotypic variation in zinc efficiency and resistance to crown rot disease (*Fusarium graminaearum* Schw. Group1) in wheat. Plant Soil, 186: 219–226.

Györi, D., Palkovics, M. (1983) The effect of borax applications on the boron uptake and yield of vines. Agrokémica és Talajtan, 32 (1–2): 88–96.

Heckman, J.R. (2007) Chlorine. In Barker, A.V., Pilbeam, D.J. (Eds.), Handbook of Plant Nutrition, CRC Press, Taylor & Francis, Boca Raton, FL, pp. 279–291.
Hotz, C., Brown K.H. (2004) Assessment of the risk of zinc deficiency in populations and options for its control. UNU Food and Nutrition Bulletin, 25(Suppl. 2): S91–S203.
Humphries, J.M., Stangoulis, J.C.R., Graham, R.D. (2007) Manganese. In Barker, A.V., Pilbeam, D.J. (Eds.), Handbook of Plant Nutrition, CRC, Taylor & Francis, Boca Raton, FL, pp. 351–374.
ISSS Working Group R.B. (1998a) World Reference Base for Soil Resources: Introduction (1st Edn.) Deckers, J.A., Nachtergaele, F.O., Spaargaren, O.C. (Eds.), ISRIC-FAO-ISSS, Acco, Leuven.
ISSS Working Group R.B. (1998b) World Reference Base for Soil Resources: Atlas, Bridges, E.M., Batjes, N.H., Nachtergaele, F.O. (Eds.), ISRIC-FAO-ISSS, Acco, Leuven.
Jewell, A.W., Murray, B.G., Alloway, B.J. (1988) Light and electron microscope studies on pollen development in barley (*Hordeum vulgare* L.) grown under copper-sufficient and deficient conditions. Plant, Cell and Environment, 11: 273–281.
Kabata-Pendias, A. (2001) Trace Elements in Soils and Plants (3rd Edn.), CRC Press, Boca Raton, FL.
Loneragan, J.F., Webb, M.J. (1993) Interactions between zinc and other nutrients affecting the growth of plants. Chap. 9. In Robson, A.D. (Ed.), Zinc in Soils and Plants, Kluwer, Dordrecht, The Netherlands, pp. 119–134.
Loué, L. (1986) Les Oligo-Elements en Agriculture. SCPA, Agri Nathan International, Paris, p. 339.
Marschner, H. (1995) Mineral Nutrition of Higher Plants (2nd Edn.), Academic, London, p. 889.
Martens, D.C., Westerman, D.T. (1991) Fertilizer applications for correcting micronutrient deficiencies, In Mortvedt, J.J., Cox, F., Shuman, L.M., Welch, R.M. (Eds.), Micronutrients in Agriculture (2nd Edn.), Soil Science Society of America Book Series No. 4, Madison, WI, pp. 549–582.
Moraghan, J.T., Mascagni, H.J. Jr. (1991) Environmental and soil factors affecting micronutrient deficiencies and toxicities. In Mortvedt, J.J., Cox, F., Shuman, L.M., Welch, R.M. (Eds.), Micronutrients in Agriculture (2nd Edn.), Soil Science Society of America Book Series No. 4, Madison, WI, pp. 371–425.
Neumann, G., Kohls, S., Landsberg, E., Stock-Oliveira Souza, K., Yamada, T., Römheld, V. (2006) Relevance of glyphosate transfer to non-target plants via the rhizosphere. Journal of Plant Diseases and Protection, 20: 963–969.
Prasad, R., Power, J.E. (1997) Soil Fertility Management for Sustainable Agriculture. CRC Lewis Publishers, Boca Raton, FL, 356 pp.
Reddy, T.Y., Reddi, G.H.S. (1997) Mineral nutrition, manures and fertilizers. In Principles of Agronomy, Kalyani Publishers, Ludhiana, India, pp. 204–256. Available at www.plantstress.com/Articles/min_deficiency_i/impact.htm.
Reith, J.M. (1968) Copper deficiency in plants and effects of copper on crops and herbage. Paper in Symposium on Copper in Farming. Copper Development Association, Potters Bar, UK.
Rerkasem, B., Jamjod, S. (2004) Boron deficiency in wheat: A review. Field Crops Research, 89: 173–186.
Scharpenseel, H.W., Eichwald, E., Hauptenthal, Ch., Neue, H.U. (1983) Zinc deficiency in a soil toposequence grown to rice at Tiaong, Quezon Province, Philippines. Catena, 10: 115–132.
Shorrocks, V.M., Alloway, B.J. (1985) Copper in plant, animal and human nutrition. Copper Development Association, TN 35, Copper Development Association, Potters Bar, UK, p. 84.
Shorrocks, V.M. (1997) The occurrence and correction of boron deficiency. Plant and Soil, 193: 121–148.
Sillanpää, M. (1982) Micronutrients and the nutrient status of soils: A global study. FAO Soils Bulletin No. 48, FAO, Rome.
Sillanpää, M. (1990) Micronutrient assessment at the country level: An international study. FAO Soils Bulletin No. 63, FAO, Rome.

Sinclair, A.H., Withers, P.J.A. (1995) Copper deficiency in UK cereal crops: Occurrence, significance and treatment. HGCA Research Review No. 31, Home Grown Cereals Authority, London.

Soil Survey Staff (1999) Soil Taxonomy: A basic system of soil classification for making and interpreting soil surveys (2nd ed.). Agricultural Handbook 436, Natural Resources Conservation, USDA, Washington DC. USA, pp. 869.

Srivistava, P.C., Gupta, U.C. (1996) Trace Elements in Crop Nutrition. Science Publishers, Lebanon, NH, 356 pp.

Tills, A. (1981) Causes and Detection of Copper Deficiency in Arable Crops. Unpublished Ph.D. thesis, University of London.

Tills, A.R., Alloway, B.J. (1981) Subclinical copper deficiency in crops on the Breckland in East Anglia. Journal of Agricultural Science, Cambridge, 97: 473–476.

United States Environmental Protection Agency (USEPA) (1993) Standards for the use or disposal of sewage sludge. Federal Register, 58: 47210–47238.

Vetter, H., Grummer, H.J., Brosch, K.I. (1985) Magnesium, manganese and copper for soil, plants and animals. Booklet 333, West German Ministry of Agriculture, AID, Oldenburg, 30 pp.

Walter, K.H. (1988) Manganese fertilizers. In Graham, R.D., Hannam, R.J., Uren, N.C. (Eds.,) Manganese in Soils and Plants, Kluwer, Dordrecht, The Netherlands.

Wild, A., Jones, L.H.P. (1988) Mineral nutrition of crop plants. In Wild, A. (Ed.), Russell's Soil Conditions and Plant Growth (11th Edn.), Longman, Harlow, UK.

Xu, G., Magen, H., Tarchitzky, J., Kafkaf, U. (2000) Advances in chloride nutrition of plants. Advances in Agronomy, 28: 97–150.

Chapter 2
Micronutrient Deficiencies in Crops and Their Global Significance

Robin D. Graham

Abstract While we now know that a spectrum of genetically controlled adaptations to soils low in available micronutrients exists in the germplasm of our major staple crops, these appear insufficient for the high yields demanded of agriculture. Thus, we have a demand for both macronutrient and micronutrient fertilisers that is growing and that must be supplied with increased efficiency via ever improving technologies. Once the macronutrient deficiencies of soils are treated, Sillanpää (1990) estimated that of the important agricultural soils of the world, 49% are deficient in zinc (Zn), 31% deficient in boron (B), 15% deficient in molybdenum (Mo), 14% deficient in copper (Cu), 10% deficient in manganese (Mn) and 3% deficient in iron (Fe). These figures may be compared with corresponding figures for the human population that depends on these same soils. In the same broad terms, it appears that more than half the human population is deficient in Fe, at least half is deficient in Zn, 25% in iodine (I) and 20% in selenium (Se). Selenium and I are not known to be required by plants and the extent of B deficiency in soils does not lead to the same high priority in human nutrition as it does for crop growth. It appears that there is more than enough Fe in food, but its bioavailability is poor. Of the micronutrients, only Zn is directly linked in the food chain such that deficiency is extensive in both humans and their food crops. Zinc deficiency is therefore the highest priority among micronutrients for agriculture to address. An agricultural solution to Zn deficiency in humans is all the more compelling because mild to moderate Zn deficiency in humans is very difficult to diagnose, so the blanket approach, justified by production gains derived through greater tolerance to a broad spectrum of stresses on the crop itself, is the surest and safest way to proceed, and is likely to yield important advances in human welfare globally.

2.1 Introduction

Since the dawn of agriculture, the land surface of planet Earth has for crop growth become widely deficient in nitrogen (N) and phosphorus (P) and to a lesser extent, potassium (K) and sulphur (S). A century ago, agriculture was focused on adding

B.J. Alloway (ed.), *Micronutrient Deficiencies in Global Crop Production.*

these nutrients to solve production problems of otherwise good soils exhausted from many hundreds of years of cultivation and of new, nutritionally marginal lands being brought into production. Rotations, green and barnyard manures and resting a field were insufficient or costly relative to the emerging use of mineral fertilisers such as Chilean saltpetre, superphosphate and potash. These mineral elements together with the other then-known essential minerals, calcium (Ca) and magnesium (Mg), were used to good effect in bringing production up to expectations, but to experienced eyes, anomalous results hinted at limitations to production as yet unknown. Agricultural science was then ready to embrace the discovery of a suite of new essential elements required for all living things in smaller amounts; they came to be known collectively as the trace elements. Later, the essential trace elements were known collectively as the 'micronutrients' (to human nutritionists this term also includes the vitamins). The micronutrients contributed greatly to 20th-century agriculture. Of course, the first of these, Fe, was already well known in the 19th century, but was always a poor fertiliser and, in any case, ubiquitous on our red planet. Serious problems of its deficiency tended to come later as the boundaries of production were stretched by attempting to breed for adaptation of crops to a wider set of more marginal and high-pH soils.

The essential microelements (micronutrients) for growth of higher plants are Fe, Mn, B, Cu, cobalt (Co), Mo, nickel (Ni), Zn and chlorine (Cl), but for animals and man, these and the additional elements, Se, I, chromium (Cr), tin (Sn), fluorine (F), lithium (Li), silicon (Si), arsenic (As) and vanadium (V) are required (Table 2.1) (Welch and Graham, 2005), leading to the question in the minds of plant nutritionists, 'how many of these additional elements will eventually be found necessary to plants as well?'. In a recent study in Xinjiang Province of China (Graham et al., 2007, in preparation), on some soils with the lowest I concentrations on earth, no response to I could be measured, even when the supply of all other essential elements was sufficient to support near-record rice yields of 14 t ha^{-1}. Therefore, the probability that I is yield-limiting in crops anywhere on earth is quite low, at least for rice. On the other hand, current work in the University of Adelaide is close to establishing that Se is essential for higher plants (Graham et al., 2005).

Ecologists argue that the sustainable population of Earth is about two billion humans, but the effect of the mass production of antibiotics during World War II has been said to have decreased death rates so much that the population exploded post-war to a current population in excess of 6 billion, with a projected 8–9 billion

Table 2.1 The known essential nutrient elements for plants, animals and humans (For the organic nutrient requirements see Table 12.1 of Chap. 12)

Nutrient group	Mineral elements
Macro elements	N, P, K, Ca, Mg, S, Na, Cl
Micro elements (plants)	B, Zn, Cu, Mn, Fe, Co (for N-fixation), Ni, Mo
Additional micro elements for animals and humans	Li, I, Se, Cr, V, Si, F, As, Sn

before the numbers stabilize and hopefully begin to decline. The contribution to production of food for such a large population made by the use of micronutrients added to compound fertilisers is undoubtedly significant but as yet far from the optimal that must be reached to achieve sustainability. This is because increases in productivity on land already in cultivation are needed to relieve the pressure on clearing more forest land. Because micronutrients are required in such small amounts, the economics of their use is generally highly favourable, as in one case of Cu that contributes to pollen fertility in wheat leading to increases in wheat production valued at $287/ha for 93 cents worth of Cu invested (Graham et al., 1987). While the economics of micronutrient use is compelling in most cases, the challenge is to get both the diagnosis and the delivery right.

Principles for use of micronutrient fertiliser were well developed in the latter part of the last century. In the same period, new information came to light that will challenge the agronomy of micronutrient use well into the new century. By 1990 it was clear that micronutrient deficiencies in humans were already major and still increasing problems, especially in developing countries, but also an enormous health cost in developed countries. The nutrients most widely deficient are Fe (affecting at least half of the human population), Zn (less well quantified, but possibly as widely deficient as is Fe), Se, I and vitamin A (an estimated one billion people affected by each). Addressing these problems has been the responsibility of the medical fraternity using their methods of food fortification and supplementation. However, the problem continues to grow, so a more sustainable approach is being advocated: a food systems strategy wherein a range of agricultural measures are combined to ensure the essential nutrients are delivered to people through the food system they depend on. These measures include biofortification by which more nutrients are packed into staple foods by selective breeding for micronutrient-dense varieties, by use of Zn, Se and I fertilisers, and by addition to the food system of new and acceptable foods richer in the missing micronutrients. A solution to this problem is urgent and world peace will depend on the equity of adequate nutrients for all, which is the goal because nutrition is fundamental to health and well-being. These in turn are fundamental to a sense of security in old age that leads to lower population growth which in turn will lead to environmental stability.

2.2 The Micronutrients

There is not yet complete agreement on what are the micronutrients under discussion, and without doubt, more remain to be established as essential either to plants or animals and humans. The trace elements are a grouping of the micronutrient minerals with other non-essential minerals of low abundance and having effects on biological systems; this chapter deals only with those trace elements known as the essential micronutrients. Table 2.1 identifies 25 mineral elements: the first eight are the essential macronutrients, required in large amounts by all living things. The second group of eight elements are widely recognised as the micronutrients of agriculture, being essential for higher plants. In addition, there are a further nine mineral elements essential for animals and humans, while two are in transition: B is only now graduating to essential status for humans and Co is recognised as an essential element for N-fixing higher plants only. However, in addition to these 25 essential mineral elements, human nutritionists recognise another 14 essential organic micronutrients, the vitamins (see Table 12.1 in Chap. 12). It is therefore important to realise in dealing with research and development at the interface of agricultural and health sciences that plant and human nutritionists refer to quite different suites of essential nutrients when they use the term, 'micronutrients'.

2.3 The Global Distribution of Micronutrient-deficient Soils

Sillanpää (1990) published a valuable assessment of the micronutrient status of Earth's soils from the specific point of view of modern agriculture. While he sampled only a small, though agriculturally representative fraction of the soils of some 15 developing countries together with some samples from Finland, the 190 field trials he conducted is still a large number considering that he carried out 'minus-one' fertiliser experiments on each of these soils, and reported the productive yields, soil and plant analyses for several crops. In addition, he earlier sampled a much larger number of soils (3,538) from 30 countries and assessed their nutrient status in the glasshouse (Sillanpää, 1982). Together these two reports represent an astonishing amount of work and leave us a rich heritage of global data that will be a benchmark for future works of such scale. Sillanpää's later contribution is widely considered to be of great value because of its thorough assessment of nutrient status for crops. In comparison, the maps of micronutrient deficiency in global soils, also useful in their own right, are based mostly on soil analysis only and this has a comparatively low predictive value for response to micronutrient fertilisers.

Table 2.2 shows the summary of Sillanpää's results. While deficiencies of NPK dominate, the figures for the micronutrients are for expression of responses to each micronutrient or combination once the macronutrient deficiencies have been corrected, as we might expect to be the case in most productive farming soils. Zinc is the most widely deficient micronutrient for crop growth, and as discussed in later sections, it may well be in humans and grazing animals as well. But there the similarity ends. Boron, the second most widely deficient micronutrient for crop production is only currently in the process of being recognised as a micronutrient for humans. While little idea of the extent of B deficiency in humans currently exists, B deficiency in crops is prominent in tropical soils of the humid zone, where a very large part of the human population lives and it is highly likely that this is important when we come to look at food systems and the diets of resource-poor peoples of the wet tropics. If Sillanpää's contribution is likely to be challenged in the future,

Table 2.2 Percentage of nutrient-deficient soils among 190 soils worldwide (Sillanpää, 1990)

Deficiency	N	P	K	B	Cu	Fe	Mn	Mo	Zn
Acute	71	55	36	10	4	0	1	3	25
Latent	14	18	19	21	10	3	9	12	24
Total	85	73	55	31	14	3	10	15	49

it may well be for an underestimate of the extent and importance of B deficiency overall and especially in the densely populated humid areas. Here, the friable, permeable, generally acidic, deep and highly leached lateritic soils (Ferralsols) easily lose soluble B of both native and fertiliser origin.

2.4 Agronomy of Micronutrient Fertiliser Use

2.4.1 Fertiliser Forms and Strategies of Use

The main challenge of fertilising with micronutrients is to deliver small amounts evenly over the target area; meeting this challenge is often a bigger cost factor than is the product itself because of the small amounts required. Micronutrients are most commonly delivered by mixing with macronutrient fertiliser destined to be used at the same time. Overall, this strategy is good but not ideal. It is widely found that dry or 'cold' mixing is inefficient as differences in particle size and/or hardness of the macro- and micronutrient components lead to segregation during handling and transport, so the distribution is not even, often seriously decreasing the overall response. Mixing during manufacture, 'hot' mixes, are better from this point of view but, in some cases, availability may be lower because of chemical reactions between components, particularly phosphorus and the micronutrients. In more recent approaches, finely powdered micronutrients are coated on the surface of macronutrient fertilisers with a sticker. With this technology, compound fertilisers can be stored separately and mixtures created on demand to client specifications. Acidifying N fertilisers tend to create transient acidic micro-sites around the granule, helping to keep micronutrient cations soluble for longer in alkaline soils where their deficiency is most common. On the other hand, high concentrations of P and micronutrients together with Ca in alkaline soils, may cause precipitates (Lombi et al., 2005) that outweigh the acidifying effect of the N in such soils.

Fluid fertilisers have advantages in these alkaline soils by keeping concentrations lower than in the case of dissolution of granules by soil moisture levels attendant on sowing. However, considerable technology and machinery are needed for effective use of fluid fertilisers, and cost effectiveness depends on this infrastructure. An effective alternative to compound mixtures, whether fluid or granular, is spraying the micronutrient on to the soil surface during preparations for sowing (ideally before cultivation). Larger amounts can be delivered to the soil surface than can be applied directly to growing plants owing to the phytotoxicity of most micronutrient fertilisers. Therefore, depending on nutrient and soil properties, repeat applications may not be necessary for several seasons. Applications of micronutrients directly to soil and subsequent mixing through the soil during cultivation, gives better distribution of micronutrients to root surfaces than do granular products. This improved distribution can be worth the extra cost of a separate application, because micronutrient cations such as Zn and Cu are less effective in the first season when delivered with N and P fertilisers in bands, or broadcast as solids (Brennan, 1994). In the latter case, root contact is better in the second season owing to further mixing from pre-seeding cultivation, and in effect the residual value can be >100%. On the other hand, foliar applications to the leaves of the crop itself after establishment also has advantages in particular circumstances, notably when unexpected symptoms appear in the crop. Moreover, some micronutrients are compatible with biocides and it is possible to apply the former in solution with certain herbicidal, fungal or insecticide sprays, thus avoiding separate application costs.

Iron and Mn are two nutrients that are soluble in the reduced state but readily oxidised in high pH soils, and so rendered highly unavailable in such soils. Generally, they are both poor fertilisers and often are best sprayed on leaves in chelated form. In the case of Mn, soluble $MnSO_4$ can be coated on the seed with considerable effect provided the seed is sown soon after coating to avoid toxicity to the seed itself, and the soil is reasonably moist at sowing (McEvoy et al., 1988). Microbial activity induced by residual carbohydrates from the endosperm of the seeds produces organic acids that seem to help to keep seed-coated Mn soluble. Other micronutrients have been successfully applied via coatings on seed.

Boron on the other hand is too soluble for efficient use as a fertiliser based on the commonly available forms of borax and boric acid, being readily leached from the soil in high rainfall conditions. This is the case in many of the extensive band of lateritic soils in the wet tropics, and its deficiency seriously limits production, especially as it is involved in pollen fertility and pollen-tube growth, both of which affect seed yield directly. A more effective fertiliser is needed, as discussed in the previous section.

2.4.2 *Interactions*

Like the macronutrients, micronutrient fertilisers are subject to many and profound interactions that both complicate the diagnosis of nutritional problems and their correction. However, these interactions can greatly increase the rewards for persistence. It is fundamental that two or more fertiliser nutrients will interact with each other (and also with components of the environment) to affect yield and quality, where deficiencies of both or all nutrients exist in the soil. Where two nutrients are deficient for plant growth, supplying both together will enhance yield to an extent much greater than the sum of the independent effects of each (synergy). In practice, however, where two nutrients such as Cu and Zn are deficient, as is common in South Australia, supply of both or of the one in relatively shorter supply will increase yield, often greatly relative to cost. However, to add only the one less deficient will give no benefit and may decrease yield below what it would have been if nothing were done. Similarly, when only one micronutrient cation is deficient, the effect of supply of a second is mainly one of antagonism when the abundance in the soil of the second element is much better. It follows from these two cases that for maximal growth rate, all nutrients must be supplied at near optimal levels, or yield will be lost, and the loss will accelerate if any other nutrient approaches high concentrations. These examples underline the risks to producers of complex interactions, and dictate that micronutrient fertilisation must be introduced with the help of experienced agronomists and backup from plant analysis.

Nutrients interact with other factors in the environment in somewhat similar ways. Any stress factor is likely to interact with a nutritional deficiency such that the crop may tolerate one stress to a considerable degree but when two stresses occur together, yield will collapse. This is common experience in South Australia. Thus, a crop may yield well in spite of mild Mn deficiency or a moderate burden of disease that would cause serious yield loss to a nutrient-deficient crop under the same conditions (e.g., see Wilhelm et al., 1985; Graham and Webb, 1991. Other interactive stresses are drought, herbicide damage (Dodds et al., 2002), poor subsoil nutrient status, salinity, direct drilling, topsoil drying, and seasonal differences. With so many interacting factors, the underlying deficiency problems are often misdiagnosed, and it is important to get the diagnosis right as many farmers lose interest after just one failure.

One of the most challenging interactions is with season. If the effort made by the farmer and his/her advisor is started in a season likely to yield a large dividend the farmer's interest is assured, whereas if in the first season the benefit is at the low end of the possible spectrum of season-dependent yield increases, the producer may be discouraged and not persist with the effort.

Other important interactions of micronutrients include those with variety of the crop. Generally, best yields are achieved in chronically micronutrient-deficient soils by the use of micronutrient-efficient crops together with the micronutrient in question as fertiliser, what is called the 'belt and braces approach' (Graham, 1984), owing to deficiencies in the subsoil that is difficult to fertilise but which can also limit yield (see also Chap. 3).

2.4.3 Diagnosis

In South Australia, soil analysis is considered not to be precise enough for diagnosis of micronutrient deficiencies except in the extreme, although it is often used for fertiliser recommendations because it can be done before a crop is sown. Conclusive diagnosis of micronutrient deficiencies generally requires plant analysis. This is because symptoms, though useful, are somewhat variable and dependent on a number of factors: symptoms vary with variety and species, age, rate of growth, cold weather, the presence of infectious disease, the levels of other nutrients, selective herbicide use, and many others in particular cases. Plant analysis for micronutrient deficiencies is generally based on plant tissue samples from younger plant parts that are reasonably free of contamination, including that from soil, metal utensils and dust, and a competent laboratory is vital.

2.5 The Global Distribution of Micronutrient-deficiencies in Human Populations

2.5.1 Iron Deficiency in Humans

The human population is astonishingly Fe-deficient (Fig. 2.1). The World Health Organisation (WHO) on its web site (WHO, 2005) estimates the global incidence of low Fe status to be between four and five billion, 60–80% of all people. Not all of this is due to Fe deficiency in the diet (see later): however, it appears that about half of the total problem is dietary in origin. The prevalence of Fe deficiency in humans stands in stark contrast to the frequency of Fe deficiency in the crops that we all depend on for staple foods, which at 3% (Table 2.1) is quite low, as we might expect of a planet with a lithosphere strongly pigmented with the colours of Fe oxides and hydroxides. Our crop plants have evolved extensively researched and publicised mechanisms for mobilising Fe from forms in the soil made extremely insoluble by atmospheric oxygen that has accumulated in our atmosphere since plants themselves evolved photosynthetic capability. Indeed, the mechanisms of Fe mobilisation via synthesis and excretion into soil of specific ferric Fe-binding siderophores by bacteria (hydroxamate group) and by the grass and cereal family (mugeneic acid group) together provide the main portals for Fe from the regosol into the biosphere. But this appears not enough: we face the paradox of an Fe-rich planet, a lithosphere, regosphere and biosphere rich in Fe, yet WHO figures suggest the majority of the human population is deficient and the problem has been getting worse.

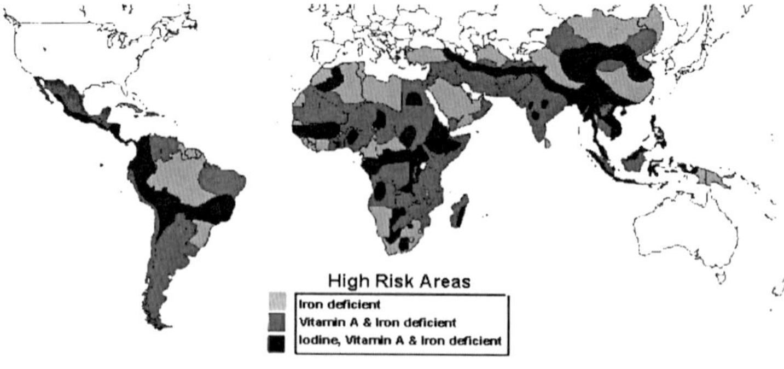

Fig. 2.1 The extent and distribution of iron, iodine and vitamin A deficiencies in the human population (Sanghvi, 1996)

Iron deficiency is most severe and widespread among growing children and premenopausal women, as adult males until old age are reasonably resistant to anaemia despite poor diets in resource-poor countries. Thus most women of child bearing age and children are debilitated to some degree in both physical and mental work capacity. In severe cases, this results in morbidity, complications in childbirth and mortality for both mothers and children. Iron deficiency is an epidemic in humans that exists in spite of relatively few problems in crop plants. An exception is deficiency of Fe in the soybean crop in the USA that has resulted from expanding the crop from its US origins on the generally acidic-soil eastern seaboard westward to the low rainfall high-pH soils of the toe-slopes of the Rocky Mountains. Remarkable progress has, however, been made in breeding for this extreme change in adaptation, utilising some 20 minor genes for increased Fe uptake and translocation efficiency in the soybean (Fehr, 1982).

Iron deficiency in humans is severe in the lateritic soil areas of the wet tropics where Fe deficiency in crops is rare, and if anything, it is Fe toxicity that is better known, especially in rice leaves. There is another apparent paradox to note here. Iron deficiency in grazing animals is rare. In ruminants, a little soil may be ingested with the pasture and its insoluble Fe, mobilised by the microbial mass of the rumen, becomes thus more available from these bacterial cells during catabolism in lower parts of the gut. In common with humans, the monogastric domesticated animals, chickens and pigs, are susceptible to Fe deficiency, especially in very early life. Piglets have been bred with such high early growth rates that successful production is not possible without supplements of Fe in the first days of life.

For humans, at least five times our needs of Fe are ingested daily in cereal products (that dominate the diets of resource-poor populations), but the bioavailability of that Fe is low. The reason for the low bioavailability of cereal Fe, largely in the form of soluble monoferric phytate, is thought to be the precipitation by dietary Ca of complex phytates and other insoluble forms in the small intestine, making it highly unavailable (see Chap. 12). As little as 1% of ingested cereal-Fe has been measured by isotopic methods as absorbed and utilised. Haem and ferritin forms of Fe are more readily absorbed by humans from both animal and plant sources. In the 'HarvestPlus' Program, that aims to increase the nutritive value of common staple foods to reduce the prevalence of nutritional Fe-deficiency in the world by a food-based system, increasing Fe in cereals by selecting Fe-dense genotypes is the main strategy. The efficacy of this strategy has yet to be established. However, it appears more logical to increase the absorbability (bioavailability) of the non-haem Fe in staples than to increase further its concentration, as bioavailability promoters are likely to be more easily enhanced genetically than Fe itself, and may also enhance the absorption of Zn and other minerals.

There are many non-nutritional causes of Fe deficiency in humans. Malaria, hookworm and thalassemia are examples. In spite of these causes, WHO still estimates that poor diet is the primary cause of Fe deficiency. It is hard to grasp the notion that the human species has eaten grains and seeds for so long and has not evolved suitable mechanisms to utilise this Fe, with the consequence of physical morbidity and intellectual impairment. It seems that another explanation is possible.

What else could cause Fe deficiency? Both Se and I deficiencies are common, with a billion people or more at risk by virtue of living in regions where soils are low in either or both of these elements. Both I and Se have been reported to interact in metabolism in ways that could lead to low Fe status in I and Se-deficient areas. Another nutritional Fe deficiency could be due to Zn deficiency. Older texts of human nutrition identify Fe deficiency anaemia as a symptom of Zn deficiency (Prasad et al., 1963). While studies in humans that gave supporting results have been deemed of poor design, this does not disprove the notion, and studies with animal models including chimpanzees have, under more controlled conditions, supported the hypothesis of Zn deficiency as a potential cause of Fe deficiency anaemia. Recent studies suggest that improved dietary Zn facilitates the synthesis of the divalent metal transporter DMT1 in the mucosal cell membrane (Yamaji et al., 2001), and that the effect may be mutual and synergistic when both nutrients are deficient in the diet. If this were a widespread phenomenon, it could explain much of the current extent of anaemia and nutritional Fe deficiency, and the failure of the gut to absorb the Fe ingested. Additionally, vitamin A deficiency, also widespread in humans (see Fig. 2.1), can also aggravate both Fe and Zn deficiencies, and supplying small amounts of these nutrients together can have larger effects than expected from single element treatments (Thurlow et al., 2005). Carotenoid pigments have been deliberately bred out of wheat and other staples during the 20th century. Similarly, vitamin C can strongly promote absorption of non-haem Fe in the gut. Finally, vitamin B12 deficiency can cause anaemia (known as megaloblastic anaemia), and although there are no extensive maps of Cobalt-deficient soils (vitamin B12 contains cobalt), the known extent of deficiency of vitamin B12 is increasing as more testing of B12 status is conducted (R.M. Welch, 2007, personal communication). The extent of these deficiencies of I, Se, Zn, vitamin A and vitamin B12 may be sufficient to explain at least some of the nutritional anaemia quantified by WHO.

2.5.2 *Zinc Deficiency and Its Importance*

The agricultural perspective on Zn is much clearer than is the human nutritional perspective. Zinc is a fertiliser that is remarkably effective, yet half of the world's soils are intrinsically deficient. As mentioned earlier, Zn deficiency occurs in all the major cropping environments and major soil types. Copper, Fe, Mo, chlorine (Cl) and Mn have more than one oxidation state, and so are easily manipulated by biological systems to reduce and release soluble ions of the elements even in the presence of a generally unfavourable pH. On the other hand, Zn, Ni, Co and B rely on coordination chemistry for changes in solubility, movement through soil and plant. Therefore, these elements tend to function in stable systems such as structural molecules like DNA and metabolic and regulatory enzymes. Zinc has been identified to bind with 925 proteins in humans and over 500 proteins in plants (Table 2.3), ten times more than does Fe (while the concentration of Zn in the lithosphere and soil

Table 2.3 Metal-containing and metal-binding selenoproteins in two species identified by proteomic techniques (Gladyshev et al., 2004)

Genome	Total proteins	Zn	Cu	Mg	Fe	Ca	Ni	Co	Mo
Homo sapiens	25,319	925	31	74	86	59	0	4	6
Arabidopsis thaliana	27,243	536	19	51	81	14	1	4	6

on average is perhaps 100 times less than that of Fe). It is not surprising therefore that the occurrence of Zn deficiency is widespread and, in both plants and humans, causes a wide range of symptoms, depending on allelic variation in genes controlling hundreds of enzymes. As such, Zn participates in almost all processes and pathways in living organisms. It can be deemed the most important metabolic promoter among the 51 essential nutrients (see Chap. 12). Because Zn interacts with such a vast number of proteins, symptoms of Zn deficiency in humans may be many and indiscriminate, and consequently many disease states are not associated with its deficiency when they should be, and in these respects, it is not surprising that Zn deficiency is exceedingly difficult to diagnose in humans and animals. Zinc deficiency is the ultimate hidden hunger.

A more important point still, Zn is a 'type 2' nutrient element, that is, its concentration does not noticeably decline in the blood stream as severity of deficiency increases, as is the case with Fe, a 'type 1' nutrient. Rather, growth is decreased and excretion of Zn reduced to conserve Zn and then Zn is mobilised from the exchangeable pool and levels in the blood and tissues fall. Stunting is a characteristic of Zn deficiency in children. Thus, unless an individual has been monitored for height and weight over several months, there is no quick and easy diagnosis as there is for Fe deficiency, but a survey by Brown et al. (1999) of Zn in global diets resulted in a map of putative Zn deficiency (dietary Zn deficiency), and these authors came up with an estimate of 2.6–3 billion people at risk. A map from this seminal work (Fig. 2.2) shows a distribution of low Zn diets that is heavily concentrated on South Asia, South-East Asia and Africa, and while not perfectly, it does clearly mimic the map of Zn-deficient soils (Fig. 2.3 in Colour Section). Until the release of the map of Zn-deficient human diets, Zn deficiency was low on the WHO list of important nutritional problems and this may be a reason why Zn deficiency has not been identified as a potential cause for some of the nutritional anaemia.

2.5.3 *Selenium and Iodine*

Recent unpublished studies in Xinjiang Province of China (Graham et al., 2007, in preparation), have attempted to prove that I could increase rice production in soils of very low I status. Iodine deficiency in humans and livestock is endemic and severe in the Taklamakan Desert. In spite of 14 t ha^{-1} yields, excellent for rice anywhere in the world, no response in yield to I fertiliser could be demonstrated.

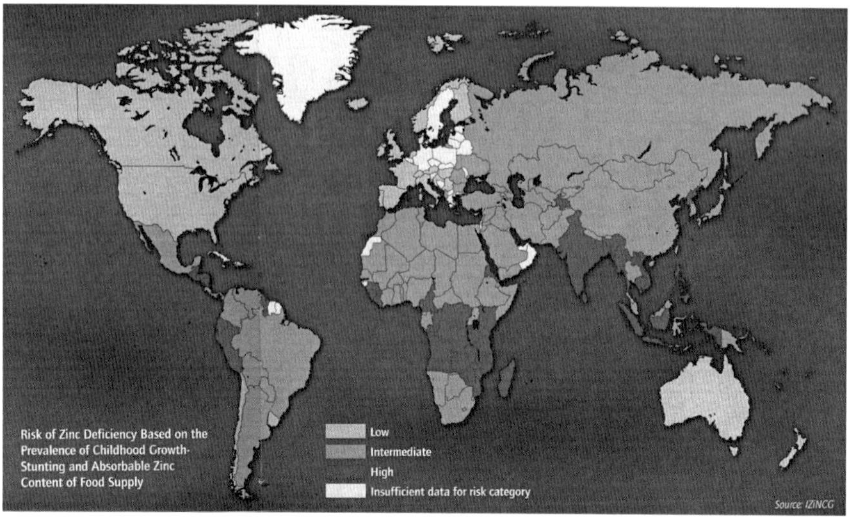

Fig. 2.2 Global distribution of diets low in zinc (Hotz and Brown, 2004)

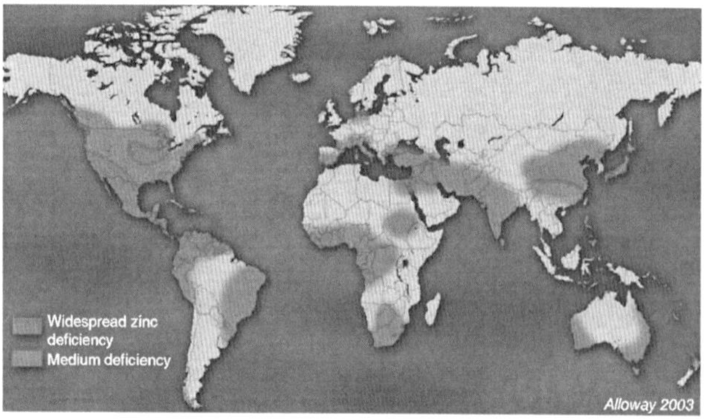

Fig. 2.3 Major areas of reported zinc deficiency in world crops (Alloway, 2004) (*See Colour Plates*)

At these near-record yields, obviously, all known limiting factors to production were satisfied, so any putative requirement for I should have been demonstrable here, if anywhere on earth. This could not be shown and we conclude that it is unlikely that rice at least, and other higher plants by implication, will respond in yield to I fertiliser anywhere. Soil I status tends to increase from high continental deserts towards the sea in every direction, and the Taklamakan Desert is further from the sea than anywhere and, as expected, its soils are exceptionally low in I. However, what did respond was the iodine concentration of both foliage and grain. Grain I concentration was approximately doubled by an application of I of several kg ha^{-1}. This is deemed from previous work in the region (Cao et al., 1994) to be sufficient to alleviate most symptoms of I deficiency in humans, including cretin-

Table 2.4 Effect of iodinating irrigation water in Xinjiang province of China. There were benefits also in animal production (Cao et al., 1994)

System component	I_2 level before mg kg^{-1} (SE)	I_2 level after mg kg^{-1} (SE)
Iodine in water	1.3	7.5
Iodine in soil	7.0	23.3
Iodine in wheat	32	65
Iodine in meat, eggs, milk	33	103
Iodine in urine (2–6 years old)	18	49

ism (Table 2.4). The use of I in the irrigation water, while eliminating symptoms in the human population, was also claimed to increase animal production (Cao et al., 1994). There is a case for use of I fertiliser to address human I deficiency wherever it is widespread, but also to treat susceptible grazing animals, as is currently done in the UK and New Zealand, for example (Morton et al., 1999). In Xinjiang Province, China, I was added to irrigation canals from which, in this desert, all production is derived. Elsewhere, where significant rainfall occurs, I could be added to urea by legislation, and all rice or wheat could be so treated to use agriculture as a public health intervention. The returns on this investment are expected to be very large. In an original I trial in the Himalayas (Hetzel, 1989), where the I was administered to a village by subcutaneous injection, the benefits to the population as a whole were so large that in addition to the elimination of cretinism and goitre, overall vigour was improved to such an extent that novel agricultural practices were adopted in the treated village. The incidence of I deficiency symptoms is increasing in many countries in which iodised salt programs were successful in the past and then were wound back.

Selenium deficiency is well known in many countries, being like I and B, a deficiency strongly promoted by leaching. Over one billion people are at risk of frank Se deficiency that affects thyroid function and muscle development, white muscle disease being common in animals grazing low Se pastures. Moreover, in the USA, a map of soil Se (Kubota and Allaway, 1972) was recognised by an oncologist to be similar in an inverse sense to a map of cancer rates. This led to a series of large feeding trials over 10 years that demonstrate that super-nutritional intakes of Se significantly decreased risk-rates of certain cancers in humans and experimental animals. Moreover, two recent studies have found that avirulent retro viruses can become much more virulent when they pass through a Se-deficient animal, underlining a vital role of Se in immune function. People infected with HIV–AIDS appear to have low Se status (Lyons et al., 2004a). Selenium fertilisers can markedly increase Se levels in plants, including grains (Lyons et al., 2004b). It is quite practical for farmers to produce high-Se wheat to sell at a premium that already exists on the Chicago grains market (Welch and Graham, 2004). Because of a role of Se enzymes in thyroid metabolism, I treatment of I deficiency is not effective where Se is also low and dual treatment is essential for a positive health outcome; this is achievable with a fertiliser strategy. Because the amounts of these two ultra-micronutrients are so small, it will be practicable for governments wishing to deal with these debilitating nutritional diseases on a population basis to spread the required amounts to target areas by transport aircraft. Toxicities due to over-application are unlikely in such a strategy properly researched and developed because the element must first pass through a plant and/or animal before being consumed by humans. In the case of these two widely deficient elements, as opposed to Zn, there is no conclusive benefit to the producer in yield and so no incentive for the farmer to use these fertilisers if supplying an unsophisticated market in a developing country where there is no price premium. Hence, a strategy founded on government policy and legislation to add these two elements to macronutrient fertilisers, as has been done in Finland for Se, should be considered where the use of urea or other macronutrient fertiliser is reasonably extensive.

2.6 Fertiliser Strategies for Enhancing Nutrient Density

2.6.1 The Problem

The use of micronutrients in agriculture has lagged well behind the science. To begin with, analysis of micronutrients in soils and plants is technically difficult, and it was not until the widespread use of inductively coupled plasma optical emission spectrometry (ICP-OES) that the availability and cost became satisfactory. Even then, diagnosis by soil analysis for most micronutrients is rarely calibrated to the specific soil in question sufficiently well to be useful for more than cases of severe deficiency. On the other hand, a careful diagnosis by plant analysis often can take the first part of a growing season and that does not suit fertiliser salesmen nor farmers. Misdiagnosis and incomplete diagnosis, on the other hand, resulting in poor returns on the first investment in micronutrients by the farmer, are likely to lead to retreat and acceptance of the status quo.

2.6.2 The Need for New Strategies for Subsistence and Other Small Areas

2.6.2.1 Farmers

New fertiliser strategies are needed for deploying micronutrient fertilisers in developing world contexts for several reasons. Firstly, they are not being used sufficiently to address primary deficiencies that currently exist and are widespread. Additionally, new and more efficient micronutrient fertiliser products and agronomic backup are needed to ensure more economic use and then there is the new imperative to address micronutrient deficiencies in the farmers themselves and other consumers they may supply. It is imperative that new strategies be worked out for extending micronutrient fertilisation technology and agronomy to such farmers who work at least half of the world's productive land.

The complexities outlined in Sects. 2.4.1, 2.4.2 and 2.4.3 in large part explain the current poor use of micronutrient fertilisers compared to that of N & P, especially by resource-poor farmers on small land holdings. Nevertheless, it remains true that such farmers will not get good value for money spent on NPK fertilisers if their crops are deficient in any essential micronutrient. As an example, in the highlands of northern Vietnam, plant tissue sampling revealed deficiencies of B, Ca and Zn in crops that farmers were unaware of, and this is the likely reason they were not using NPK fertilisers although they knew of their availability. Currently, products are not optimised for these food systems.

The widespread deficiency of B in soils of tropical and subtropical lands with high population density underlines a priority for new boron fertiliser products with better agronomic performance. This may mean slow-release products combined in some cases with macronutrients and with lower risk of under and over supply during the growing season.

2.6.3 Seed Nutrient Content

A micronutrient strategy, not dealt with above, is that of improving the micronutrient content of the seeds themselves, an important factor in production as well as quality. Indeed, high nutrient content is one reason for the advantage of certified seed over farmers' seed when it is produced by seed companies on better soils with better management than is given to the subsequent farmers' crops. Plant breeders can select for higher micronutrient content of seeds but greater enhancement in most cases can be achieved by fertilisers, both soil-applied or sprayed on the flowers, seedpods or ears, one to three times during seed development. Nutrient concentrations can be increased greatly, from less than double for Zn in rice to a hundred times in the case of Se in wheat (Lyons et al., 2004a; Fig. 2.4).

The increments in content of critical micronutrients can materially increase the vigour, disease resistance and grain yield of the subsequent crop on soils deficient in the treatment nutrient. In Bangladesh, in comparison to farmers' seed, yields in responsive soils over 4 years averaged 24% higher in wheat-growing soils by using seeds enhanced by foliar sprays on the mother plants (Duxbury et al., 2005). Studies of the genetics of seed nutrient loading traits suggest several to many genes involved, so the genetic approach, though it has much potential, is not easy. On the other hand, Fe salts in particular are very poor fertilisers even when foliar-applied, so the breeding strategy is a more viable option to enhance Fe levels. In contrast, Zn in seeds is easily enhanced: Fig. 2.5 shows the Zn content of wheat tripled by post-anthesis foliar application of zinc sulphate solution (Genc et al., 1999).

While the objective of the farmer in increasing seed nutrient content is a yield benefit, nutrient-dense seeds also have value in animal and human nutrition that until recently has been overlooked, but increasingly, may be more important economically than the extra yield, especially to subsistence farmers and their families, and to their animal production. In Xinjiang Province of China, adding I to

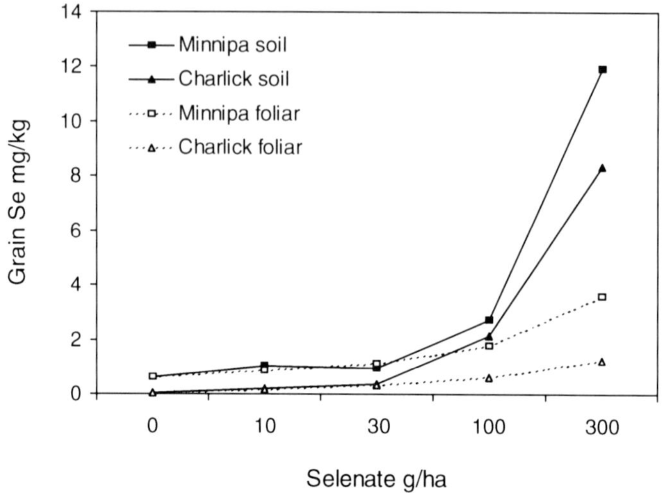

Fig. 2.4 The increase in selenium concentration in wheat grains by application of selenate to the soil before sowing (Lyons et al., 2004)

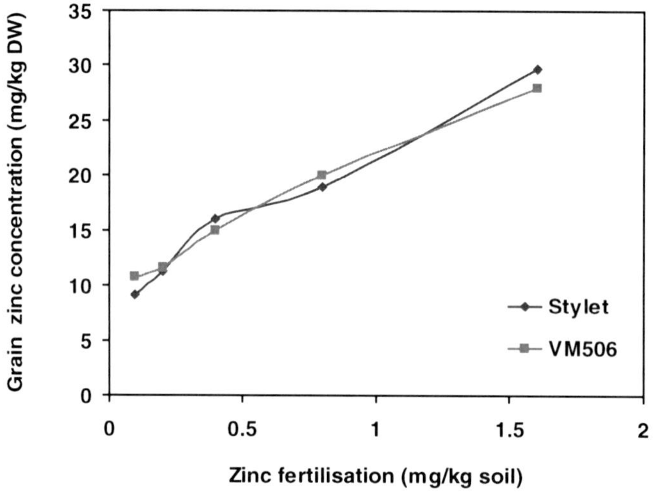

Fig. 2.5 The increase in grain zinc concentration by application of zinc sulphate to the soil before sowing (Genc et al., 1999)

the irrigation water was aimed at eliminating I deficiency disorders in the human population, but it also increased animal production to an extent estimated to cover the cost of the whole program so the spectacular health and subsequent economic benefits to human health are free and can drive a major lift in the economy. In I studies by medical intervention in the 1960s, Hetzel observed a trend in the treated village to adopt new agricultural technologies faster than in the control village (Hetzel, 1989).

2.6.4 *The Need for More Advances in Zinc Fertiliser Technologies*

As outlined by Alloway (2004), there are however several reasons to justify greater use of Zn fertilisers. Not only the yield advantage as demonstrated in well-conducted classical fertiliser trials, but Zn can enhance yields in less than favourable conditions, including those of subclinical disease burdens, water stress, salinity, excess radiation, presence of toxicities and damage from selective herbicides. However, the primary purpose of agriculture is the production of food for the human population of the planet, and although the yield advantage is important, it is probably more important that Zn fertilisation has been shown to increase the concentration of Zn in all edible parts, notably in grains. If, as argued earlier, Zn may also enhance the absorption of iron from plant food sources where both are limiting to humans, then it follows that addressing zinc deficiencies in all countries is the issue of paramount importance in micronutrient nutrition. It could be further argued that addressing complex micronutrient deficiencies, notably of Zn and B, is an important component of efforts to establish an economic, N- and P-responsive, productive food system. This would reward farmers sufficiently to stabilise their agricultural system at a higher level of productivity, allowing a move away from shifting cultivation.

An important situation where Zn fertiliser is not fully effective is where subsoils are severely deficient, and in these situations, Zn-efficient cultivars derived through plant breeding efforts contribute to significant yield increases, a situation true for other relatively immobile nutrients as well. Another concern is that the weakest effect of Zn fertiliser on grain Zn concentration is in wet (paddy) rice. Here it seems that the ferric hydroxide plaque formed on the roots of rice plants under saturated conditions binds Zn (Kirk and Bajita, 1995), and limits the increase to as little as 20%, where in wheat it may be 200% or more. Breeding will also be important in producing high-Zn rice.

2.6.5 *Vitamin A Deficiency and Its Significance*

Vitamin A is widely deficient in humans (Fig. 2.1) but is not a nutrient for plants as they can synthesise for themselves the carotenes that the animal body converts into vitamin A. Its importance here is that its deficiency can cause anaemia in humans, and solving the problem of vitamin A is important to eliminating anaemia. As carotenes are not nutrients for plants, there is no fertiliser strategy, and new foods must be added to vitamin A-deficient food systems or existing staples enriched with pro-vitamin A carotenes by plant breeding. These strategies combined with a Zn strategy thereby address not only the vitamin A deficiency problem in humans but may also address more effectively the Fe deficiency in humans than any Fe fertiliser is likely to do.

To sum up, increasing carotenoids in staples by plant breeding and increasing Zn in diets by fertiliser use (and breeding where appropriate) is advocated. High seed Zn, while vital to human consumers (and feedlot animals), is also vital to the next generation of the crop as high-Zn seeds are more vigorous and resistant of other stresses in the seedling environment. This means a need for outstanding innovations in delivering Zn to the food chain across a wide range of environments, farm sizes and farming systems. Finally, while the emphasis has been placed on Zn fertilizers among the micronutrients for both agronomic and human health reasons, in both cases, nutrient balance is the continuing challenge, and an emphasis on Zn must not be pursued to the point of imbalance with other nutrients.

2.7 Food Systems Strategies

The HarvestPlus Global Challenge Program, seeks to improve diets in all countries, especially for resource-poor populations by using staple food crops as a vehicle for delivering more micronutrients (principally Fe, Zn, Se, I and vitamin A) through plant breeding. This is a huge task and the challenge is to minimise the number of genes involved. One way to help meet the challenge is supplemental use of fertilisers where they have a comparative advantage. So far, we have seen little prospect of breeding for high Se or I, so fertiliser strategies are appropriate for these, and for Zn in a number of situations for which a case has already been made.

To combine effectively the HarvestPlus strategy with the resources of the fertiliser industry, we need to work within individual food systems that support the bulk of the populations at risk of micronutrient deficiencies. It is clear that fertiliser strategies will not solve Fe or vitamin A deficiencies. These can be solved by breeding more Fe-dense staples, a primary HarvestPlus strategy, but also by use of more Zn, I and Se fertilizers where the soils of the food system are deficient in them. Vitamin A must be addressed by breeding, biotechnology (such as golden rice) or by introduction into the food system of an additional food crop rich in pro-vitamin A carotenoids. Often, where a food system is struggling to meet basic expectations for calories to avoid starvation, an additional food requires that land be allocated for it and this in turn requires greater productivity on existing land. Therefore, emphasis on macronutrients must be considered an important component of an holistic approach to functional food systems. Other nutrients that are likely to be limiting for humans and that may require additional crops are vitamin C (vegetables, cassava, and potatoes) and Ca (vegetables and pulses), and a stable, economic food system must be capable of including the preferred crops and providing at the same time sufficient calories, and be both economic and socially acceptable.

Integrating all this requires successful deployment of expertise in several disciplines, and includes agronomy, fertiliser and nutritional expertise.

2.8 Conclusions

Currently, products and technologies are not optimised for the market of the small landholder in resource-poor countries. Optimal fertiliser efficiency for these food systems will probably be achieved by that part of the private sector that is willing to bankroll competent agronomists with the necessary experience and the supporting infrastructure and with new, user-friendly products that can solve the problems of delivery, application and distribution. Such enterprise will share with these farmers the benefits of increased productivity. Known technologies will likely include and improve on coating micronutrients onto the seeds themselves, coating onto urea and other macronutrient fertilisers, fluid fertilisers, including foliar sprays, and slow release products containing B, S and N. They will also provide access to competent analytical services and help with interpretation and product selection, and additionally, in many cases, arrange access to credit. In the right environments, the increase in use of and benefit from macronutrient fertilisers may underpin the economic driver for these efforts in achieving balanced crop nutrition through micronutrients. The economic drivers are, however, not just higher yield, but increased health and vigour of the human productive community itself. In such small-scale systems, however, there is no potential for exploitation so the private sector must enter such markets for the long haul.

References

Alloway, B.J. (2004). Zinc in Soils and Plant Nutrition. International Zinc Association, Brussels, 116 pp. Available at www.zincworld.org.

Brennan, R.F. (1994). The residual effectiveness of previously applied copper fertiliser for grain yield of wheat grown on soils of south-west Australia. Fert. Res., 39: 11–18.

Cao, X.Y., Jiang, X., Amette, K., Rakeman, M.A., Dou, Z., O'Donnell, K., Zhang, M., Ma, T., DeLong, N., DeLong, G.R. (1994). Iodination of irrigation water as a method of supplying iodine to a severely iodine deficient population in Xinjiang Province, China. Lancet, 344: 107–110.

Dodds, D.M., Hickman, M.V., Huber, D.M. (2002). Micronutrient uptake by isogenic glyphosate tolerant and normal corn. Proc. Weed Sci. Soc. Amer., 42: 2.

Duxbury, J.M., Bodruzzaman, Md., Johnson, S., Mayer, A.B.M., Lauren, J.G., Meisner, C.A., Welch, R.M. (2005). Impacts of increased mineral micronutrient content of rice and wheat seed/grain on crop productivity and human nutrition in Bangladesh. In Li, C.J., Zhang, F.S., Doberman, A. (Eds.), Plant Nutrition for Food Security, Human Health and Environmental Protection. Tsinghua University Press, Beijing, China, pp. 30–31.

Fehr, W.R. (1982). Control of iron deficiency chlorosis in soybeans by plant breeding. J. Plant Nutr., 5: 611–621.

Genc, Y., McDonald, G.K., Graham, R.D. (1999). Effect of seed zinc content on early growth of barley (*Hordeum vulgare* L.) under low and adequate soil zinc supply. Aust. J. Agric. Sci., 51: 37–46.

Genc, Y., Humphries, J.M., Lyons, G.H., Graham, R.D. (2005). Exploiting genotypic variation in plant nutrient accumulation to alleviate micronutrient deficiency in populations. J. Trace Elem. Med. Biol., 18: 319–324.
Gladyshev, V.N., Kryukov, G.V., Fomenko, D.E., Hatfield, D.L. (2004). Identification of trace-element containing proteins in genomic databases. Annu. Rev. Nutr., 24: 579–596.
Gopalan, C. (1999). The changing epidemiology of malnutrition in a developing society. The effect of unforeseen factors. Bull., Nutr. Found. India, 20: 1–5.
Graham, R.D. (1984). Breeding for nutritional characteristics in cereals. Adv. Plant Nutr., 1: 57–102.
Graham, R.D., Webb, M.J. (1991). Micronutrients and resistance and tolerance to disease. Chapter 10. In Mortvedt, J.J., Cox, F.R., Shuman, L.M., Welch, R.M. (Eds.), Micronutrients in Agriculture (2nd edn.). Soil Science Society of America, Madison, WI, pp. 329–370.
Graham, R.D., Ascher, J.S., Ellis, P.A.E., Shepherd, K.W. (1987). Transfer to wheat of the copper efficiency factor carried on rye chromosome arm 5RL. Plant Soil, 99: 107–114.
Graham, R.D., Stangoulis, J.C.R., Genc, Y., Lyons, G.H. (2005). Selenium increases growth and fertility in higher plants. Extended abstract. In Proceedings, 15th International Plant Nutrition Colloquium, Beijing, China, September 2005.
Hetzel, B.S. (1989). The Story of Iodine Deficiency. Oxford Medical Publications, New York.
Hotz, C., Brown, K.H. (Eds.) (2004). Assessment of the risk of zinc deficiency in populations and options for its control. Food Nutr. Bull., 25: S91–S204.
Kirk, G.J.D., Bajita, J.B. (1995). Root-induced iron oxidation, pH changes and zinc solubilisation in the rhizosphere of lowland rice. New Phytol., 131: 129–137.
Kubota, J., Allaway, W.H. (1972). Geographic distribution of trace element problems. In Mortvedt, J.J., Giordano, P., Lindsay, W. (Eds), Micronutrients in Agriculture. Soil Science Society of America, Madison, WI, pp. 525–554.
Lombi, E., McLauchlin, M.J., Johnston, C., Armstrong, R.D., Holloway, R.E. (2005). Mobility, solubility and lability of fluid and granular forms of P fertiliser in calcareous and non-calcareous soils under laboratory conditions.
Lyons, G.H., Lewis, J., Lorimer, M.F., Holloway, R.E, Brace, D., Stangoulis, J.C.R., Graham, R.D. (2004a). High selenium wheat: Agronomic biofortification strategies to improve human nutrition. Food Agric. Environ., 2: 171–178.
Lyons, G.H., Stangoulis, J.C.R., Graham, R.D. (2004b). Exploiting micronutrient interaction to optimize biofortification programs: The case for inclusion of selenium and iodine in the HarvestPlus program. Nutr. Rev., 62: 247–252.
McEvoy, J., Ascher, J.S., Graham, R.D., Hardy, P. (1988). Seed coating with manganese sulphate is a method of reducing manganese deficiency and improving early seedling vigour. In Webb, M.J., Nable, R.O., Graham, R.D., Hannam, R.J. (Eds.), International Symposium on Manganese in Soils and Plants: Contributed Papers. Manganese Symposium, 1988, Adelaide, Australia.
Morton, J.D., Grace, N.D., O'Connor, M.B. (Eds.) (1999). The Principles and Practice of Meeting Trace Element Requirements of Pastures and Animals, New Zealand Fertiliser Manufacturers' Research Association, Newmarket, New Zealand, 67 pp.
Moussavi-Nik, M., Rengel, Z., Hollamby, G.J., Ascher, J.S. (1997). Seed manganese (Mn) content is more important than Mn fertilisation for wheat growth under Mn-deficient conditions. In Ando, T., Fujita, K., Mae, T., Matsumoto, H., Mori, S., Sekiya, J. (Eds.), Plant Nutrition for Sustainable Food Production and Environment. Kluwer, Dordrecht, The Netherlands, pp. 267–268.
Prasad, A.S., Miale, A., Farid, Z., Schulert, A., Sandstead, H.H. (1963). Zinc metabolism in patients with the syndrome of iron deficiency anemia, hypogonadism, and dwarfism. J. Lab. Clin. Med., 61: 537–549.
Sanghvi, T.G. (1996). Economic Rationale for Investing in Micronutrient Programs. A Policy Brief Based on New Analyses. Office of Nutrition, Bureau for Research and Development, United States Agency for International Development, Washington, DC.

Sillanpää, M. (1982). Micronutrients and the nutrient status of soils: A global study. Soils Bulletin 48, FAO, Rome, 444 pp.

Sillanpää, M. (1990). Micronutrient assessment at country level: An international study. Soils Bulletin No. 63, FAO, Rome, 208 pp.

Thurlow, R.A., Winichaoon, P., Green, T., Wasantwisut, E., Pongcharoen, T., Bailey, K.B., Gibson, R.S. (2005). Only a small proportion of the anemia in North-east Thai schoolchildren is associated with iron deficiency. Am. J. Clin. Nutr., 82: 380–387.

Welch, R.M., Graham, R.D. (2000). A new paradigm for world agriculture: Productive, sustainable, nutritious, healthful food systems. Food Nutr. Bull., 21: 361–366.

Welch, R.M., Graham, R.D. (2004). Breeding for micronutrients in staple food crops from a human nutrition perspective. J. Exp. Bot., 55: 353–364.

Welch, R.M., Graham, R.D. (2005). Agriculture: The real nexus for enhancing bioavailable micronutrients in food crops. J. Trace Elem. Med. Biol., 18: 299–307.

WHO (2005). Available at http://www.who.int/nut.

Wilhelm, N.S., Fisher, J.M., Graham, R.D. (1985). The effect of manganese deficiency and cereal cyst nematode infection on the growth of barley. Plant Soil, 85: 23–32.

Yamaji, S., Tenant, J., Tandy, S., Williams, M., Srai, S.K.S., Sharp, P. (2001). Zinc regulates the function and expression of iron transporters DMT1 and IREG1 in human intestinal Caco-2 cells. FEBS Lett. 507:137–141.

Chapter 3
Micronutrient Deficiencies in Australian Field Crops

Robert E. Holloway, Robin D. Graham, and Samuel P. Stacey

Abstract Australia's ancient landscape has soils of exceptionally low fertility and deficiencies of all known nutrients have been recorded. Deficiencies of Mo and Zn are most widespread, being common on acid and alkaline soils respectively. Zinc deficiency is notable for being the most widely distributed micronutrient problem globally as well as in Australia, occurring on all soil classes, acid and alkaline, sandy and clayey, humid and arid, and in hot and cold growing seasons. Many Australian soils are affected by the presence of fine, free lime in the form of shell-grit blown up over the continent when sea levels were low during the last ice age; such soils, especially the more sandy types are low in micronutrient cations, Fe, Zn, Mn, Cu and/or Co. Multiple nutrient deficiencies are common, giving rise to a wealth of nutrient interaction effects. Interactions between two or more micronutrients and between micro- and macro-nutrients are agronomically and economically important. The classical micronutrient sensitivities reported elsewhere are also seen in Australia, but importantly, breeding has been carried out for tolerance to deficient soils in the major cereal crops, as well as tolerance to the common nutrient toxicities, the latter in common with activities in many parts of the world. The first deliberately bred cereal variety (barley) tolerant to Mn deficiency was released in South Australia in 2004. An important feature of the agronomy of micronutrients is the yield benefit in micronutrient-deficient soils of sowing seeds with a high micronutrient density. The use of plant analysis for diagnosis is almost always warranted as some crop varieties may lose much yield potential before symptoms of some micronutrient deficiencies appear. Recent research in South Australia has demonstrated that on calcareous soils, multi-nutrient fluid fertilisers have provided more efficient responses to both macronutrients and micronutrients than granular fertilizers.

3.1 Introduction

The greater part of Australia's ancient and eroded agricultural landscape is predictably low in micronutrients. Extensive tracts of previously unproductive land in South Australia (SA) and Western Australia (WA) were rendered arable and able to

B.J. Alloway (ed.), *Micronutrient Deficiencies in Global Crop Production.*

support healthy field crop and pasture growth by scientists who, with primitive equipment and scant resources by today's standards, first deduced the connection between the impoverished state of the soil and the lack of micronutrients. It is in these two states where the greater part of soils naturally low in plant-available micronutrients are found and where much of the early work was done in establishing the need for micronutrients in vast areas incapable of supporting crops, pastures or livestock. As Welch et al. (1991) pointed out, WA has the largest single contiguous area of naturally zinc (Zn)-deficient soil in the world.

3.2 Early Micronutrient Research in Australia

The first record of a micronutrient deficiency in Australia was of manganese (Mn) deficiency in oats in the lower south-east of SA (Samuel and Piper, 1928). The disorder was colloquially referred to as "road take-all" in SA because of the deleterious effect of limestone dust blown from the road (Donald and Prescott, 1975). The same disorder in wheat and oats was referred to in Western Australia (WA) and also in the Netherlands and Germany as "grey speck disease" (Carne, 1927). This occurred only a few years after Mn was shown to be an essential nutrient for higher plants (McHargue, 1922). Samuel and Piper (1929) were able to reproduce the deficiency in solution culture, the first laboratory demonstration of a known micronutrient deficiency (Donald and Prescott, 1975).

Similarly, studies of the poor growth of crop plants and pastures in extensive regions of aeolian, calcareous sands along the south-eastern coastline of SA led to the identification of acute copper (Cu) deficiency (Piper, 1938; Riceman and Donald, 1938) accompanied by less severe, but clinically important, Zn deficiency (Donald and Prescott, 1975). In a more extensive region of podzolised sands in the upper south-east of SA (known originally as the "90 mile desert", although Donald and Prescott (1975) preferred the less derogatory term "90 mile plain"), extending into western Victoria (a total area of 26,000 km^2 – Reuter et al., 1988) applications of Cu, Zn and superphosphate had resulted in spectacular increases in pasture growth and crop yields by 1942 (Riceman and Anderson, 1943). The first report of wheat yield responses to the application of Zn in WA occurred on calcareous sandy soils in 1940–1941 (Riley et al., 1992). Millikan (1938) reported large responses to the application of Zn sulphate ($ZnSO_4$) (with superphosphate) at sowing in wheat afflicted with cereal root eelworm disease (*H. schachtii* (sic)) (now *Heterodera avenae*) in western Victoria. The disease was still present on the cereal roots but was much less severe in those plants treated with Zn and superphosphate than on those treated with superphosphate alone. This is one of the earliest reports of the positive effects of a micronutrient in reducing the severity of a root disease. The emergence of Zn deficiency in the late 1950s and early 1960s was reported in wheat, linseed and maize on black earths (Vertisols) of the Darling Downs in Queensland (Duncan, 1961).

Copper deficiencies were first recognised in WA in 1939 and by 1948 were found to be widespread on sandy and gravelly (ironstone or "buckshot") soils and calcareous coastal soils. In the following decade, 1.2 Mha of sandy and gravelly soils in WA had been treated with Cu and Zn, again with large responses, although it was noted that livestock could still be deficient in those nutrients where no yield responses were recorded in crops (Toms, 1958).

Anderson (1942) reported the first known response to molybdenum (Mo) in the field, with subterranean clover (*Trifolium subterraneum* L.) on a lateritic ironstone soil in the Adelaide Hills of SA. The first lead to the possible contributing factor to the poor growth of pasture was a farmer's observation that plants grew well where a tree stump had been previously burned and that ash added to superphosphate elicited an immediate and spectacular response. Interestingly, Arnon and Stout (1939), who first established the essentiality of Mo for higher plants, considered that it was of no practical importance and that no soils in the world could be deficient in this element, which was required in such low amounts (Stout, 1972). Piper (1940) first demonstrated in SA that oats (*Avena sativa* L.) suffered a 60% loss of grain yield in the absence of Mo when grown in solution culture. The first recorded field response of a cereal to Mo was in Tasmania where the application of Mo, cured "blue chaff disease" (Donald and Prescott, 1975). Widespread reports of field responses of broadacre crops to Mo were delayed until Gartrell (1966) reported Mo responses in wheat and oats in WA and Doyle et al. (1965) recommended the use of Mo on acidic "scrub-plain soils". These are highly weathered, sandy and gravelly soils developed on gneiss and granite in the northern and eastern wheat-belt of south-western WA. As an immediate consequence, the percentage of superphosphate-based fertilisers containing Mo increased from 3.5% in 1963–1964 to 70% in 1966–1967 (Williams and Andrew, 1970). Gartrell and Glencross (1968) also recommended the addition of Mo to Cu and Zn applications to sandy and gravelly soils in the south-west of the agricultural zone of WA. By this time, Mo had become the second most extensively used fertiliser in Australia on the basis of area (Donald and Prescott, 1975). Responses to Mo were more recently recorded in sunflower grown on sand over clay soils of the south-east of SA, in which the soil pH had declined from 6.0 in the virgin soil to 4.5 after 25 years of pasture growth (McFarlane et al., 1980). A large area of poor grazing land, mainly podzols, on the southern tablelands of NSW and in central Victoria was transformed by the addition of Mo and superphosphate (Newman, 1955, 1962).

Iron (Fe) was shown to be essential for plants as early as about 1860 (Tiller, 1983). Iron deficiency was first reported in flax at Westmere in south-western Victoria in 1934 on a black self-mulching clay soil (Vertisol) over limestone (Millikan, 1951 cited in McFarlane, 1999). According to Donald and Prescott (1975), the first published account of widespread Fe deficiency in field and horticultural crops, which occurred on irrigated, solonised brown soils (Calcic Luvisols; Calciorthids) of northern Victoria, was that of Baxter (1957), although the deficiencies referred to are horticultural rather than field crops. These soils are often referred to as "mallee" soils, after the colloquial name of the predominant *Eucalyptus oleosa* vegetation. They are now classified generally in Australia as

Calcarosols (Calcisols) (Isbell, 1996) because of the presence of calcium carbonate ($CaCO_3$) throughout the profile. Iron deficiency in cereal crops on these mallee soils generally appears in isolated areas of regrowth in field crops, which have been cut for hay.

Legumes require cobalt (Co) for N fixation where it is used by symbiotic root-dwelling *Rhizobia* to produce cobamides (e.g., deoxyadenosinecobalamins – Vitamin B12 coenzymes). Ahmed and Evans (1959) first established the essential nature of Co for *Rhizobia* in the laboratory and in the following year demonstrated a Co response in soya beans (*Glycine max* L.) (Ahmed and Evans, 1960). Concurrently, Powrie (1960) demonstrated the first plant response in the field through responses in subterranean clover on a grey siliceous sand over clay soil in the south-east of SA, followed by Ozanne et al. 's (1963) response on siliceous sand in WA. Cobalt was reported to have positive effects on the growth and N fixation of sweet lupins, (*Lupinus angustifolius*) on an infertile sandy soil in WA by Chatel et al. (1978).

Boron (B) was shown to be essential for higher plants by Warington (1923). Donald and Prescott (1975) cited Atkinson's (1935) important work in apples in New Zealand and shortly afterwards Carne and Martin (1937) demonstrated the successful treatment in Australia of "corky pit" or "internal cork" in apples with B solution applied to the trees.

Nickel (Ni) and chlorine (Cl) deficiencies have not to this point been recorded in field situations in Australia. Nickel has an important role in a storage protein in hydrogenase, which is associated with nitrogenase in some energy-efficient species of *Rhizobia* that exist symbiotically in nodules associated with the roots of grain legumes.

3.3 Micronutrient Deficiencies in Cool Temperate Cropping Zones

The area of grain production in Australia is generally divided into three regions, the Western, Southern and Northern Zones. The south-western corner of WA together with the irrigated tropical region centred around the Ord River Scheme in northern WA make up the Western Zone; the Southern zone extends from western SA into Victoria and into mid-NSW. The Northern Zone includes the northern cropping region in central NSW and extends into central Queensland (Qld). The cool temperate regions are confined to the south-east corner of WA, SA, Victoria and NSW. Micronutrient deficiencies in field crops have been recorded in each of these areas. In broad-acre field crops, WA and SA constitute the major areas of potentially low soil micronutrient status and hence are most liable to suffer loss of yield through sub clinical or unrecognised deficiencies. Donald and Prescott (1975) listed the red brown earths (Calcic, Chromic and Albic Luvisols) of Australia as the only important soil group to which micronutrients were not being applied. However, Reuter et al. (1988) later referred to reports of Zn deficiency on these soils in SA and on black earths and grey heavy textured soils (Vertisols) in the northern cropping zones of the State. For detailed descriptions of the incidence of micronutrient disorders and treatment in field crops, pasture and livestock in SA in particular, see Reuter (1975) and Reuter et al. (1988).

3.3.1 *Zinc*

Zinc deficiency is the most ubiquitous potential micronutrient deficiency in Australia, particularly in the sandy podzolic soils in the central south and west of the continent. Potential areas of Zn deficiency identified by Donald and Prescott (1975) Reuter et al. (1988) and are outlined in the Soil Atlas of Australia (Fig. 3.1).

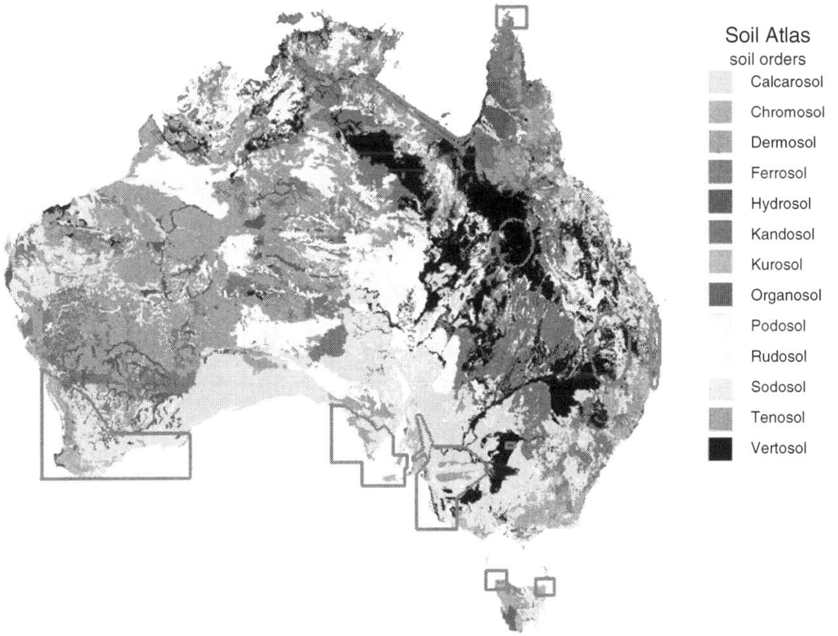

Fig. 3.1 Soil Atlas of Australia showing the major soil orders described in the Australian Soil Classification System (Isbell, 1996). The areas outlined in red indicate areas of potential Zn deficiency[1] (Commonwealth of Australia 2001, National Land and Water Resources Audit 2001; Bureau of Rural Sciences after Commonwealth Scientific and Industrial Research Organisation (1991). Digital Atlas of Australian Soils (ARC/INFO vector format). Available at html: http://www.brs.gov.au/data/datasets) (*See Colour Plates*)

[1] This hierarchical soil classification system was developed specifically for Australian Soils (described in Isbell, 1996), and follows the broad concepts of the USDA Soil Taxonomy (Soil Survey Staff, 1975) and begins with soil Orders, followed by Suborders, Great groups, Subgroups and Families. In the ASC there are 14 soil orders.

For further information see http://www.clw.csiro.au/aclep/asc_re_on_line/soilkey.htm

Applications of soil-applied Zn fertiliser for decades and the general awareness by farmers of the marginal Zn status of many arable soils have had a major effect on reducing at least overt catastrophic Zn-deficiency in field crops.

Besides the large area of 8 Mha of contiguous Zn-deficient soils in WA referred to by Welch et al. (1991) (which includes podzolic sands, lateritic podzolic sands, yellow earths and calcareous sands), Reuter et al. (1988) indicated that an estimated 70–80% of SA's cropping region was potentially subject to Zn deficiency. Since then, the susceptible area has probably increased with more intense cropping and the use of high analysis NP fertilisers. Zinc deficiency in crops was originally encountered on many virgin sandy soils with naturally low fertility, such as on acid sands of the south-east of SA, western Victoria and WA and on sand over clay duplex soils where the basic infertility of the soil was the predisposing cause of the deficiency. Large areas of alkaline cracking clay soils, or Vertosols, (Isbell, 1996) in NSW and Qld are also subject to Zn deficiency in field crops. Low availability of Zn on alkaline soils has also led to Zn deficiency on a wide range of alkaline soil types, including grey and red-brown calcareous sands and sandy loams, shallow grey and red calcareous soils, alkaline grey, red and black clays (which occur in the Wimmera district of western Victoria) and heavy grey and black cracking clays which are also found in NSW and Qld. The heavy use of lime and gypsum as ameliorants for acid soils can also decrease the availability of Zn and other micronutrients other than Mo.

Reuter (1975) tabulated the estimated consumption of trace element fertilisers in each State between 1962 and 1968, with 70% of Zn used in WA, 14% in SA, 15% in Victoria and 1% in Qld. In the 1980s superphosphate usage began to decline in favour of high analysis NP fertilisers – monoammonium phosphate (MAP) and diammonium phosphate (DAP). Until then, the 0.04% Zn impurity in superphosphate manufactured from Nauru, Christmas Island and Moroccan rock phosphate (Williams, 1974; Brennan, 1991) was sufficient to prevent Zn deficiency in superphosphate-fertilised crops and pastures. The incidence of Zn deficiency began to increase when highly refined MAP and DAP from Florida became more widely used. By the late 1980s, symptoms of Zn deficiency had begun to appear in crops in calcareous soils on the Upper Eyre Peninsula in SA and by 1990, annual applications of foliar $ZnSO_4$ to crops, or less frequent but heavier (e.g., 2.5 kg Zn ha^{-1}) applications to the soil surface, had become common practice. Zinc is also applied to the soil with the seed at sowing, integral with MAP or DAP as in 17:19:0 Zn 2.5. Rainbow (1997, unpublished data), reported responses to Zn applied as Zn-coated DAP in split placement (50% of fertiliser with the seed at sowing, 50% banded below the seed), with urea also banded below the seed, on a range of soils on the upper Yorke Peninsula and the mid north of SA.

In many countries of the world, Zn deficiency is a major human nutrition problem because of the predominance of cereal grains in the diet (Welch and House, 1984). In his review of the requirements of humans for Zn in plants, Welch (1993) indicated that marginal deficiency was endemic in a sufficient number of human population groups and livestock as to be cause for concern. The recommended daily intake (RDI) of Zn in Australia is 12–16 mg day^{-1} but 67% of males and 85% of females consume less than this (Baghurst et al., 1991). Hence, producing grain with high concentrations of Zn (and other trace elements such as Fe and Se) is likely to assume increasing importance, given that Australia exports a high proportion of its grain production.

While grains contribute only a small proportion of the dietary Zn uptake in some developed countries, it is likely that plant foods will constitute a higher proportion of Zn supply in future diets.

In an experiment with fluid suspension fertilisers on a grey highly calcareous soil, (Holloway et al., 2004), granular fertiliser containing a mixture of 17:19:0 Zn 2.5% and 13:15:0 Mn 6% was applied at sowing. An identical amount of the mixture was converted into a suspension by mixing with water, bentonite clay and sulphuric acid. Other granular fertilisers such as DAP, TSP and MAP were also compared in the two forms. Granular $ZnSO_4.7H_2O$, $MnSO_4.nH_2O$ and urea with 5% Zn were also added as basal nutrients where necessary. Concentrations of Zn in the whole shoots of plants treated with the mixed suspension fertiliser were 25.2 mg Zn kg^{-1} and in the granular fertiliser 20.9 mg Zn kg^{-1} (LSD $\leq 0.01 = 2.6$). Zinc uptake in shoots with the suspension-treated plants was more than twofold that in the granular treated plants. Zinc concentrations in grain were similar with the two fertilisers but Zn uptake in grain was 19.6% higher with the suspension than with the granular treatment. It has been hypothesised (M. J. McLaughlin and E. Lombi, 2004, personal communication) that the clay in suspensions may adsorb cationic micronutrients and hence avoid or, at least, reduce their rapid immobilisation in calcareous soils. This would be particularly important in the case of Mn and it is interesting that in the experiment of Holloway et al. (2004) above, Mn concentrations in whole shoots were higher ($P < 0.001$) in the mixed suspension containing Zn and Mn than the granular fertiliser (data not shown). Manganese uptake in shoots treated with the suspension was almost threefold higher than those in plants treated with the granular mix.

On the highly calcareous soils of the Eyre Peninsula in SA, P deficiency is endemic, despite many decades of application of P fertiliser, and the limitation imposed by poor availability of P often attenuates or prevents responses to micronutrients, particularly Zn. Granular triple superphosphate (TSP) on these soils was shown by Wilhelm and Growden (1999) to be a highly inefficient source of P. On the other hand, fluid NP fertilisers in solution with Zn were 4–15 times more effective than granular fertiliser (MAP) in supplying P to spring wheat on highly calcareous soils on the Eyre Peninsula in SA (Holloway et al., 2001b) when the solutions were applied to the seedbed at sowing. Fluid solutions have also been shown to increase Zn uptake in shoots and grains, compared with granular NP fertilisers containing Zn, on these soils (Holloway et al., 2001b, 2002). The chemical bases for the differences in efficiency of the two forms of fertiliser on calcareous and some other soils has been investigated by Bertrand et al. (2002) and Lombi et al. (2004). Fluid ammonium polyphosphate (APP) has performed well as a fluid source of P on calcareous soils but the amount of Zn that can be added as $ZnSO_4.7H_2O$ without precipitation is limited to about 3% by weight (Mortvedt et al., 2001).

Holloway et al. (2002) acidified APP by adding 40% of the total P as ortho-phosphoric acid in a P rate-response trial (0–16 kg P ha^{-1}) on a red-brown calcareous soil

(Calcarosol – Australian Soil Classification System) (Calcixerolic Xerochrept – USDA Soil Taxonomy) with 2.2 kg Zn ha^{-1} applied as $ZnSO_{4.}7H_2O$ in solution with the APP. Fluid APP was also applied alone with a separate solution of acidified $ZnSO_{4.}7H_2O$ applied simultaneously through a second delivery system. Granular fertiliser treatments consisted of mixtures of 16:18:0 Zn 2.5% and Urea Zn 5%. At a commercial application rate of 8 kg P ha^{-1}, APP and acidified APP produced 36% more grain than the granular fertiliser. Overall, the mean Zn concentration in the grain with acidified APP + Zn was 20 mg Zn kg^{-1}, compared with 16.8 mg kg^{-1} with the APP and separate Zn treatment and 14.4 mg kg^{-1} with the granular fertiliser ($P < 0.001$). Similarly, fluid mixtures of P with N and Zn have produced significant responses to both N and Zn on red brown calcareous soils when granular mixtures have produced no response (Holloway et al., 2000).

3.3.2 Manganese

In their review of SA micronutrient deficiencies, Reuter et al. (1988) argued that Mn deficiency was widespread in the State's cropping districts, because of the poor residual value of soil-applied Mn fertilisers and the need for persistent fertilisation. Response to Mn has been reported in most of SA's major cropping regions including the upper and lower South-East, the Eyre and Yorke Peninsulas, the Mallee districts, Kangaroo Island, north and south of Adelaide and the Mount Lofty Ranges (Tiver, 1955; Reuter et al., 1973b; Hannam et al., 1984).

Chronic Mn deficiencies are common on the highly calcareous soils of the Eyre and Yorke Peninsulas, many of which have $CaCO_3$ contents of over 80% and the ability to rapidly adsorb free Mn (Graham et al., 1985). In fact, up until the 1950s, agricultural development on the southern Yorke Peninsula was limited by severe Mn deficiencies (Higgs and Burton, 1955; Carter and Heard, 1962). Early attempts to crop these lands, or establish pasture, had failed. Manganese fertilisers have helped to remediate the problem and crop and pasture responses have been enormous. For example, Carter and Heard (1962) obtained a 20-fold increase in barley yields with Mn fertilisers on the southern Yorke Peninsula.

In early pot experiments with SA soils, Samuel and Piper (1928) measured oat responses to $MnSO_4$ at application rates of up to 84 kg ha^{-1}. Field applications have generally been smaller. However, on the Yorke Peninsula, Higgs and Burton (1955) found that an application of just over 22 kg $MnSO_4$ ha^{-1} was adequate but had very little residual value into the following year (Carter and Heard, 1962). In WA, Brennan (1998) recommended 15 kg of $MnSO_4$ ha^{-1} for cereals and 30 kg/ha for narrow-leafed lupins (*Lupinus angustifolius* L.). These Mn application rates would be considered extremely high in many regions of the world, but were found to be necessary in southern Australia due to the high sorptive capacities of the soils being cropped.

On some SA soils, foliar application of Mn has been more effective than soil application (Hannam et al., 1984). The poor effectiveness of the Mn fertilisers could be due to precipitation with P (when coated onto, or incorporated within fertiliser granules) and sorption of free Mn in the soil (Reuter et al., 1973a; Walter, 1988). The residual effectiveness of Mn fertilisers is also known to be poor and annual applications are often required on southern Australian soils (Reuter et al., 1973a, 1988; Hannam et al., 1984). As yet there are no truly effective Mn chelates that can be used in the soils of southern Australia because Mn is rapidly exchanged by Ca, Fe and other cations according to the Irving–Williams order (Irving and Williams, 1948). Nevertheless, chelation does prevent the formation of insoluble Mn and P precipitates in fertiliser formulations, thereby allowing these elements to be applied to the soil together (S. P. Stacey, unpublished data).

Severe Mn deficiencies have been reported in southern WA on both calcareous and acidic sandy soils. Deficiencies in cereals have caused yield losses of up to 1.3 t ha^{-1} and yield losses in lupins have been even higher (Brennan, 1998). Recent trial work has shown that considerable yield responses to Mn are still obtainable in some WA cropping systems (Brennan, 1999; Brennan and Longnecker, 2001; Brennan and Bolland, 2003). Specifically, Brennan (1999) showed that lupin grain yields could be increased by 190–1,870 kg ha^{-1} with Mn fertiliser. The Mn responses were highly significant, as the average grain yield in WA was only 1 t ha^{-1} for lupins (Brennan, 1999). These results were obtained from 31 field experiments run in four regions, both north and south-east of Perth. All the soils had pHs between 5 and 6.5, which shows that Mn deficiencies in WA are not restricted to the alkaline, or calcareous soil types.

In a separate pot experiment, Brennan and Bolland (2003) found that Mn fertilisers almost doubled lentil dry matter production on two alkaline soils from southern WA. The fact that such large Mn responses have been obtained suggests that Mn deficiency is an ongoing problem in southern WA and may continue to have a significant deleterious impact on field crop productivity.

Responses to Mn have been recorded in citrus trees grown in NSW near Griffith and in the coastal regions north of Sydney (Levitt and Nicholson, 1941; Connor, 1953). Manganese deficiencies in NSW have more often been reported in citrus trees than in field crops. Few Mn responses have been reported in Victorian field crops. However, it would not be surprising if Mn responses were found on the alkaline soils of northwest Victoria. Deficiencies may also be present on lands bordering the south-east district of SA. However, to our knowledge, Mn responses have not been detected in that region.

3.3.3 Copper

Copper deficiencies have been recorded in all states of Australia, but the major areas of potential deficiency include 8–9 million ha in WA (Brennan and Best, 1999), the Eyre Peninsula and the upper south-east of SA and western Victoria. These areas are generally contiguous with the areas of potential Zn deficiency and the regions of potential Mn deficiency in WA, SA and Victoria lie within them. A distinguishing feature of Cu in soils is its affinity for organic matter; it is more strongly bound to organic matter than other micronutrients and 20–50% of Cu in soils occurs in complexes with organic matter (Robson, 1981). Copper deficiencies occur on acid sands and calcareous soils alike and Cu is often occluded with soil minerals or within clay lattices, Fe-Al nodules, large organic molecules or organic matter. Coastal and inland calcareous soils of marine origin are often inherently low in Cu, as are highly weathered soils based on parent materials inherently low in Cu such as acid igneous rocks (granite) and sandstones (Gartrell, 1981; Robson and Reuter, 1981). Large potential areas of Cu deficiency in SA also include the inherently infertile calcareous and siliceous sands of the south-east of SA and western Victoria.

Copper fertiliser applications to soil appear to have long lasting effectiveness and Gartrell (1981) reported that Cu applied to a soil in WA 12 years previously was still two to three times more effective than current applications banded with the seed. However, as is often the case with micronutrient fertilisers in general, Gartrell (1981) commented that regional recommendations were often adopted from elsewhere and if they "seemed to work" were prescribed for many years "without question or investigation". Given the restricted amount of site-specific field research in micronutrient nutrition over the past 20 years, it is unlikely that the situation has changed dramatically. Gartrell (1981) noted that five times as much Cu had to be applied in Cu-OSP granules >3 mm diameter drilled with wheat seed to equal the effectiveness of granules 1 mm in diameter (due to the higher probability of root contact (Gilkes, 1975). Both Gartrell (1981) and Gilkes (1981) commented separately, but prophetically, that the ultimate means of uniform distribution was likely to be through fluids, given the rapidly increasing interest in the use of fluid fertilisers in recent years, especially in WA. In the highly calcareous soils of the Eyre Peninsula in SA, endemic P deficiency is most efficiently treated with fluid P. The reasons for the increased efficiency are largely chemical rather than physical (e.g., Lombi et al., 2004) and once this major limitation to growth has been overcome, micronutrient uptake can be enhanced by applying them to soils in solution or in suspension with NP fertilisers (Holloway et al., 2001a, 2004).

3.3.4 Molybdenum

It became increasingly apparent between the early 1940s and late 1960s that potential Mo deficiency was generally found on highly weathered acidic soils throughout Australia, including the Adelaide hills, Kangaroo Island, the south-east and part of the lower Eyre Peninsula in SA, the southern tablelands of NSW, central Victoria, north western Tasmania, the south-western corner of WA and coastal lowlands in Qld (Donald and Prescott, 1975).

Because Mo normally occurs in aqueous solutions as the molybdate oxyanion (MoO_4^{2-}) (Marschner, 1995), deficiencies in crops are most often encountered on acidic soils. Hence, crops grown on alkaline soils, which are conducive to Zn, Mn and Cu deficiencies are unlikely to be deficient in Mo, unless the total soil concentration of Mo is very low. Because the availability of Mo increases with increasing alkalinity, responses to Mo on alkaline soils are so rare as to be noteworthy. Donald and Prescott (1975) referred to two records of Mo response on alkaline soils in WA, one on a calcareous sand of pH 8.2 and the other on a loam of pH 8.6, containing 70% $CaCO_3$.

Brennan et al. (2004) reported on the extent of Mo deficiencies in WA cereal crops. Out of 33 field sites located across south-west WA, 16 were Mo-deficient. Many of the sites were also deficient in K, S, P and Cu. Most sandy soils in WA had a pH ($CaCl_2$) range of 5.5–6.5 when newly cleared. Soil acidification since then has led to Mo deficiency in crops resulting from stronger retention of Mo (Brennan et al., 2004).

Soil acidification has induced Mo deficiencies under long-term legume pastures in WA (Brennan, 2002). Molybdenum is a key component of the nitrogenase enzyme, used by symbiotic *Rhizobia* for biological nitrogen fixation (Marschner, 1995). Therefore, subclinical Mo deficiency may at first present as N deficiency in leguminous species.

Molybdenum deficiency may be remedied with fertiliser applications. This element has been incorporated with superphosphate as either sodium molybdate ($Na_2MoO_4.7H_2O$), ammonium molybdate ($(NH_4)_6Mo_7O_{24}.4H_2O$) or molybdenum trioxide (MoO_3) (Williams and Andrew, 1970; Brennan, 2002). Recommended rates are 50–60 g Mo ha^{-1} in SA and 75–80 g Mo ha^{-1} in WA (Gartrell and Glencross, 1968; Reuter et al., 1988; Brennan and Bruce, 1999; Brennan, 2002). Reuter et al. (1988) also recommended 50 g Mo ha^{-1} applied as a foliar spray.

The residual value of Mo fertiliser depends strongly on physical and chemical soil characteristics and on removal, principally through plants (Brennan, 2002).

The pH of newly cleared sandy soils in WA range from pH 5.5 to 6.5 ($CaCl_2$) but subsequent acidification of these soils has begun to induce widespread Mo deficiency in pastures where 20 years previously subterranean clover and cereal crops were adequately supplied (Bolland, 1985; Brennan et al., 2004). Brennan et al. (2004) measured a 15% decrease in yield of dry clover herbage after 80 g of Mo ha^{-1} had been applied 10 years previously and concluded that Mo applied as fertiliser could not be considered long lasting in acidic soils.

3.4 Micronutrient Deficiencies in Tropical Cropping Zones

While micronutrient deficiencies in southern Australia are well known, deficiencies in tropical and subtropical regions have not been widely reported in the scientific literature. Nevertheless, micronutrient deficiencies are a common occurrence on specific soil types and continue to limit agricultural production. A limited number of journal papers have described micronutrient deficiencies in northern Qld with the majority having been published in the early 1970s. More recently, reports of micronutrient deficiencies have been made in conference papers and in some grower extension materials. However, very little data exists on the extent of micronutrient deficiencies in Australia's tropical north. More importantly, the costs of these deficiencies to agricultural production are not known. In a review published in 1975, Donald and Prescott wrote:

> The infrequency of recorded [micronutrient] deficiencies in [Australia's] dry monsoon zone is almost certainly due to the low incidence of cropping and pasture development, and thus of agronomic trials in the zone, rather than to the absence of deficiencies.

Since 1975, the cultivation of land in Australia's north has increased significantly, particularly following the construction of irrigation infrastructure in Qld and WA. However, it appears that the incidence and severity of micronutrient deficiencies in these expanded cropping regions have still not been adequately quantified.

At present, severe visual deficiency symptoms are treated in regions where no micronutrient deficiencies have been reported in the literature. For example, Cu deficiency symptoms in sugarcane ('droopy tops') have been treated on siliceous sands around Bundaberg Qld. (I. K. Dart, 2005, personal communication, even though micronutrient deficiencies have not been formally reported in this region. Because the extent of micronutrient deficiencies in tropical and subtropical Australia is largely unknown, field crops may go untreated until strong visual symptoms present.

Late treatment may be a widespread problem due to the lack of effective soil testing protocols for micronutrients such as Cu, and a poor appreciation by farmers and extension staff of the deleterious effects of subclinical micronutrient deficiencies on final yields (Schroeder et al., 1999).

In 1953, the first micronutrient-deficient soils were identified in northern Australia. McLachlan (1953) grew subterranean clover on four soils collected between Katherine and Darwin in the Northern Territory. Significant dry matter responses to Mo were found on three of the four soils. Despite these early findings, micronutrient deficiencies have never been widely studied in the agricultural zones of the Northern Territory. More recently, Ross and Calder (1990) reported dry-weight responses of Siratro (*Phaseolus atropurpureus* DC.) to Mo and Zn on sandy red earth soils in the Douglas-Daly region of the Northern Territory, where the expansion of dryland and irrigated cropping is currently underway. Micronutrient deficiencies have been reported in some horticultural crops grown in the Northern Territory Horticulture Division (Northern Territory Horticulture Division, 1999). However, the extent of these deficiencies and their cost to agricultural production in the Northern Territory appears not to have been determined.

In 1961, Egan and Whitaker first reported Cu deficiencies in north Qld sugarcane. This early work stimulated a number of studies into the extent of micronutrient deficiencies in the region. By 1969, Teitzel had reported Cu deficiencies on soils of granitic origin. A later study showed that Zn was deficient on the same soil type (Teitzel and Bruce, 1971). Teitzel and Bruce (1971) also found responses to Zn and Mo on basaltic soils, responses to Mo on soils derived from metamorphic parent material, and responses to Cu and Mo on soils of alluvial origin (Teitzel and Bruce, 1972a, b, 1973a). Soils derived from beach sand showed significant plant growth responses to Cu, Zn, Mo and B (Teitzel and Bruce, 1973b).

The same region was surveyed by Reghenzani (1993) during a study on the Zn status of north Qld sugarcane soils. The survey indicated that up to 15% of the area under sugarcane (~18,000 ha) was likely to be Zn-responsive. The survey confirmed that Zn deficiency was of significant concern on the granitic, siliceous sand and metamorphic soils. In addition, low levels of Zn were found on some organic, tidal zone and alluvial soils (Reghenzani, 1993). These results suggest that the micronutrient deficiencies first reported in the early 1970s have not been adequately remedied and may continue to have a significant deleterious impact on the production of field crops in Qld. No recent survey of the Cu and Mo status of these soils has been undertaken.

Micronutrient deficiencies have also been reported on alkaline sodic soils in the Ord River Irrigation Area (ORIA), northern WA (Regan et al., 2001). Zinc fertiliser rates up to 3 kg Zn ha^{-1} have been recommended to overcome deficiencies in chickpea crops grown on the heavy clay and sandy loam soils typical of the Ord River area (Regan et al., 2001).

Cracking clay soils in the La Grange sub-basin of northern WA are believed to be inherently low in Zn, Cu, Mo, Fe, B and Mn (Australian Cotton CRC, 2004). The La Grange sub-basin has recently been the subject of a feasibility study into the expansion of cotton cropping into the region.

Despite the prevalence of micronutrient deficiencies in northern Australia's most important irrigated and dryland cropping regions, very few studies have been undertaken to determine the extent of these deficiencies or their cost to Australia's tropical agricultural production.

3.5 Micronutrients and the Subsoil

Subsoil constraints to root growth have only recently been studied in any detail because of the difficulties involved in obtaining unequivocal data, as Graham and Ascher (1993) observed. In the lower rainfall (<350 mm per annum) cereal growing areas of southern Australia, the majority of soils on which cereals are grown are naturally infertile and this problem has traditionally been addressed by fertilising the top 0.05 m of soil or by annual foliar applications.

Zinc, Cu and Mn applied to the topsoil are likely to be retained there, even in coarse textured sandy soils (Reuter, 1975; Brennan and McGrath, 1988). The soil below this depth in Australia's temperate cropping regions (referred to hereinafter as the subsoil) often confronts plant roots with some kind of hazard, whether low nutrient availability, sodicity, salinity, B toxicity, acidity, Al toxicity, compaction, Mn toxicity, alkalinity, or excessive bicarbonate, either alone or in various combinations. In some highly alkaline subsoils (pH > 9), there appears to be a real possibility of toxicity from an anionic form of Al (aluminate $Al(OH)_4^-$) (Ma et al., 2003). Because the greater proportion of fertiliser is applied to the cultivated layer, seldom deeper than 0.05 m, nutrients become unavailable as the surface dries, which is a common occurrence in the low rainfall cereal cropping areas of Australia. Grundon and Best (1981) reported that when neutral to alkaline clay topsoils in Qld were dry just before booting, the formation of infertile pollen severely curtailed yield despite the application of $CuSO_4$ with the seed at planting. It is has been proposed that the placement of fertilisers in the subsoil would encourage root growth and enhance nutrient and water uptake from soil that remains moist as the surface dries. This was clearly demonstrated with respect to Cu by Graham (1991). Yield responses in crops and pastures have been reported with deep placement of P (Simpson and Pinkerton, 1989; Jarvis and Bolland, 1990), N (Garwood and Williams, 1967), N and P combined (Murphy et al., 1978; Slattery and Rainbow, 1995) and to placement of Mn (as $MnSO_4$) 0.08 m below the seed (0.13 m below the surface) at sowing in lupins (Brennan, 1999). The mean yield increase with the deeper placement of Mn (1,100 kg ha^{-1}) was 30% higher than with seed-placed Mn and 64% higher than with surface broadcasting.

Graham and Ascher (1993) began to address the problem of infertile, highly alkaline subsoils in SA as early as 1985 by measuring responses to added organic matter or nutrients over a period of up to 7 years. Subsoils were removed from pits in layers to a depth of 0.9 m, mixed with organic matter or fertilisers and replaced. Control treatments consisted of soil removal and replacement alone with no additions of nutrients (physical disturbance) and a no-disturbance treatment. Substantial responses to the addition of organic matter or added N, P and micronutrients in terms of vegetative growth and grain yields were recorded on a calcareous soil over a period of 7 years. Large increases in grain yield above the controls were recorded with the same treatments at most other sites (apart from more fertile red-brown earths) and in the case of the NP + micronutrient treatments, yield increases of 50–100% were still being recorded 6 years after the original placement at some sites. It is encouraging that yield increases also occurred at sites in which the subsoils contained high concentrations of B (>15 mg B kg^{-1} soil; hot 0.01 M $CaCl_2$ extraction, Cartwright et al., 1984; Holloway and Alston, 1992). It is estimated that the cropping area affected by high concentrations of B exceeds 10 Mha in the low rainfall (<350 mm per annum) cereal growing areas of southern Australia (Graham et al., 1992b). The Upper Eyre Peninsula in SA constitutes about 20% of the State's wheat producing area and is characterised by coarse textured highly calcareous soils (10–90% $CaCO_3$), with low annual rainfall. In the majority of agricultural regions, alkalinity increases with depth and B and salt concentrations are often sufficiently high to present a hazard to root growth (Holloway, 1991; Holloway and Alston, 1992). Such subsoils have been justifiably labelled "inhospitable" and "hostile" (Graham et al., 1992b). Apart from B, these subsoils are deficient in most micronutrients (and P) required for adequate plant growth.

There are few reports of where the effects of deep placement of Zn in the field have been investigated in alkaline soils, or in any other soils for that matter. There are sound reasons, however, for such an investigation. While it has been shown that, given an adequate supply of Zn in part of the root zone, Zn can be transported in the phloem to maintain root growth rates in unsupplied zones (Loneragan et al., 1987; Webb and Loneragan, 1990), it has also been demonstrated that adequate root function requires Zn in the external medium (Welch et al., 1982; Loneragan et al.,

1987; Nable and Webb, 1993). The integrity of root cell membranes is dependent on a supply of Zn in the external medium, as Zn is a component of superoxide dismutase, which prevents the peroxidation of membrane proteins and lipids by superoxide radicals (Welch et al., 1982; Cakmak and Marschner, 1988).

Glasshouse and laboratory experiments have shown that substantial benefits may accompany the correction of Zn deficiency in the subsoil. In the specific case of the effect of Zn on B toxicity, Graham et al. (1987) conducted a solution culture study in which toxic concentrations of B were measured in the tissue of Zn-deficient barley plants, although dry matter was not affected. Singh et al. (1990) treated a loamy sand soil with factorial combinations of 0, 2.5, 5.0, 7.5 or 10 mg B kg^{-1} soil and 0, 10 or 20 mg Zn kg^{-1} soil. Increasing Zn decreased Mn, Mg and P uptake and increased Cu, Fe and K uptake. Increasing B decreased the uptake of Cu, Fe, Mn, Ca, Mg, K and P, but adding Zn reduced this effect. Boron toxicity in wheat and barley is a common problem with many of the cereal producing soils in the low rainfall areas of southern Australia (Cartwright et al., 1984; Hirsch and Manton, 1989). Many of these soils are also characterised by low Zn availability (Reuter et al., 1988), particularly in the subsoil at a depth of between 40 and 60 cm, where B levels reach the highest concentrations – up to 120 mg B kg^{-1} soil (Tiller, 1983; Holloway, 1991). Similarly, Al toxicity is likely to inhibit micronutrient uptake in acid soils (e.g., Cu – Gartrell, 1981), through its effect on root growth, and possibly also in alkaline soils (Ma et al., 2003).

In an elegant pot experiment, Nable and Webb (1993) compared the performance of Zn-efficient (Excalibur) and Zn-inefficient (Gatcher) wheat cultivars with laterally split root systems in which topsoil (0.10 m deep) was supplied with Zn, but the "subsoil" (0.25 m deep) was either fertilised with Zn or was left without Zn. Withholding Zn from the subsoil had no effect on root growth in either zone or on vegetative yield, or on plant appearance before booting in either genotype. However, inefficient Gatcher plants grown in pots containing subsoil without added Zn demonstrated lower water usage in the 60-day period preceding maturity, delayed head emergence and lower grain yield, compared with plants grown in pots with Zn-fertilised subsoil. Excalibur displayed none of these effects in the Zn-deficient subsoil.

In both genotypes, Zn concentrations in the flag leaf were higher in pots containing subsoil Zn. While only 20% of the root weight occurred in the subsoil, about 50% of the Zn appeared to be taken up by these roots. Excalibur had consistently higher Zn concentrations in flag leaves and lower concentrations in grain than Gatcher, independent of the subsoil treatment. The consistently lower Zn concentrations in Excalibur are a cause for some concern; when the subsoil was not fertilised with Zn, Zn concentrations in Excalibur grain were halved compared with grain from pots in which both soil layers were fertilised.

To better understand the relationship between the placement of Zn fertiliser and transport to grain, Holloway (1996) grew Excalibur plants in pots 0.70 m deep and 0.15 m diameter. The soil was infertile white (Laffer) sand from the south-east of SA and was divided into three 0.20 m deep horizons all fertilised with basal nutrients. Zinc was added to, or withheld, from the horizons in all possible combinations. Zinc concentrations in YEBs were higher in plants grown in pots with Zn added to the top two layers. Zinc concentrations in grain indicated a general trend for Zn concentration to reflect the total amount of Zn available in the root zone. Where Zn was added to a single layer only, concentrations of Zn in grain tended to increase with the depth of placement. Plants grown in the standard field nutrient placement configuration where Zn was placed in the top horizon only, produced the lowest Zn concentration and Zn uptake (content) in grain.

While it appears that stems and leaves constitute an important source for the loading of seed with Zn, a constant or increasing supply throughout the season may be necessary to maintain Zn concentrations and hence cell membrane integrity in maturing roots. This is unlikely to occur in the field environment of much of southern Australia where, even if bio-available Zn concentrations are adequate in the top 0.05 m of soil, this is unlikely to be the case in the subsoil. While the supply of Zn later in the season may be supported by foliar applications, the problem of maintaining root function remains. A high internal concentration of Zn does not prevent leakage through cell membranes if the external solution is Zn-deficient, nor can phloem re-translocation of Zn from parts of the root zone adequately supplied with Zn compensate for lack of Zn elsewhere in the root zone (Welch et al., 1982; Graham and Rengel, 1993). For Australian agricultural cropping systems with shallow nutrient placement systems, deeper application of fertilizer containing Zn is likely to be beneficial for increasing Zn concentrations in grain.

Holloway (1996) conducted field experiments on a red brown alkaline sandy loam soil over calcareous subsoil (Regolithic Calcarosol, Isbell, 1996) at Minnipa on the Upper Eyre Peninsula in SA. Solutions of $ZnSO_4.7H_2O$ were applied with and without MAP and ammonium nitrate solution evenly down the soil profile to a depth of 0.40 m before sowing or alternatively $ZnSO_4.7H_2O$ was applied with NP granular fertiliser with the seed at sowing. Zinc concentrations in YEBs were highest with plants where Zn, P and N solution was applied to the subsoil. This improvement in Zn availability remained for at least 3 years. Consistent benefits in Zn concentrations and uptake in shoots and grain, and in water use efficiency, were confined to fluid applications of N, P and Zn in a single solution applied to 0.40 m.

3.6 Alternative Approaches in Australia to Managing Micronutrient Deficiencies

3.6.1 Micronutrient Content of Seeds

In his review on the role of seed reserves in the early establishment of crops, Asher (1987) emphasised the effect of the supply of mineral nutrients early in the seedling phase of the crop on seedling vigour and events such as floral initiation, that determine potential yield. Nutrients in the seed enable the seedling to grow until the roots are fully functional and there is also a cross-over period when essential nutrients are drawn from both the soil and seed. Consequently, the supply of adequate nutrients in the seed can be quite critical in determining the potential of the plant to produce dry matter and grain. Asher (1987) also pointed out the similarity of general seed concentrations of micronutrients to concentrations required for healthy growth in seedlings, but that any reserves exceeding those of seedling requirements could be particularly important in adverse conditions. For phloem-immobile nutrients like B and Mn, deficiencies during seed formation can be damaging to the seed itself and result in severely restricted germination and early growth. Manganese deficiency in maturing crops causes "Marsh spot" in field pea (*Pisum sativum* L.) seeds and split seed in narrow leaf (or sweet) lupins (*Lupinus augustifolia*). In some parts of Australia, micronutrient solutions are applied to ripening cereal crops to increase the concentration in seeds. In SA, the procedure is colloquially referred to as "supercharging". Frischke (1999) reported that two applications of micronutrient solutions to wheat during grain fill on calcareous soils had positive effects on the growth and grain yield in the following year where farmers keep their own seed, depending on the solution used and the wheat genotype.

Longnecker et al. (1988) reported that barley (*Hordeum vulgare* L.) seed with high Mn content (e.g., 0.64–0.80 µg $seed^{-1}$) improved dry matter production in Mn-deficient conditions compared with low Mn seed (0.14–0.36 µg $seed^{-1}$). In field plots on a highly calcareous soil (80–90% $CaCO_3$) the high Mn seed produced a higher plant density and, in years when Mn deficiency was severe, also produced grain yield responses. It was considered that to minimise losses due to Mn deficiency, both high Mn content seed and Mn fertiliser were required.

Rengel and Graham (1995a, b) evaluated the effects of Zn content in seeds of two wheat genotypes with different Zn efficiency on vegetative growth and grain yield. Zn efficiency % is defined as:

$$\frac{Yield\ (-Zn) \times 100}{Yield\ (+Zn)} \tag{1}$$

Low Zn seed (~250 ng Zn $seed^{-1}$) and high Zn seed (~700 ng Zn $seed^{-1}$) of Zn-efficient Excalibur and Zn-inefficient Gatcher wheat were sown in Zn-deficient sand treated with basal nutrients and 0, 0.05, 0.2, 0.8 or 3.2 mg Zn kg^{-1} soil. Higher seed Zn increased the vegetative and root growth of both genotypes at 6 weeks in the non-fertilised Zn-deficient soil. Zinc fertilisation at a rate of 0.2 mg Zn kg^{-1} soil was required to produce 90% of the maximum yield for plants grown from the high Zn seed, but 0.8 mg Zn kg^{-1} was required for the low Zn seed. It was concluded that the higher Zn seed content acted similarly to a starter fertiliser, improving vegetative growth and reducing the effect of the difference in Zn efficiency. Plants grown to maturity from high seed Zn produced more and larger grains in the unfertilised soil. They also produced more grain dry matter per unit of Zn absorbed by above-ground plant parts, transported a higher proportion of absorbed Zn to the grain and were able to reach close to the maximum harvest index at a lower fertilisation rate (0.05 mg Zn kg^{-1} soil) than was required for the plants grown from low seed Zn (0.2 mg Zn kg^{-1} soil). The Zn-efficient genotype Excalibur also exhibited greater fertiliser efficiency when provided with 0.05 mg Zn kg^{-1} soil, a low level of fertilisation, and a greater harvest index in the nil-Zn soil.

3.6.2 *Nutrient Efficiency*

The identification of Zn efficiency in the South Australian wheat genotype Excalibur (Graham and Rengel, 1993) was an important development where a survey of farms in part of SA's low rainfall wheat producing area indicated that 96% of crops had foliar Zn concentrations of less than 20 mg Zn kg^{-1}, a concentration considered to be marginal at best (Hannam, 1991). Zinc efficiency is likely to be of particular benefit where Zn fertiliser is concentrated at the surface and the topsoil is subject to drying.

Graham and Rengel (1993) have identified a wide range of genetic diversity for efficiency in uptake of Zn (and other micronutrients) in wheat. The South Australian cultivar Excalibur was outstanding in this respect in field experiments. By contrast, the cultivar Gatcher was shown to be consistently Zn-inefficient. There is little information concerning root growth parameters (in soil) that may affect Zn-efficiency (Zn-efficiency in soil and nutrient culture do not appear to be related – Graham and Rengel, 1993). Dong Bei et al. (1995) indicated that in Zn-deficient soil, Excalibur is able to produce more fine roots (diameter < 0.2 mm) than Gatcher and the surface to volume ratio increases as root diameter falls. This would allow the exploration of a greater volume of soil if root mass is similar. It was considered that this difference could account to some degree for the difference in Zn-efficiency between the cultivars.

To compare the Zn-efficiency, root growth and production characteristics of the two cultivars Excalibur and Gatcher in calcareous subsoils (containing salt and B), Holloway (1996) conducted a deep pot experiment in controlled conditions designed to compare the growth of plant tops and roots of the two genotypes (*Gt* levels) grown in subsoil typical of that occurring on the Upper Eyre Peninsula in a three way factorial design. The whole soil was either treated with basal nutrients or not (*nu+ and nu−* levels) fertilised with or without Zn (*Zn+ and Zn−* levels). Overall, the time to reach maturity was delayed by 10 days in the absence of Zn.

Both Gatcher and Excalibur plants grown in *nu+Zn−* pots displayed delayed head emergence compared with plants grown in *nu+Zn+* soil. In the case of Gatcher, however, some plants in the *nu+Zn−* pots did not produce any heads at all. Dry matter production was significantly higher with Zn+ treatments than Zn− treatments, but clearly the greatest difference occurred between *nu+* and *nu−* treatments, where the fertilised subsoil increased total production by more than 20-fold.

Grain yield was highest in Excalibur and Gatcher plants grown with added nutrients and zinc (*nu+Zn+*). There was no significant difference between the performance of the two cultivars in the *nu+Zn+* treatment. However, the absence of Zn when other nutrients were added (*nu+Zn−*) had a much greater effect on Gatcher than on Excalibur. Zinc-efficiency was 74% for Excalibur compared with only 12%

for Gatcher. Excalibur produced six times more grain in the *nu+Zn−* soil than Gatcher. While Zn uptake in grain was greater with Excalibur *nu+Zn−* than with the corresponding Gatcher plants (*nu+Zn−*), there were no differences in Zn uptake in total mature tops, a suggestion that Excalibur's Zn-efficiency may be due to an ability to divert Zn into grain production when the supply of Zn is limited. The addition of Zn to the soil reduced B concentration in grain by 22%.

The relative transport of Zn to grain ("Zn harvest index") (% of zinc in grain of the total content in tops) is an indicator of the ability of plants to load seed with zinc. In this case, Excalibur (42%) was considerably more efficient than Gatcher (25.4%). Harvest index (the ratio of grain to whole tops) was eight fold higher with the *nu+Zn−* Excalibur plants than with Gatcher.

The greatest effect on rooting density over all depths was due to the addition of nutrients other than Zn. In the top layer, Gatcher produced 12% more roots and in the lower two depth intervals, Gatcher produced root length densities twice that of Excalibur. The presence or absence of added Zn had no effect on rooting density at any of the depth intervals. Overall, total root length and surface area of roots in pots was about twice as much with Gatcher as with Excalibur.

While the total uptake of Zn in tops (straw + chaff + grain) was similar in both cultivars in *nu+Zn−* soil, Excalibur produced about half the total length of roots and estimated root surface area in this soil. This indicates a considerably more efficient uptake mechanism in Excalibur.

3.7 Micronutrient Interactions

Interactions between macro and micronutrients have such a large impact on the bioavailability of micronutrients that as yet they still constitute one of the least understood aspects of plant nutrition. Some progress has been made in this field in Australia and elsewhere, but in field crops in Australia at least, the application of micronutrients is still often an ad hoc procedure, with no sound information on how micronutrient interactions are likely to affect the outcome, other than the standard insurance policy of "throwing in a bit of Mn (or Zn or Cu) as well!"

The availability of rapid plant tissue tests has allowed more detailed studies to be made of nutrient interactions in plants. Workers in this field seem to be extraordinarily prone to chastisement by colleagues. Sumner and Farina (1986) published an interesting review of interactions between P and other nutrients including micronutrients, and complained of the lack of field research with micronutrient interactions at the time. They continued:

> All too often, one finds numerous reports in the literature of the relationship between yield and the quantities of nutrients applied to the soil without any presentation of meaningful soil or plant analyses. Such an approach, aptly named "spread and measure", contributes nothing to our understanding of the [interactional] processes involved in the yield responses measured. Indeed, if anything, this approach merely adds worthless pages to an already voluminous literature and, into the bargain, probably leads to confusion.

The introduction of multi-element inductively coupled optical emission spectrometry (ICP-OES) analytical technology, not long after this was written, should have eased the concerns of the authors. However, Loneragan and Webb (1993) commented on the current state of research into P–Zn interactions at the time. Two paragraphs from their review are worthy of reproduction here:

> There is a voluminous and confusing literature on P-Zn interactions. Much of the confusion has arisen from workers who failed to identify the factor operative in an interaction or who used conditions irrelevant to zinc deficiency. Others have compounded the confusion by accepting conclusions without critical evaluation of the experimental conditions and data by looking for a single magic phenomenon, "the P-Zn interaction" or "the P-induced Zn deficiency" to explain the many phenomena now known to occur; some even look for an interaction under conditions where, since neither Zn nor P are limiting or excessive, there is no reason to expect any!

Mercifully, no examples of the offending literature are provided. Loneragan and Webb (1993) continued:

> The value of careful, critical research in the understanding of nutrient interactions is well illustrated by recent research which has provided a new insight into the long puzzling phenomenon of "P enhanced Zn requirements". It contrasts with the huge volume of uncritical and unproductive research correlating P/Zn ratios with Zn deficiency and serves as a warning against the current trend of producing "desktop" interactions from the uncritical development of relationships between every possible combination of nutrients resulting from the availability of multi-element analysers and high capacity computers.

When P and Zn are both limiting or marginal, increased growth induced by adding P alone can induce or exacerbate Zn deficiency – the "growth dilution effect" (Loneragan et al., 1979; Singh et al., 1988; Loneragan and Webb, 1993); adding both P and Zn corrects both limitations. At times, Zn concentrations are depressed below that which can be explained by dilution alone. Sharma et al. (1968) and Loneragan and Webb (1993) have suggested several mechanisms for this, based on P-induced depression of Zn absorption by roots, or on the translocation of Zn from roots to shoots. Welch et al. (1982) concluded from P accumulation studies with Zn-deficient plants and absorption studies on excised roots that Zn also played a fundamental role in the stability of cell membranes and that the effect of destabilising membranes was not easily reversed, as it is with Ca. Interestingly, these effects can occur very early in the growth cycle when the external root environment is deprived of Zn, despite adequate amounts for normal shoot growth (from seed reserves) in roots and shoots. An important conclusion from this work is that Zn is required as is Ca, P, B and Mn (Graham et al., 1992a) in the external solution to maintain root cell membrane integrity. From experiments conducted by Welch et al. (1982), it was deduced that Zn-deficient root cells, having impaired membrane integrity, could allow non selective mass flow of P into the roots which could then accumulate, particularly in actively transpiring older leaves.

Loneragan et al. (1982) confirmed that Zn deficiency in okra (*Abelmoschus esculentis* L. Moench) strongly increased P transport to tops and accumulation in leaves, interfered with P metabolism and, with high P supplies, allowed P to accumulate to toxic concentrations in leaves, inducing or aggravating symptoms resembling

Zn deficiency. Similarly, leaf symptoms could also be attributed to P toxicity in subterranean clover (Loneragan et al., 1979), cotton (Cakmak and Marschner, 1986) and wheat supplied with low Zn and high P concentrations (Webb and Loneragan, 1988). Loneragan and Webb (1993) concluded that Zn precipitation by high P concentrations within the plant (decreasing the proportion of water soluble Zn) is the principal cause of symptom development in the presence of undiminished (total) Zn concentrations.

Nitrogen fertilisers can induce or exacerbate Zn deficiency in plants growing in soils with low Zn status by dilution of Zn in the plant through large increases in total growth (Chaudhry and Loneragan, 1970). Nitrogen fertilisers may also affect Zn uptake by changing the rhizosphere pH (Marschner and Cakmak, 1986). Graham (1999) showed that applying N (as NH_4^+) in NP fertiliser to a depth of 0.40 m on a highly calcareous soil in SA substantially increased Mn concentrations in YEBs (42.1 mg Mn kg^{-1}) compared with topsoil placement (25.4 mg Mn kg^{-1}) or when micronutrients (Mn, Zn and Cu) were added to the subsoil with the NP fertiliser (33.5 mg Mn kg^{-1}). Zinc concentrations in YEBs with the same treatments were 23.5, 31.1 and 31.6 mg Zn kg^{-1}, respectively.

Nitrogen fertilisers can restrict the movement of Cu in plants with low Cu concentrations (Gartrell, 1981). Gartrell's (1981) astute observation that large areas of WA had a sufficient Cu supply for wheat crops of 1 t ha^{-1} of grain but not for those with potential yields of 2–3 t ha^{-1} has increasing relevancy nearly 25 years later. Doubtless, there are many other areas of Australia where micronutrient supplies are now limiting the full expression of yield potentials as more intensive cropping phases, improvements in technology in weed control, increasing soil water use efficiency, varietal improvements, seed and fertiliser placement and timing of sowing increase pressure on soil nutrient reserves.

Copper and Zn applied to an infertile acid loamy sand in pot culture had no effect on vegetative or grain yields unless N was also added (Chaudhry and Loneragan, 1970). There was a strong positive interaction between added Zn and N if Cu was adequate, and a negative interaction in the absence of Cu. Copper fertiliser without Zn aggravated the effects of N on Zn deficiency. In Chaudhry and Loneragan (1970) and Harry and Graham (1981), Zn was implicated in exacerbating Cu deficiency by depressing the uptake of Cu and hence had a large effect on grain yield because of the key role of Cu in grain production through pollen viability. Gartrell (1969) suggested that increasing use of Zn fertiliser was a significant factor in accentuating Cu deficiency in Australia (Gartrell, 1969).

In interesting work in WA, not often quoted in the literature, Beckwith (1966) reported that fertilisation of oats with $CuSO_4$ reduced their Mn content to below that which could be explained by dilution alone. In a calcareous soil from WA, Beckwith (1966) recorded more consistent reductions of Mn and Fe content in cereal shoots than reductions in Cu after the addition of Zn to the soil. Occasional reductions in Fe and Zn contents accompanied added Cu and Mn applications. On a soil from the south-east of SA, added Mn decreased Cu and Zn contents. Beckwith's (1966) conclusion regarding the non-dilutional reduction in some micronutrients when others were added was that the phenomena occurred only when the micronutrient inducing the reduction was added to correct a deficiency. Alternatively, oats or wheat shoots deficient in Zn translocated abnormally large amounts of other non-deficient micronutrients to the shoots if the Zn deficiency was not corrected.

Harry and Graham (1981) demonstrated the tolerance of rye (*Secale cereale* L.) to a wide range of soil pH and deficiencies of Cu and Zn, separately or together, and contrasted this with the susceptibility of wheat. Robson (1981) considered that differences within and between species in their tolerance of Cu deficiency were more a function of different abilities to absorb Cu from soil than in internal requirements. Kochian (1993) considered that the main competitive interaction for Zn^{2+} is Cu^{2+} and Rengel and Graham (1995a) confirmed this in their study with Zn-deficient plants. Kochian also referred to studies by Schmid et al. (1965), which demonstrated that, in contrast to Cu, Mn did not affect Zn uptake. Copper and Zn appear to share competitive uptake mechanisms by roots (Loneragan, 1975), possibly utilising the same transport protein (Kochian, 1993). However, as Loneragan and Webb (1993) pointed out, Zn^{2+} activity is likely to be much higher than Cu^{2+} activity at absorbing sites because a higher proportion of Cu is complexed. They were unaware of any study in which Cu unequivocally reduced growth or yield by affecting Zn uptake. Loneragan and Webb (1993) showed that wide variations in Mn concentrations and activities in chelate-buffered solutions between near deficiency and toxicity did not have pronounced effects on Zn concentrations in roots and shoots of barley. The often-recorded phenomena of inverse correlations between Mn and Zn concentrations in tissues (e.g., Holloway, 1996) is possibly due to different patterns of re-translocation and seed loading of the two micronutrients within the plant (Pearson and Rengel, 1994, 1995; Pearson et al., 1995).

3.8 Detecting and Predicting Micronutrient Deficiencies in Australia

The ability of tissue tests to reveal the micronutrient status of plants (both overt deficiency and subclinical deficiency which can also reduce grain yield) in time to take corrective action has been a major breakthrough in plant nutrition. The now common use of ICP-OES has allowed rapid tissue analysis to become a standard practice for nutrient assessment in field crops in Australia. For a complete description of plant analysis and interpretation in Australia, the reader is referred to Reuter and Robinson (1997). Recent improvements in instrumentation such as the introduction of inductively coupled plasma mass spectrometry (ICP-MS) have allowed detection limits to fall so that concentrations of elements such as Mo, Ni, Se and Co can now be routinely detected in plant samples.

At present, soil tests based on chemical extractions require more detailed calibration than is generally available to be useful as predictive tools. Because of the different abilities of various species, and even cultivars within species, to extract micronutrients from different soil types, soil tests are not able to predict yield responses from micronutrient applications without highly specific information on both sites and crops. At best, it may be possible to differentiate between the likelihood of micronutrient deficiency and adequacy in a given soil, the reliability of the test depending on the amount of previous soil x cultivar calibration (and information on other associated soil properties that may be available for multiple factor regression studies). In the case of Mn (Uren, 1999) and Fe (McFarlane, 1999), soil tests alone cannot be relied on to predict or diagnose deficiencies.

Chemical extractants tend to lack specificity in distinguishing how nutritional elements are distributed and bound among different soil mineral phases (Bertrand et al., 2002). Mid Infrared (MIR) spectroscopy can be used to identify soil minerals and organic matter species. Janik et al. (1995, 1998) and Janik and Skjemstad (1995) used MIR diffuse reflectance (DRIFT) spectroscopy with a partial least square (PLS) regression technique to successfully predict the major elemental composition, pH, organic carbon, N, CEC, carbonate and clay concentrations of a wide range of Australian soils. Bertrand et al. (2002) tested the ability of this technology to determine chemical characteristics, and micronutrient concentrations along with major mineral phases associated with the micronutrients on a range of solid phase alkaline soils from the Upper Eyre Peninsula in SA and in western Victoria. The soils from the Eyre Peninsula were calcareous, with $CaCO_3$ concentrations ranging from 3% to 87%. Mid Infra Red spectrometry initially predicts total metal concentrations but Bertrand et al. (2002) used other information from the MIR output (e.g., pH and soil organic matter), combined with the partitioning model of Sauvé et al. (2000), to estimate the mobile pools of micronutrients in these soils. In the alkaline soils tested, retention of Cu, Mn, Zn and Fe were closely associated with the characteristics of the clay fraction; positively with the amount of illite/smectite in the soil and negatively with kaolinite. It was thus possible to rapidly determine the potential bioavailability of several metal micronutrients. Further development of this technology is required specifically for micronutrients, but in terms of increasing the speed, simplicity and practical value and reducing the costs of soil testing for micronutrient management in agriculture, MIR is one of the most promising developments in many years.

3.9 Summary

It is often said that Australia is a micronutrient desert, and largely this remains true in spite of a significant body of research on micronutrients carried out in this country, and corresponding benefits in crop productivity. Much more could be done to further enhance the productivity of the inherently infertile soils of a land mass that has been changed relatively little by tectonic, volcanic and glacial activity over much of geological time. New crops, new varieties of crops, new fertiliser techniques and strategies, new markets, new perceptions of agriculture as an instrument of public health, and new attitudes to soil and water as priceless natural resources mean that the need to re-optimise our approaches to plant nutrition for the production of food for humans and animals is as great as it ever was.

Despite the sizable body of research in the past century identifying the critical nature and extent of serious micronutrient deficiency in Australia, much remains to be done to understand precisely where and when micronutrients are needed in topsoil and subsoil, how they interact with macronutrients like N and P in fertilisers, how to improve their plant availability and longevity in the soil, and how to predict the ability of the soil to provide micronutrients in a balanced form. Sadly, micronutrient research in broadacre crops in Australia appears to have diminished since it has been possible to correct other deficiencies while the less visible impediments such as subclinical deficiency, infertile subsoils, and the inability of fertilisers to provide micronutrients that remain in plant-available forms on alkaline soils remain.

References

Ahmed, S., Evans, A. A. (1960). Cobalt: A micronutrient element for the growth of soybean plants under symbiotic conditions. Soil Science, 90: 205–210.

Ahmed, S., Evans, H. J. (1959). Biochemistry & Biophysiology Research Communications, 1: 271–275.

Anderson, A. J. (1942). Molybdenum deficiency on a South Australian ironstone soil. Journal of Australian Institute of Agricultural Science, 8: 73.

Arnon, M. B., Stout, P. R. (1939). Molybdenum as an essential element for higher plants. Plant Physiology, 14: 599–602.

Asher, C. J. (1987). Crop nutrition during the establishment phase: Role of seed reserves. Paper Presented at the Crop Establishment Problems in Queensland. Recognition, Research and Resolution, Queensland Agricultural College.

Baghurst, K. I., Dreosti, I. E., Syrette, J. A., Record, S. J., Baghurst, P. A., Buckley, R. A. (1991). Zinc and manganese status of Australian adults. Nutrition Research, 11: 23–32.

Baxter, P. (1957). Trace element deficiencies in fruit trees on Mallee and Wimmera soils. Journal of Agriculture Victoria, 55: 11–26.

Beckwith, R. S. (1966). The availability of micronutrient metals to plants with special reference to copper, zinc and manganese. Unpublished Ph.D. thesis, University of Western Australia, Perth, Australia.

Bertrand, I., Janik, L. J., Holloway, R. E., Armstrong, R. D., McLaughlin, M. J. (2002). The rapid assessment of concentrations and solid phase associations of macro-and micronutrients in alkaline soils by mid-infrared diffuse reflectance spectroscopy. Australian Journal of Soil Research, 40: 1339–1356.

Bolland, M. D. A. (1985). Effect of soil acidity and nutrient deficiencies on the growth and persistence of subterranean clover in pastures grown on sandy soils near Esperance, WA. Australian Journal of Experimental Agriculture, 25: 893–901.

Brennan, R. (1998). Manganese. In Moore, G. (Ed.), Soil Guide. A Handbook for Understanding and Managing Agricultural Soils (pp. 197–199). Agriculture Western Australia, Bulletin No. 4343.

Brennan, R. F. (1991). Effectiveness of zinc sulfate and zinc chelate as foliar sprays in alleviating zinc deficiency of wheat grown on zinc-deficient soils in Western Australia. Australian Journal of Experimental Agricultural Research, 31: 831–834.

Brennan, R. F. (1999). Lupin grain yields and fertiliser effectiveness are increased by banding manganese below the seed. Australian Journal of Experimental Agriculture, 39: 595–603.

Brennan, R. F. (2002). Residual value of molybdenum trioxide for clover production on an acidic sandy podzol. Australian Journal of Experimental Agriculture, 42(5): 565–570.

Brennan, R. F., Best, E. (1999). Copper. In Peverill, K. I., Sparrow, L. A., Reuter, D. J. (Eds.), Soil Analysis: An Interpretation Manual (pp. 303–307). CSIRO, Melbourne, Australia.

Brennan, R. F., Bolland, M. D. A. (2003). Application of fertilizer manganese doubled yields of lentil grown on alkaline soils. Journal of Plant Nutrition, 26(6): 1263–1276.
Brennan, R. F., Bruce, R. C. (1999). Molybdenum. In Peverill, K. I., Sparrow, L. A., Reuter, D. J. (Eds.), Soil Analysis: An Interpretation Manual (pp. 303–307). CSIRO, Melbourne, Australia.
Brennan, R. F., Longnecker, N. E. (2001). Effects of the concentration of manganese in the seed in alleviating manganese deficiency of *Lupinus angustifolius* L. Australian Journal of Experimental Agriculture, 41: 1199–1205.
Brennan, R. F., McGrath, J. F. (1988). The vertical movement of zinc on sandy soils in southern Western Australia. Australian Journal of Soil Research, 26: 211–216.
Brennan, R. F., Bolland, M. D. A., Bowden, J. W. (2004). Potassium deficiency, and molybdenum deficiency and aluminium toxicity due to soil acidification, have become problems for cropping sandy soils in south-western Australia. Australian Journal of Experimental Agriculture, 44: 1031–1039.
Cakmak, I., Marschner, H. (1986). Mechanism of phosphorus-induced zinc deficiency in cotton 1. Zinc deficiency-enhanced uptake rate of phosphorus. Plant Physiology, 68: 483–490.
Cakmak, I., Marschner, H. (1988). Enhanced superoxide radical production in roots of zinc deficient plants. Journal of Experimental Botany, 39: 1449–1460.
Carne, W. M. (1927). Grey speck disease of wheat and oats. Journal of Agriculture Western Australia, 45: 515–519.
Carne, W. M., Martin, D. (1937). Journal of the Council of Scientific & Industrial Research of Australia, 10: 47.
Carter, E. D., Heard, T. G. (1962). Land development on southern Yorke Peninsula. Journal of Agriculture, South Australia, 65: 392–395; 448–457.
Cartwright, B., Zarcinas, B. A., Mayfield, A. H. (1984). Toxic concentrations of boron in a red-brown earth at Gladstone, South Australia. Australian Journal of Soil Research, 22: 261–272.
Chatel, D. L., Robson, A. D., Gartrell, J. W., Dilworth, M. J. (1978). The effect of inoculation and cobalt applications on the growth of and nitrogen fixation by sweet lupins. Australian Journal of Agricultural Research, 29: 1191–1202.
Chaudhry, F. M., Loneragan, J. F. (1970). Effects of nitrogen, copper and zinc fertilizers on the copper and zinc nutrition of wheat plants. Australian Journal of Agriculture Research, 21: 865–879.
Connor, J. (1953). Effect of cultural practices on the manganese status of soil and citrus trees under irrigation. Australian Journal of Agricultural Research, 5: 31–38.
Donald, C. M., Prescott, J. A. (1975). Trace elements in Australian crop and pasture production. In Nicholas, D. J., Egan, A. R. (Eds.), Trace Elements in Soil-Plant-Animal Systems (pp. 7–37). Academic, New York.
Dong, B., Rengel, Z., Graham, R. D. (1995). Effects of herbicide chlorsulfuron on growth and nutrient uptake parameters of wheat genotypes differing in Zn-efficiency. Plant and Soil, 173: 275–282.
Doyle, R. J., Parkin, R. J., Smith, J. A. C., Gartrell, J. W. (1965). Molybdenum increases cereal yields on wheat belt scrub plain. Journal of Agriculture of Western Australia, 27: 118–120.
Duncan, O. W. (1961). Correction of zinc deficiency in wheat on the Darling Downs, Queensland. Queensland Journal of Agricultural and Animal Sciences, 24: 287–291.
Eagan, B. T., Whitaker, G. H. (1961) Copper deficiency in the Mossman area. Proceedings of the Queensland Society of Sugar Cane Technologists, 28.
Frischke, A. J. (1999). Supercharging wheat seed. In Eyre Peninsula Farming Systems: 2000 Summary (pp. 93–94). PIRSA (SARDI), Minnipa, Australia.
Gartrell, J. W. (1966). Nature (London), 209: 1050.
Gartrell, J. W. (1969). Technical Bulletin No. 3. Western Australian Department of Agriculture, South Perth, WA, Australia.
Gartrell, J. W. (1981). Distribution and correction of copper deficiency in crops and pastures. In Loneragan, J. F., Robson, A. D., Graham, R. D. (Eds.), Copper in Soils and Plants (Chapter 14, pp. 313–349). Academic, Sydney, Australia.
Gartrell, J. W., Glencross, R. N. (1968). Copper, zinc and molybdenum fertilisers for new land crops and pastures – 1969. Journal of Agriculture Western Australia, 9: 517–529.

Garwood, E. A., Williams, T. E. (1967). Growth, water use and nutrient uptake from the subsoil by grass swards. Journal of Agricultural Science, Cambridge, 69: 125–130.
Gilkes, R. J. (1975). Some properties of granulated superphosphate and its behaviour in the soil. Australian Journal of Soil Research, 13: 203–215.
Gilkes, R. J. (1981). Behaviour of Cu additives – fertilisers. In Loneragan, J. F., Robson, A. D., Graham, R. D. (Eds.), Copper in Soils and Plants (Chapter 5, pp. 97–117). Academic, Sydney, Australia.
Graham, R. D. (1991). Breeding wheats for tolerance to micronutrient deficient soil: Present status and priorities. In Saunders, D. A. (Ed.), Wheat for the Non-traditional Warm Areas (pp. 315–332). CIMMYT, Mexico.
Graham, R. D. (1999). Overcoming Subsoil Nutrient Limitations to Cereal Production (Final report). The University of Adelaide, Adelaide, Australia.
Graham, R. D., Ascher, J. S. (1993). Nutritional limitations of subsoils. Paper Presented at the Proceedings of the Twelfth International Plant Nutrition Colloquium, Perth, Australia.
Graham, R. D., Ascher, J. S., Hynes, S. C. (1992a). Selecting zinc efficient cereal genotypes for soils of low zinc status. Plant and Soil, 146: 241–250.
Graham, R. D., Davies, W. J., Ascher, J. S. (1985). The critical concentration of manganese in field grown wheat. Australian Journal of Agricultural Research, 36: 145–155.
Graham, R. D., Rengel, Z. (1993). Genotypic variation in zinc uptake and utilisation by plants. Paper Presented at the Proceedings of the International Symposium on Zinc in Soils and Plants, University of Western Australia, Perth, Australia.
Graham, R. D., Turner, N. C., Ascher, J. S. (1992b). Evidence for subsoil constraints and potential benefits from amelioration. Paper Presented at the Proceedings of the National Workshop on Subsoil Constraints to Root Growth and High Soil and Water use by Plants, Tanunda, South Australia.
Graham, R. D., Welch, R. M., Grunes, D. L., Carey, E. E., Norvell, W. A. (1987). Effect of zinc deficiency on the accumulation of boron and other mineral nutrients in barley. Soil Science Society of America Journal, 51: 652–657.
Grundon, N. J., Best, E. K. (1981). Rainfall pattern and response of wheat to soil dressings of copper sulphate (Poster). Proceedings of the Golden Jubilee International Symposium of "Copper in Soils and Plants", Perth, Australia.
Hannam, R. J. (1991). Nutrition and zinc update. In: Agronomy Technical Conference (pp. 90–94). SARDI, Adelaide, Australia.
Hannam, R. J., Davies, W. J., Graham, R. D., Riggs, J. L. (1984). The effect of soil- and foliar-applied manganese in preventing the onset of manganese deficiency in Lupinus angustifolius. Australian Journal of Agricultural Research, 35: 529–538.
Harry, S. P., Graham, R. D. (1981). Copper-zinc interactions in wheat, rye and triticale growing on soils of contrasting pH (Poster). Proceedings of the Golden Jubilee International Symposium of "Copper in Soils and Plants", Perth, Australia.
Higgs, E. D., Burton, M. V. (1955). Land development on southern Yorke Peninsula. Journal of Agriculture, South Australia, 59: 114–121.
Hirsch, M., Manton, R. (1989). Barley leaf diseases survey, 1988 (Technical report no. 143). Department of Agriculture SA, Adelaide, Australia.
Holloway, R. E. (1991). Factors affecting the growth of wheat roots in the subsoils of Upper Eyre Peninsula, South Australia. Unpublished M.Ag.Sci. thesis, University of Adelaide, Adelaide, Australia.
Holloway, R. E. (1996). Zinc as a subsoil nutrient for cereals. Unpublished Ph.D. thesis, University of Adelaide, Adelaide, Australia.
Holloway, R. E., Alston, A. M. (1992). The effects of salt and boron on growth of wheat. Australian Journal of Agricultural Research, 43: 987–1001.
Holloway, R. E., Bertrand, I., Frischke, A. J., Brace, D., McLaughlin, M., Shepperd, W. (2001a). Fluid fertilisers – an efficient source of P for calcareous soils. Plant Nutrition – Food Security and Sustainability of Agro-ecosystems, 832–833.
Holloway, R. E., Bertrand, I., Frischke, A. J., Brace, D., McLaughlin, M., Shepperd, W. (2001b). Improving fertiliser efficiency on calcareous and alkaline soils with fluid sources of P, N and Zn. Plant and Soil, 236: 209–219.

Holloway, R. E., Frischke, A. J., Brace, D. (2000). Multi-nutrient fluid fertiliser experiments. In Eyre Peninsula Farming Systems: 2000 Summary. PIRSA (SARDI), Minnipa, Australia.
Holloway, R. E., Frischke, A. J., Brace, D. (2002). How much fluid P fertiliser is enough? In Eyre Peninsula Farming Systems: 2002 Summary (pp. 84–89). PIRSA (SARDI), Minnipa, Australia.
Holloway, R. E., Frischke, B., Brace, D., Richter, I. (2004). Suspension fertilisers. In Eyre Peninsula Farming Systems: 2004 Summary (pp. 91–93). PIRSA (SARDI), Minnipa, Australia.
Irving, H., Williams, R. J. P. (1948). Order of stability of metal complexes. Nature (London), 162: 746–747.
Isbell, R. F. (1996). The Australian soil classification. CSIRO, Melbourne, Australia.
Janik, L. J., Skjemstad, J. O. (1995). Characterization and analysis of soils using mid-infrared partial least-squares. 2. Correlations with some laboratory data. Australian Journal of Soil Research, 33(4): 637–650.
Janik, L. J., Merry, R. H., Skjemstad, J. O. (1998). Can mid infrared diffuse reflectance analysis replace soil extractions? Australian Journal of Experimental Agricultural Research, 38: 681–696.
Janik, L. J., Skjemstad, J. O., Raven, M. D. (1995). Characterization and analysis of soils using mid-infrared partial least-squares. 1. Correlations with XRF-determined major element composition. Australian Journal of Soil Research, 33(4): 621–636.
Jarvis, R. J., Bolland, M. D. A. (1990). Placing superphosphate at different depths in the soil changes its effectiveness for wheat and lupin production. Fertilizer Research, 22: 97–107.
Kochian, L. V. (1993). Zinc absorption from hydroponic solutions by plant roots. In Robson, A. D. (Ed.), Zinc in Soils and Plants (pp. 45–57). Kluwer, Dordrecht, The Netherlands.
Levitt, E. C., Nicholson, R. I. (1941). Manganese deficiency in citrus in N.S.W. coastal districts. Agricultural Gazette of New South Wales, 52: 283–286.
Lombi, E., McLaughlin, M. J., Johnston, C., Armstrong, R. D., Holloway, R. E. (2004). Mobility and lability of phosphorus from granular and fluid monoammonium phosphate differs in a calcareous soil. Soil Science Society of America Journal, 68: 682–689.
Loneragan, J. F. (1975). The availability and absorption of trace elements in soil–plant systems and their relation to movement and concentration of trace elements in plants. In Nicholas, D. J. D., Egan, A. R. (Eds.), Trace Elements in Soil-Plant-Animal Systems (pp. 109–134). Academic, New York.
Loneragan, J. F., Webb, M. J. (1993). Interactions between zinc and other nutrients affecting the growth of plants. In Robson, A. D. (Ed.), Zinc in Soils and Plants (pp. 119–134). Kluwer, Dordrecht, The Netherlands.
Loneragan, J. F., Grimes, D. L., Welch, R. M., Aduayi, E. A., Tengal, A., Lazar, V. A. (1982). Phosphorus accumulation and toxicity in leaves in relation to zinc supply. Soil Science Society of America Journal, 46: 345–352.
Loneragan, J. F., Grove, T. S., Robson, A. D., Snowball, K. (1979). Phosphorus toxicity as a factor in zinc–phosphorus interactions in plants. Soil Science Society of America Journal, 43: 966–972.
Loneragan, J. F., Kirk, G. J., Webb, M. J. (1987). Translocation and function of zinc in roots. Journal of Plant Nutrition, 10: 1247–1254.
Longnecker, N. E., Graham, R. D., Marcar, N. E. (1988). The effect of seed manganese on barley (Hordeum vulgare L.) growth and yield. Paper Presented at the International Symposium on Manganese in Soils and Plants, Adelaide, Australia.
Ma, G., Rengasamy, P., Rathjen, A. J. (2003). Phytotoxicity of aluminium to wheat plants in high-pH solutions. Australian Journal of Experimental Agriculture, 43: 497–501.
Marschner, H. (1995). Mineral Nutrition of Higher Plants (2nd ed.). Academic, London.
Marschner, H., Cakmak, I. (1986). Mechanism of phosphorus induced zinc deficiency in cotton. II. Evidence for improved shoot control of phosphorus uptake and translocation under zinc deficiency. Plant Physiology, 68: 491–496.
McFarlane, J. D. (1999). Iron. In Peverill, K. I., Sparrow, L. A., Reuter, D. J. (Eds.), Soil Analysis: An Interpretation Manual (Chapter 20, pp. 295–300). CSIRO, Melbourne, Australia.
McFarlane, J. D., Potter, T. D., Lewis, D. C. (1980). Correction of molybdenum deficiency in dryland sunflowers on acid sands in the south east of South Australia. Sunflower, 4: 8.

McHargue, J. S. (1922). The role of manganese in plants. Journal of the American Chemical Society, 44: 1592.

McLachlan, K. D. (1953). Phosphorus, sulphur and molybdenum deficiencies on some soils of the northern territory. Journal of the Australian Institute of Agricultural Science, 19: 197–199.

Millikan, C. R. (1938). A preliminary note on the relation of zinc to disease in cereals. Journal of Agriculture, Victoria (August), 409–416.

Mortvedt, J. J., Murphy, L. S., Follett, R. H. (2001). Fertilizer Technology and Application (2nd ed.). Meister, Willoughbuy, OH.

Murphy, L. S., Keikam, D. R., Lamond, R. E., Gallagher, P. J. (1978). Applying N & P at the same time into the same soil shows promise for winter wheat. In Martin, S. (Ed.), Better Crops with Plant Food (pp. 16–24). Potash and Phosphate Institute, GA.

Nable, R. O., Webb, M. J. (1993). Further evidence that zinc is required throughout the root zone for optimal plant growth and development. Plant and Soil, 150: 247–253.

Newman, R. J. (1955). Molybdenum deficiency in central highlands and upper Goulburn Region and adjoining districts. Journal of Agriculture, Victoria, 53: 451.

Newman, R. J. (1962). (Bulletin 1): Grassland Society, Victoria. Northern Australia Scoping Study. (2004). Available at http://cotton.pi.csiro.au/AboutUS/CrcNASs.htm.

Northern Territory Horticulture Division Horticulture Technical Annual Report (1999). Available at http://kakadu.nt.gov.au/pls/portal30/docs/folder/dbird_pi/horticulture/research_development/tb280.pdf

Ozanne, P. G., Greenwood, E. A. N., Shaw, T. C. (1963). The cobalt requirement of subterranean clover in the field. Australian Journal of Agricultural Research, 14: 39.

Pearson, J. N., Rengel, Z. (1994). Distribution and remobilization of Zn and Mn during grain development in wheat. Journal of Experimental Botany, 45: 1829–1835.

Pearson, J. N., Rengel, Z. (1995). Uptake and distribution of ^{65}Zn and ^{54}Mn in wheat grown at sufficient and deficient levels of Zn and Mn 1. During vegetative growth. Journal of Experimental Botany, 46: 833–839.

Pearson, J. N., Rengel, Z., Graham, R. (1995). Transport of zinc and manganese to developing wheat grains. *Physiologia Plantarum*, 95: 449–455.

Piper, C. S. (1938). The occurrence of "Reclamation Disease" in cereals in South Australia (Pamphlet). Council of Scientific & Industrial Research, Australia.

Piper, C. S. (1940). Molybdenum as an essential element for plant growth. Journal of Australian Institute of Agricultural Science, 6: 162–164.

Powrie, J. K. (1960). A field response by subterranean clover to cobalt fertilizer. Australian Journal of Science, 23: 198.

Regan, K., Siddique, K., Shackles, B. (2001). Kabuli chickpea production in the Ord River Area (Technical report no. 99/2001). Agriculture Western Australia, Perth, Australia.

Reghenzani, J. R. (1993). A survey of the nutritional status of north Queensland sugarcane soils with particular reference to zinc. Proceedings of the Australian Society of Sugar Cane Technologists, 15: 298–304.

Rengel, Z., Graham, R. D. (1995a). Importance of seed Zn content for wheat growth on Zn-deficient soil 1.Vegetative growth. Plant and Soil, 173: 259–266.

Rengel, Z., Graham, R. D. (1995b). Importance of seed Zn content for wheat growth on Zn-deficient soil 2.Grain yield. Plant and Soil, 173: 267–274.

Reuter, D. J. (1975). The recognition and correction of trace element deficiencies. In Nicholas, D. J. D., Egan, A. R. (Eds.), Trace Elements in Soil-Plant-Animal Systems. Academic, New York.

Reuter, D. J., Robinson, J. B. (1997). Plant Analysis: An Interpretation Manual (2nd ed.). CSIRO, Melbourne, Australia.

Reuter, D. J., Cartwright, B., Judson, G. J., McFarlane, J. D., Maschmedt, D. J., Robinson, J. B. (1988). Trace Elements in South Australian Agriculture (No. 139). South Australian Department of Agriculture, Adelaide, Australia.

Reuter, D. J., Heard, T. G., Alston, A. M. (1973a). Correction of manganese deficiency in barley crops on calcareous soils 2. Comparison of mixed and compound fertilizers. Australian Journal of Experimental Agriculture and Animal Husbandry, 13: 440–445.

Reuter, D. J., Heard, T. G., Alston, A. M. (1973b). Correction of manganese deficiency in barley crops on calcareous soils I. Manganous sulphate applied at sowing and as foliar sprays. Australian Journal of Experimental Agriculture and Animal Husbandry, 13: 434–439.

Riceman, D. S., Anderson, A. J. (1943). Copper and zinc deficiency in pasture and crops in South Australia. Journal of Agricultural Science, Australia, 47: 16–29.

Riceman, D. S., Donald, C. M. (1938). A crop deficiency in plants at Robe, South Australia. Preliminary investigations of the effect of copper & other elements on the growth of plants in a calcareous sand at Robe, South Australia. (Pamphlet): Council of Scientific & Industrial Research, Australia.

Riley, M. M., Gartrell, J. W., Brennan, R. F., Hamblin, J., Coates, P. (1992). Zinc deficiency in wheat and lupins in Western Australia is affected by the source of phosphate fertiliser. Australian Journal of Experimental Agriculture, 32: 455–463.

Robson, A. D. (1981). Conclusion: Copper in soils and plants – an overview. In Loneragan, J. F., Robson, A. D., Graham, R. D. (Eds.), Copper in Soils and Plants. Academic, Sydney, Australia.

Robson, A. D., Reuter, D. J. (1981). Diagnosis of copper deficiency and toxicity. In Loneragan, J. F., Robson, A. D., Graham, R. D. (Eds.), Copper in Soils and Plants (Chapter 13, pp. 287–312). Academic, Sydney, Australia.

Ross, B. J., Calder, G. J. (1990). Nutrient studies on sandy red earths in the Douglas-Daly area, northern territory. Tropical Grasslands, 24: 121–123.

Samuel, G., Piper, C. S. (1928). Grey speck (manganese deficiency) disease of oats. Journal of Agriculture in South Australia, 31: 696–705.

Samuel, G., Piper, C. S. (1929). Manganese as an essential element for plant growth. Annals of Applied Biology, 16: 493.

Sauvé, S., Hendershot, W., Allen, H. E. (2000). Solid-solution partitioning of metals in contaminated soils: Dependence on pH, total metal burden and organic matter. Environmental Science and Technology, 34: 1125–1131.

Schmid, W. E., Haag, H. P., Epstein, E. (1965). Absorption of zinc by excised barley roots. Plant Physiology, 18: 860–869.

Schroeder, B. L., Webster, K., Davies, B., Wood, A. W. (1999). Fertiliser recommendations. Paper Presented at the Sustainable Nutrient Management in Sugarcane Production Short Course, Brisbane.

Sharma, K. C., Kranzt, B. A., Brown, A. L., Quick, J. (1968). Interaction of zinc and phosphorus in top and root of corn and tomato. Journal of Agronomy, 60: 453–456.

Simpson, J. R., Pinkerton, A. (1989). Fluctuations in soil moisture and plant uptake of surface applied phosphate. Fertilizer Research, 20: 101–108.

Singh, J. P., Dahiya, D. J., Narwal, R. P. (1990). Boron uptake and toxicity in wheat in relation to zinc supply. Fertilizer Research, 24: 105–110.

Singh, J. P., Karwasna, S. P. S., Singh, M. (1988). Distribution and forms of copper, iron, manganese and zinc in calcareous soils of India. Soil Science, 146: 359–365.

Slattery, M. G., Rainbow, R. W. (1995). Investigation of seeder related factors on seed placement, crop establishment, growth and yield of wheat (Technical research report 1990–1993). SARDI & University of South Australia, Adelaide, Australia.

Stout, P. R. (1972). In Trace elements in Soil-Plant-Animal systems: Trace elements in Australian crop and pasture production. In Mortvedt, J. J., Giordano, P. M., Lindsay, W. L. (Eds.), Micronutrients in Agriculture (1st ed.). Soil Science Society of America, Madison, WI.

Sumner, M. E., Farina, N. P. W. (1986). Phosphorus interactions with other nutrients and lime in field cropping systems. In Stewart, B. A. (Ed.), Advances in Soil Science (Vol. 5). Springer, Heidelberg.

Teitzel, J. K. (1969). Responses to phosphorus, copper and potassium on a granite loam of the wet tropical coast of Queensland. Tropical Grasslands, 3(1): 43–48.

Teitzel, J. K., Bruce, R. C. (1971). Fertility studies of pasture soils in the wet tropical coast of Queensland 2. Granitic soils. Australian Journal of Experimental Agriculture and Animal Husbandry, 11: 77–84.

Teitzel, J. K., Bruce, R. C. (1972a). Fertility studies of pasture soils in the wet tropical coast of Queensland 3. Basaltic soils. Australian Journal of Experimental Agriculture and Animal Husbandry, 12: 49–54.

Teitzel, J. K., Bruce, R. C. (1972b). Fertility studies of pasture soils in the wet tropical coast of Queensland 4. Soils derived from metamorphic rocks. Australian Journal of Experimental Agriculture and Animal Husbandry, 12: 281–287.

Teitzel, J. K., Bruce, R. C. (1973a). Fertility studies of pasture soils in the wet tropical coast of Queensland 5. Mixed alluvial soils. Australian Journal of Experimental Agriculture and Animal Husbandry, 13: 306–311.

Teitzel, J. K., Bruce, R. C. (1973b). Fertility studies of pasture soils in the wet tropical coast of Queensland 6. Soils derived from beach sand. Australian Journal of Experimental Agriculture and Animal Husbandry, 13: 312–318.

Tiller, K. G. (1983). Micronutrients. SOILS: An Australian viewpoint (Chapter 25, pp. 365–387). CSIRO, Melbourne, Australia and Academic, London.

Tiver, N. S. (1955). Deficiencies in South Australian soils. Journal of Agriculture, South Australia, 59: 100–113.

Toms, J. (1958). The use of copper and zinc in the cereal-growing districts of Western Australia. The Journal of Agriculture of Western Australia, 197–203.

Uren, N. C. (1999). Manganese. In Peverill, K. I., Sparrow, L. A., Reuter, D. J. (Eds.), Soil Analysis: An Interpretation Manual (Chapter 19, pp. 287–294). CSIRO, Melbourne, Australia.

Walter, K. H. (1988). Manganese fertilizers. In Graham, R. D., Hannam, R. J., Uren, N. C. (Eds.), Manganese in Soils and Plants (pp. 225–241). Kluwer, Dordrecht, The Netherlands.

Warington, K. (1923). The effect of boric acid and borax on the broad bean and certain other plants. Annals of Botany (London), 37: 629–672.

Webb, M. J., Loneragan, J. F. (1988). Effect of zinc deficiency on growth, phosphorus concentration and phosphorus toxicity of wheat plants. Soil Science Society of America Journal, 52: 1676–1680.

Webb, M. J., Loneragan, J. F. (1990). Zinc translocation to wheat roots and its implications for a phosphorus/zinc interaction in wheat plants. Journal of Plant Nutrition, 13: 1499–1512.

Welch, R. M. (1993). Zinc Concentrations and forms in plants for humans and animals. In Robson, A. D. (Ed.), Zinc in Soils and Plants (pp. 183–195). Kluwer, Dordrecht, The Netherlands.

Welch, R. M., House, W. A. (1984). Factors affecting the bioavailability of mineral nutrients in plant foods. In Welch, R. M., Gabelman, W. H. (Eds.), Crops as Sources of Nutrients for Humans (pp. 37–54). American Society of Agronomy, Madison, WI.

Welch, R. M., Allaway, W. H., House, W. A., Kubota, J. (1991). Geographic distribution of trace element problems. In Mortvedt, J. J., Cox, F. R., Shuman, L. M., Welch, R. M. (Eds.), Micronutrients in Agriculture (2nd ed.) (Chapter 2, pp. 31–57). Soil Science Society of America, Madison, WI.

Welch, R. M., Webb, M. J., Loneragan, J. F. (1982). Zinc in membrane function and its role in phosphorus toxicity. Paper Presented at the Proceedings of the 9th International Plant Nutrition Colloquium, Warwick University, England.

Wilhelm, N. S., Growden, B. (1999). Phosphorus fertilisation for field crops in difficult soils. Paper Presented at the Proceedings of the 91st Annual meeting American Society of Agronomy, Salt Lake City, UT.

Williams, C. H. (1974). Heavy metals and other elements in fertilizer-environmental considerations. In Leece, D. R. (Ed.), Fertilizers and Environment (pp. 123–130). Proceedings of the Symposium of the Australian Institute of Agricultural Science, Australian Institute of Agricultural Science, Sydney, Australia.

Williams, C. H., Andrew, C. S. (1970). Mineral nutrition of pastures. In Moore, R. M. (Ed.), Australian Grasslands (Chapter 21, pp. 321–338). Australian National University Press, Canberra, Australia.

Chapter 4
Micronutrient Deficiencies in Crops and Soils in India

Maha V. Singh

Abstract This chapter describes the micronutrient status of soils in India, the occurrence of deficiencies, responses of different crops to them, the suitability of various sources of the elements, and techniques for their application. Crops grown in most soils in India suffer from deficiencies of one or more micronutrients, even though the soils often contain apparently adequate total amounts of the respective elements. The nature and extent of deficiencies varies with soil type, crop genotype, management and agro-ecological situations. With the intensive cropping of high yielding varieties of rice and wheat, deficiency of zinc (Zn) initially, and subsequently deficiencies of iron (Fe) in rice, and manganese (Mn) in wheat, emerged as threats to sustaining high levels of food crop production. Micronutrient deficiencies are now frequently observed in intensively grown cereals, oilseeds, pulses and vegetable crops. With widespread and regular application of Zn fertilizers, the occurrence of Zn deficiency has declined in recent years, but multi micronutrient deficiencies are now becoming an increasing problem. Analysis of soil and plant samples has indicated that 49% of soils in India are potentially deficient in Zn, 12% in Fe, 5% in Mn, 3% in copper (Cu), 33% in boron (B) and 11% in molybdenum (Mo). Basal application to soil and/or foliar sprays of Zn, B and Mo, and foliar sprays of Fe and Mn have been recommended as the most suitable methods for correcting such deficiencies in crops.

4.1 Introduction

Indian agriculture during the past 50 years has achieved a fourfold growth in food production by adopting modern agricultural practices. The availability of fertilizer-responsive, high yielding varieties (HYV) of rice, wheat and maize has made it possible to produce 15–20 t ha^{-1} $year^{-1}$ of biomass drymatter. Initially, this became possible by using nitrogenous (N) fertilisers alone, as the soil could provide all other essential nutrients needed by plants. However, within a few years, the nutrient reserves in soil were gradually exhausted and it was no longer possible to sustain higher yields even by applying both N and phosphorus (P) (Kanwar and Randhawa,

B.J. Alloway (ed.), *Micronutrient Deficiencies in Global Crop Production.*

1974; Singh, 1988). Thus, in areas of high cropping intensity, deficiencies of secondary nutrients and micronutrients were frequently observed in cereal, oilseed, pulse and vegetable crops, which became critical in obtaining and sustaining higher crop production over the years (Fig. 4.1 in Colour Section) (Singh, 2001a; Singh et al., 2004a).

India is endowed with a diverse climate. The rainfall ranges generally between 200 and 2,200 mm per annum. The country has 142 Mha of cultivable land out of a total area of 329 Mha. The cropping intensity is 145%. Cropping systems based on rice (*Oryza sativa* L.), wheat (*Triticum aestivum* L.), maize (*Zea mays* L.), sorghum (*Sorghum vulgare* L.), cotton (*Gossypium hirsutum* L.), soybean (*Glycine max* L.), sugarcane (*Saccarum officinarum* L.), potato (*Solanum tuberosum* L.), mustard (*Brassica juncea* Coss.), groundnut (*Arachis hypogaea* L.), green gram (*Phaseolus aureus* Roxb), black gram (*Phaseolus mungo* L.) and chick pea (*Cicer arietinum* L.) occupy most of the area. These crops show high responses to the application of both macro and micronutrients.

Among the major soil groups, the predominant soils are red soils (USDA Soil Taxonomy: Inceptisols) (105.5 Mha), black clayey, swell-shrink soils (Vertisols) (73.8 Mha), alluvial soils (Inceptisol/ Entisols) (58.4 M ha), lateritic soils (Alfisols) (7 Mha), desert soils (Aridisols) (30 Mha) and foothill tarai soils (Mollisols) (26.8 Mha). Most Indian soils are inherently low in N fertility. About 50, 53 and 41% of soils are deficient in phosphorus (P), potash (K) and sulphur (S), respectively (Singh, 2001a). Crops grown in about half of the country's soils suffer from one or more micronutrient disorders (Singh, 1991a; Takkar et al., 1990). Field-scale zinc (Zn) deficiency was first noticed in rice in tarai soils (Mollisols) of the foot hills of the Himalayas, which caused the complete failure of the rice crop (Nene, 1965).The problem is more acute in alkali soils (Singh and Abrol, 1986b).

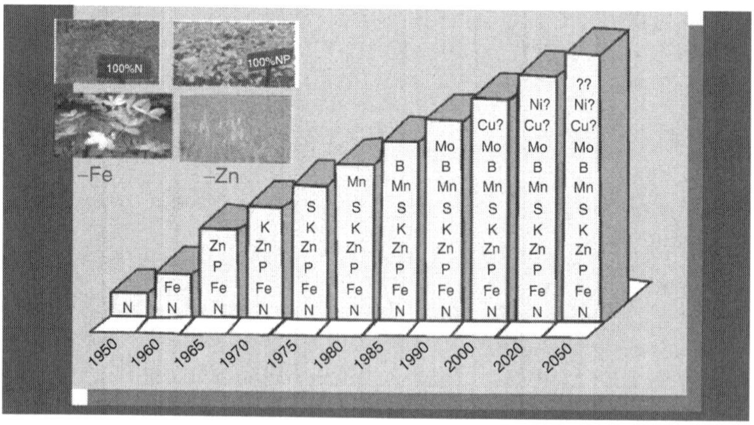

Fig. 4.1 Impact of the green revolution on the emergence of micronutrient deficiencies in crops in India (*See Colour Plates*)

India is the third largest fertiliser-consuming country in the world, its average NPK usage was 106 kg ha^{-1} in 2004–2005. However, the consumption of fertilizer is highly uneven. Among 466 districts, 37 districts consumed 25% of the total NPK, 102 districts 50% and 202 districts consumed the remaining 25% of the total 18.1 Mt of fertilizer nutrients (FAI, 2002).

With a predicted population of 1.5 billion (1.5×10^9) by the year 2050, India needs to increase its current food production by 5 Mt as compared to 3.1 Mt $year^{-1}$ achieved during the past 4 decades. This is a daunting task as there is very little scope for expanding the area under cultivation. Therefore, future food, fuel and fibre requirements will have to be met through increasing agricultural productivity. Such practices will further accentuate the deficiencies of micronutrients in crops and soils and may pose a threat to the nutritional security and environmental safety of the country. In view of this, research carried out on micronutrients in crops and soils, summarized in this chapter is of major importance.

4.2 Boron

4.2.1 Total Boron Content in Soils

Total B contents in Indian soils varied from 7 to 630 mg B kg^{-1}. Soils in arid and semi-arid tropical regions had the highest contents of B. Satyanarayan (1958) reported the influence of the geology of soil parent materials on the total B content as 7 mg B kg^{-1} in granite (gneiss), 22 mg B kg^{-1} in lavas, 36 mg B kg^{-1} in basalt, 13 mg B kg^{-1} in limestone, 62 mg B kg^{-1} in alluvium, 44 mg B kg^{-1} in slate and 371 mg B kg^{-1} in laterite. Certain young calcareous alluvial soils (Calciorthents/Calcifluvents) had B contents of up to 83 mg kg^{-1}. Coarse textured, leached soils (Haplustalfs) showed lower B contents and its accumulation was dependant upon the calcium carbonate ($CaCO_3$) content in the profile (Sakal and Singh, 1995). On the other hand, Vertisol and Alfisol soils of the Deccan trap formed from granite, basalt, shale, slate and lime stone had 28–57 mg B kg^{-1}. Raychaudhary and Datta Biswas (1964) found that Xerochrepts developed on granite and crystalline gneiss contained 8.5 mg B kg^{-1}, but lateritic (Alfisol) soils from more or less similar geologic parent materials contained 25 mg B kg^{-1}.

4.2.2 Available Boron Contents in Soils

Concentrations of 'available' (hot water soluble – HWS) B in Indian soils ranged from 0.75 to 8.0 mg B kg^{-1} (Singh, 2001b). Availability of soil B to plants is often related to the total B content as well as other properties such as pH, $CaCO_3$ and organic matter contents, nutrient interactions, plant type or variety and environmental factors which strongly influence the emergence of B deficiency or toxicity in plants (Sakal et al., 1996; Saha and Singh, 1997).

Deficiencies of B in Indian soils ranged from 2% in alluvial soils (Ustipsamments) of Gujarat, to 68% in red soils (Calciorthhents, Haplustalfs) in Bihar, with a mean of 33% for the whole country (Singh, 1999a; Singh et al., 2006). A maximum occurrence of B deficiency (54–86%) was recorded in Alfisol soils of Assam and West Bengal, due to a decrease in water soluble B with increase in rainfall (Sakal and Singh, 1995). A significant negative correlation exists between soluble B and $CaCO_3$ content (Saha and Singh, 1997). An increase in organic carbon (OC) content from 0.50 to 0.75% enhances the fixation of B in soils by 48–60%. Thus, association of B with OC prevents its leaching and thereby ensures its higher availability to crop plants (Katyal and Vlek, 1985). High concentrations of B were recorded in the saline soils of the Indo-Gangetic plain and moderate level in Vertic Ustochrepts of Rajasthan and Madhya Pradesh (Mathur et al., 1964; Saha et al., 1998).

4.2.3 Crop Responses to Boron

Boron is neither a constituent of enzymes, nor directly involved in the enzymatic activities of plants. It forms very stable organic compounds with a cis-diol configuration and it is strongly complexed in cell walls. Its concentration in cell walls reflects the differences in B requirement among the plant species. Boron is also important for transport of K into guard cells, thus stomatal opening increases with B deficiency. The role of B in pollen germination and pollen tube growth is very important for enhancing crop production.

Responses of different crops to B fertilisation are shown in Table 4.1 (Singh, 2001c). In B-deficient soils, the average yield response of maize, wheat and rice to the application of 1.0 to 1.5 kg B ha^{-1} together with NPK was 570, 390 and 320 kg ha^{-1} (32.5%, 15.1% and 16.6%), respectively over NPK without B (Sakal and Singh, 1995). Chickpea, pigeon pea (*Cajanus cajan* Milsp.), and groundnut gave economic yield responses of 430, 340, 370 kg ha^{-1} (69%, 29.6%, 30%), respectively, to B fertilisation. Seed yields of sesame (*Sesamum indicum* L.) and mustard increased by 90 and 210 kg ha^{-1} (15.9% and 14.0%) whereas raya (*Brassica compestris* L.) showed a 48% yield increase over the B control plots (Singh, 2001c). The yield of sunflower (*Helianthus annuus* L.) increased significantly by 320–480 kg ha^{-1} (24–38%) when 1–2 kg B ha^{-1} was applied, either prior to sowing, or when 1–2 kg B ha^{-1} was dusted on the crop or the plants sprayed with 0.2% borax solution, twice on the flower head during the ray floret opening stage (growth stage, Zadoks 30).

Application of B significantly enhanced ‘pegging’ in groundnuts and increased the pod yield by 20.9%. Boron fertilisation increased the linseed (*Linum ustatissinum* L.) oil yield by 59–68%, but it had no effect on the oil content of seeds.

Table 4.1 Crop response to boron fertilisation in Indian soils (Compiled by Singh, 1999b, 2001c)

Crop	No. of field trials	Responses to B added kg ha^{-1} over NPK		Percent response over NPK controls
		Range	Mean	
Rice	107	10–1,670	320	16.6
Wheat	35	30–1,170	390	15.1
Maize	5	170–1,050	570	32.5
Chickpea	7	90–900	350	44.1
Lentil	4	40–490	240	18.6
Groundnut	11	50–420	120	9.9
Sesame	5	24–168	90	23.9
Mustard	15	120–500	268	32.8
Sunflower	3	30–850	550	35.2
Cotton	2	60–450	260	11.6
Onion	4	3,870–7,300	4,470	29.5

Crops grown in Alfisol soils of Jharkhand and north West Bengal are more vulnerable to B deficiency. Lentil (*Lens esculenta* Moench) yields increased by 50, 300 and 350 kg ha^{-1} (4%, 24%, 28%), respectively, with the application of 0.5, 1.0, 2.0 kg B ha^{-1}. Application of B enhanced growth and fruiting in chickpea, and gave yield responses 250–750 kg ha^{-1} (16.6–48.5%) over controls (Singh and Saha, 1994, Singh, 2000, Singh et al., 2006). The response of crops to B depends upon the inherent fertility status of soils. Chickpea gave 35–47.7% extra yield in calcareous soils having hot water soluble B concentrations below 0.5 mg kg^{-1}, but yields were reduced by 3–24% in similar soils having hot water soluble B concentrations above 0.5 mg B kg^{-1}. The yield of onion (*Allium cepa* L.) increased by 4,470 kg ha^{-1} (29.5%) and that of sugarcane by 8,700 kg ha^{-1} (12.2%) with the application of 2 kg B ha^{-1} in highly B-deficient Calciorthents (Sakal et al., 1996; Singh, 2005).

Spectacular yield responses in cereals, pulses and oilseed crops were observed with the application of 0.5–1.0 kg B ha^{-1} largely in calcareous soils (Calciorthents) of Bihar, swell-shrink (Typic Haplusterts /Vertic Ustochrepts) soils of Gujarat and Madhya Pradesh, acidic leached soils (Rhodolstults, Ochracults) and red and lateritic soils (Haplustalfs) of Assam, and in several leached coastal soils (Sakal et al., 1997, 2001c) (Table 4.2). Red lateritic soils (Alfisols) are highly deficient in B, so application of 0.5–2 kg B ha^{-1} increased the rice yield by 460–1,500 kg ha^{-1} (12.5–37.2%) over controls. Response to B was greater in rabi (winter season) rice than in kharif (monsoon season) rice (Ali, 1992; Nandi, 1992; Datta et al., 1992). In 19 field trials in deltaic alluvial (Ustorthent) soils in Assam, application of 2 kg B ha^{-1} to rice increased the grain yield significantly by 180–460 kg B ha^{-1} (25.2–39%) and that of the following wheat crop by 230–320 kg ha^{-1} (23.1–35.8%) (Sakal et al., 1997). Vegetable, cereal and coconut crops also responded significantly to B fertilisation in the north eastern hill region.

Table 4.2 Cumulative response (grain yield) of different crops as influenced by rates and frequency of boron application in three cropping cycles in calcareous soil (Singh et al., 2006)

B rate kg ha^{-1} and use frequency	Total B added kg ha^{-1}	Rice–wheat (3 + 3 = 6 crops) kg ha^{-1}	kg kg^{-1} B	%	Maize–mustard (3 + 3 = 6 crops) kg ha^{-1}	kg kg^{-1} B	%	Sesame–chickpea (3 + 3 = 6 crops) kg ha^{-1}	kg kg^{-1} B	%
0.8 A	2.4	2,710	1,129	14.2	1,590	663	13.2	377	157	7.2
0.8 C	4.8	3,380	704	17.8	2,650	553	22.1	955	199	18.1
1.6 A	4.8	4,130	860	21.7	2,730	569	22.7	1,162	242	22.1
1.6 C	9.6	2,540	265	13.3	2,530	264	21.1	1,117	116	21.2
32 I, 8C	72	2,850	39	15.0	2,310	32	19.2	719	10	13.6
F.S. 2	9.0	1,330	147	7.0	1,520	168	12.7	555	62	10.5
LSD ($P < 0.05$)		378	–	–	469	–	–	182	–	–

B added I = Once to first crop, A = Alternate, C = All crops, F.S. = Foliar spray of 0.25% H_3BO_3

4.2.4 *Correction of Boron Deficiency*

Dicotyledon legume crops absorb more B than monocotyledon cereal crops and so the former require higher levels of B. Studies show that plants of the *Crucifereae, Papilionaceae* and *Leguminaceae* families have relatively high B requirements and, therefore, plants of these families contain generally high concentrations of B (>25 mg kg^{-1}) in their foliage. Boron deficiency occurs in many plants when its concentration in the upper fully matured leaves is found to be below 15 mg B kg^{-1}. The sufficiency range of B is between 20 and 100 mg kg^{-1} (Singh, 1998). Boron concentrations are higher in upper leaves than lower leaves and decrease with the age of plants.

Boron is generally applied through broadcasting and mixing into the soil prior to sowing or before transplanting the crop (Singh et al., 2006). Band placement of 0.5–2.0 kg B ha^{-1} resulted in higher B concentrations in cauliflower (*Brassica oleracae* var *botrytis* L.) plants than broadcasting (Singh, 2005). However, band placement can also lead to B toxicity in plants, when it is applied in excess, or if it is placed too close to seedlings or shoots. Therefore, B should neither be placed in contact with seed or seedlings, nor should excessive doses be used because of potential toxicity problems. Regular use of higher doses of B can lead to its toxicity in crop plants (Sakal et al., 1996; Singh, 2005, 2006b).

Boron deficiency is one of the important nutritional problems limiting crop production in calcareous soils (Calcifluvents, Calciorthents) of Bihar and Saurashtra. Optimum doses of B for calcareous and heavy textured Vertisols ranged between 1 and 2 kg B ha^{-1} and 0.5 and 1.0 kg B ha^{-1} in light textured Entisols. Optimum requirements of B for maize, wheat and rice were found to be between 0.75 and 1.5 kg B ha^{-1} (Dangarwala et al., 1994) (Table 4.2) whereas in chickpea, pigeon pea, groundnut, sunflower and mustard, productivity increased significantly with the application of 1–2 kg B ha^{-1}. Sesame and linseed responded significantly to 0.5–0.75 kg B ha^{-1} broadcast prior to sowing (Singh, 2001c). Onion, cotton and sugarcane crops require more B, with applications of between 1 and 2 kg B ha^{-1} (Sakal and Singh, 1995).

Frequency of application depends upon doses of B applied and the nature of the crop. Studies showed that application 0.8 kg B ha^{-1} to each crop, or 1.6 kg B ha^{-1} to alternate crops gave higher yields in rice–wheat, maize–wheat and sesame–chickpea cropping systems and were found optimum for sustaining higher productivity (Table 4.2). Regular applications of more than 2 kg B ha^{-1} caused adverse effects on the growth and yield of crops. Therefore, one should apply the optimum dose of B at the desired frequency (Singh et al., 2006).

Crop species and their cultivars differ significantly in their relative response to B, so its fertilisation has to be practiced accordingly. Sakal and Singh (1995) reported the tolerance of sesame cultivars to B stress in the order: RT-54 > OMT-11-63 > OMT-11-6-5 = Krishna > TC 25 and that of mustard cultivars as: Pusa Bold > RH-30 > = Kranti > RAURD = BR-40 = Varuna in B-deficient Psamment soils in Bihar. Among the various sources of B, the efficacy of borax, Granubor-II and boric acid was found to be almost equal in increasing the productivity of groundnuts, soybeans, rice, cauliflower, chickpea, and maize. Basal application of B, through broadcasting, gave the best response. However, if a basal application is missed, foliar sprays of 2.0–2.5 g l^{-1} of boric acid or Solubor can be used for correcting the B deficiency quite efficiently.

4.3 Copper

4.3.1 Total Copper Contents in Soils

Total Cu contents of Indian soils varied between 1.8 and 285 mg Cu kg^{-1} as compared to 2–100 mg Cu kg^{-1} in agricultural soils of the world (Swaine, 1955). Raychaudhary and Datta Biswas (1964) reported mean Cu contents of 197, 70 and 20 mg Cu kg^{-1} in soils derived from shale, igneous rocks and limestone, respectively. Surface soils in the semi-arid tropics, contained more Cu than those in humid and sub-humid regions (Katyal et al., 1982). Total Cu contents in 28 surface samples of normal cultivated soils in Punjab ranged between 6.6 and 36.4 mg Cu kg^{-1} compared to 52–63 mg Cu kg^{-1} in swell-shrink (Vertisol) soils of Madhya Pradesh (Rathore et al., 1995). Total Cu contents increased significantly with increase in clay content in soils. Copper contents in Vertisols in central India varied from 13 to 167 mg Cu kg^{-1}. These soils are better endowed with Cu than coarse textured Alfisols and Entisols (Katyal and Vlek, 1985; Singh, 1999b).

4.3.2 Available Copper Contents in Soils

Available Cu content, extracted by the diethylene triamine penta acetic acid (DTPA) method (Lindsay and Norvell, 1978), in Indian soils ranged between 0.1 and 8.2 mg Cu kg^{-1} with a mean of 1.90 mg Cu kg^{-1} (Singh, 1999a, 2000). Analysis of 252,000 soil samples indicated that most Indian soils have sufficient contents of available

Cu (Kanwar and Randhawa, 1974; Singh, 2001b). Copper availability is dependant upon soil characteristics. It increases with increase in organic matter and clay content, but decreases with increase in pH and $CaCO_3$ content of the soil (Nayyar et al., 1990). Studies on the redistribution of Cu in different chemical forms in a calcareous soil indicated that water soluble plus exchangeable, complexed, organically bound, occluded and residual Cu fractions amounted to 0.22%, 0.80%, 9.8%, 15.4% and 68.5%, of the total Cu content, respectively. Among the different chemical forms, only 10.8% of the total Cu in these soils was found in pools available to plants and the rest remains in unavailable forms (Sakal et al., 1996). Application of manure decreased the $CaCO_3$-bound Cu fraction and increased the organically bound Cu, which showed a beneficial effect on maize yield (Patel et al., 2005).

Organic peat soils (Histosols) of Kerala, hill (Alfisols) and submontane soils (Mollisols) of the Himalayan tarai zone, alluvial (Inceptisol) soils in Uttar Pradesh, Bihar and north Madhya Pradesh all showed Cu deficiency. Forage and fodder crops gave a high response to Cu fertilisation in the sand dunes of Rajasthan due to their low Cu status (Patel and Singh, 1995). Regular sprays of Cu-containing fungicides result in a build up of Cu in orchards, which poses a potential Cu toxicity problem in grapes and citrus (Singh, 1998, 2001b).

4.3.3 Crop Responses to Copper

Crop responses to Cu fertilisation have not been observed very often in India, due to the generally high available-Cu status of soils. Continuous use of Cu-containing fungicides leads to an improvement in the Cu status of soil. Application of 5 kg Cu ha^{-1} had an adverse effect on green gram, whereas its application to groundnut had no beneficial effect on pod yield or root nodulation in Alfisol soils of Tamil Nadu (Krishnasamy et al., 1992). Application of 5 kg Cu ha^{-1} or foliar sprays of 0.25% $CuSO_4$ solution in 60 field trials on rice and 32 trials on wheat gave yield increases of 440 kg ha^{-1} (14.7%) and 390 kg ha^{-1} (15.2%) in foothill tarai (Hapludalfs) soils due to better grain filling and seed setting (Singh et al., 1979). Rice showed a higher response to Cu fertilisation in Haplustalf soils of the Kokan region, peat soils (Histosols) of Kerala and acidic leached soils (Rhodoustult/Hapludalfs) of West Bengal (Singh, 2001a). Recently, soybean responded to Cu fertilisation in sandy loam soils of Punjab (Nayyar et al., 2005).

4.3.4 Correction of Copper Deficiency

Copper sulphate pentahydrate ($CuSO_4.5H_2O$, 25% Cu) is the most common, cheap and easily available source of Cu employed for correcting Cu deficiencies in field crops in India. However, copper sulphate monohydrate ($CuSO_4.H_2O$, 35% Cu), cupric oxide (CuO, 75% Cu), Cu chelates (Cu EDTA, 12%Cu), organic manure or

Cu-containing fungicides are also used. (Kanwar and Randhawa, 1974). Application of Cu-containing fungicides like Bordeaux Mixture on foliage meets the Cu requirement of crops (Singh, 2001b). Nevertheless, soil application of 5 kg Cu ha^{-1} to wheat increased yields to a greater extent than three foliar sprays of 0.4% $CuSO_4$. Copper application in soils also leaves a residual effect for succeeding crops (Takkar et al., 1990). In Kerala, soaking of rice seed for 24 h in 0.25% $CuSO_4$ solution or foliar sprays of 500 g $CuSO_4$ in 100 L water ha^{-1} have been found to be useful in increasing rice yields. Seed-soaking of potato (*Solanum tuberosum* L.) tubers was found least effective, while soil application and foliar sprays remained comparable and gave additional potato tuber yields of 0.9–1.0 t ha^{-1} (10–12%) (Grewal and Trehan, 1990). Among wheat cultivars, UP 370, UP 262 and UP 2001 were found to be relatively more susceptible to Cu deficiency than WL 711, UP 368 and UP 115. Addition of P fertiliser had a positive effect, while addition of Zn, Mo and S showed antagonistic effects on the availability and absorption of Cu by various crops (Patel and Singh, 1995; Singh, 2006).

4.4 Iron

4.4.1 Total Iron Contents in Soils

Iron is one of the commonest elements in the earth's crust and in soils. Total Fe contents of Indian soils varied from 0.46% to 27.3% with an average of 3% (Kanwar and Randhawa, 1974). Iron is generally abundant in Indian soils. Average Fe contents are higher in igneous rocks than in limestones. Oxisols have the highest Fe contents followed by Vertisols and Alfisols. Ultisol soils had the lowest total Fe contents (Katyal et al., 1982). In the arid and semi-arid tropics, Fe distribution in soil profiles was found to be uniform compared to limed soils which had a high Fe accumulation in the lower layers (Lal and Biswas, 1973).Variation in Fe distribution in soil is dependant upon the parent material from which the soil is formed. The Fe content increases with an increase in organic matter and finer soil fractions (Lal and Biswas, 1974).

4.4.2 Available Iron Contents in Soils

Only a small fraction of the total Fe reserve in soil becomes available to growing plants. Total Fe in soils is considered to be of little significance from the plant availability point of view. Available concentrations of Fe are very low in semi-arid regions of Haryana, followed by alkali soils in the Indo-Gangetic alluvial plains. Analysis of 256,000 surface soil samples revealed that 12% of Indian soils are deficient in Fe (Singh, 2001b). Available (DTPA-extractable) Fe contents ranged between 21 and 352, 14.4 and 371, and 5.9 and 386 mg kg^{-1} in calcareous (Orthent) soils of Bihar, alluvial (Ochrept) soils of West Bengal and alluvial (Psamment) soils of Uttar Pradesh and Punjab, respectively (Singh, 1998; Takkar et al., 1990).

Iron deficiency chlorosis is one of the common nutritional problems in highly alkaline soils (Natrustalfs), calcareous soils (Calciorthents), compact soils with restricted aeration, and sandy soils (Ustipsamments) with low organic matter and high permeability (Nayyar et al., 1990). Redistribution of Fe among different forms in old alluvial soils (Haplusterts) indicated that the water soluble plus exchangeable, insoluble ferric (Fe^{3+}), crystalline sesquioxide-bound and residual Fe fractions constituted about 0.06%, 9.9%, 22.3% and 76.8% of the total Fe content, respectively. Among these, the water soluble plus exchangeable fractions represented the most readily available form of Fe to growing plants (Sakal et al., 1996).

4.4.3 Responses of Crops to Iron

Crops grown in calcareous, coarse textured and alkaline upland soils showed higher responses to Fe fertilisation than those cultivated under submerged conditions. Iron chlorosis is a conspicuous problem in groundnut, sorghum, upland rice and gram. Wheat grown in alkaline, calcareous, swell-shrink (Typic Haplustert/Vertic Ustochrept) soils of Madhya Pradesh and sugarcane in swell-shrink clay soils (Haplusterts) of Maharashtra showed severe chlorosis and a high response to Fe fertilisation (Malewar and Randhawa, 1978; Singh, 2001b).

The acidity of ferrous sulphate ($FeSO_4.7H_2O$) solution markedly influenced its efficiency in mitigating Fe chlorosis in rice. Grain yields increased by 74.2%, 54.8% and 33.3% when the pH of spray solution was adjusted to pH 2.30, 2.60 and 3.15, respectively (Takkar et al., 1990). Rice suffered from Fe chlorosis to the greatest extent when transplanted in coarse textured upland soils which were brought into rice cultivation for the first time. Puddling of such soils helped to increase rice yield by 4,130 kg ha^{-1} (283%), compared with an unpuddled field, by mitigating Fe deficiency and prolonging submergence. Foliar sprays of 2% $FeSO_4$ solution increased the rice yield by 198% in unpuddled and by 50% in puddled plots, compared to non-sprayed (control) plots (Nayyar et al., 1990). These researchers also reported that green manuring, plus three foliar sprays of 2% $FeSO_4$ solution, increased the rice yield response by 2,670, 2,140 and 1,980 kg ha^{-1} (195%, 76.4%, 47.1%) over the non-sprayed controls during the first, second and third years, respectively, of rice–wheat cropping in coarse textured soils of the northern Indo-Gangetic alluvial plains.

4.4.4 Correction of Iron Deficiency

Several inorganic, synthetic and other sources of Fe compounds are available to combat Fe chlorosis in plants. Since Fe applied to soil through inorganic Fe carriers is susceptible to rapid transformation into unavailable forms, Fe deficiency in crops is one of the most difficult micronutrient problems to manage. Ferrous sulphate ($FeSO_4.7H_2O$, 19–20.5% Fe) is the most common and cheapest source of Fe to correct its deficiency in growing crops. Various chelates (e.g., Fe-EDTA), farm yard manure (FYM) and green manure are also commonly recommended to correct Fe chlorosis. Among various other sources, the minerals pyrite (FeS_2) and biotite proved inferior to $FeSO_4$ in increasing rice yield and Fe uptake in Vertisols (Deore et al., 1994). Fe-EDTA or $FeSO_4$ were found to be equally efficient in increasing rice yield when seeds were coated with a 2% slurry of these compounds. Singh and Chaudhary (1991) reported that mixing citric acid with $FeSO_4$ enhances its efficacy in increasing groundnut pod yield (16–24%), compared to sprays of $FeSO_4$, Fe-citrate and Fe-EDTA separately. Foliar sprays of $FeSO_4$ were found to be more efficient than soil application in correcting Fe-chlorosis in groundnut, rice and sugarcane (Nayyar et al., 1990; Krishnasamy et al., 1992). Soil application of 20 kg Fe ha^{-1} increased the average yield of chickpea by 22% over the NPK controls in Vertisols, whereas three foliar sprays of 0.5%, 1.0% and 2% $FeSO_4$ solution gave yield responses of 4%, 14% and 27%, respectively (Rathore et al., 1995). Mixing of 1.0 t ha^{-1} pyrite (FeS_2) with poultry manure enhanced the Fe use-efficiency as well as yield response by 8% over that from application of manure alone.

Rates of Fe application, through broadcasting $FeSO_4$ as a basal treatment, are very high (10–40 kg Fe ha^{-1}) compared with 2.5–7.5 kg ha^{-1} required for foliar sprays. Therefore, soil application is not economical. Similarly, the high cost of synthetic carriers (Fe-chelates) discourages farmers from using these sources for correcting Fe deficiency despite their high agronomic efficiency (Nayyar et al., 1990).

Deficiency of Fe in rice is encountered in upland soils or in highly permeable sandy soils, due to low mobilization of native soil Fe into the plant-available (Fe^{2+}) form, as the desired degree of reduction does not occur. Puddling and addition of FYM, green manure and submergence enhances Fe availability and thus markedly reduces the extent of Fe-chlorosis in rice. A combination of green manure and a foliar spray of 1% $FeSO_4$ solution gave the maximum rice grain yield, followed by green manuring or foliar sprays in loamy sands of Ludhiana. Green manuring helped in the mobilization of the native soil Fe reserve during its decomposition (Nayyar et al., 1990).

Fitting plants to the soil is an alternative approach, rather than ameliorating soil to support normal plant growth. The relative tolerance of sorghum cultivars to Fe stress was found to be in the order: TNS-3 > CO-23 > CSC-541 > TNS-294 > CO-24 > SPY-881. Krishnasamy et al. (1992) reported that the sorghum genotypes SPY-881, CO-4 and TNS-31-1 were highly susceptible to Fe chlorosis. The groundnut cultivar 59-3 was found to be least responsive to Fe spray and could be grown successfully on Fe-deficient soils, whereas other cultivars SL 44, JS 20, and JS 263 failed to produce comparable yields without Fe application. The spreading-type of groundnut varieties (GAU 6-10, Punjab-1, GAU 6-11) were more tolerant to Fe chlorosis than erect ones: MH-1, MH-2, JL-24, PKV-68 (Singh and Chaudhari, 1991).

4.5 Manganese

4.5.1 Total Manganese Contents in Soils

The distribution of Mn in soils depends upon the geological parent material, weathering processes, soil pH, age and contents of organic matter, clay and $CaCO_3$ contents. Leaching and submergence also modify the distribution of Mn in soils (Lal and Biswas, 1973). The total Mn content of Indian soils ranges between 2 and 11,500 mg kg^{-1} (Singh, 2001b). The majority of soils contain 200–2,000 mg Mn kg^{-1}. Soils derived from sandstones and acid igneous rocks had the least Mn, but those derived from basalts, shales and limestones are richer in Mn. Raychaudhary and Datta Biswas (1964) reported average Mn contents of 1,270, 805, 500 and 368 mg kg^{-1} in swell-shrink clay soils (Vertisols), lateritic (Alfisols), red (Inceptisols) and alluvial soils (Entisols and Inceptisols), of India, respectively. In surface soils of the semi-arid tropical regions, where leaching is generally limited, the average Mn content was found to be 718 mg kg^{-1} compared to 371 mg Mn kg^{-1} in soils of the humid tropics where leaching occurs due to heavy precipitation (Katyal et al., 1982). Among Indian soils, Vertisols of the Deccan Trap contained more Mn than acid Ultisols. Total Mn in alluvial soils of arid regions varied from 247 to 760 mg kg^{-1} with an average of 439 mg Mn kg^{-1}. The concentration of total Mn in generic soil horizons was principally linked with the montmorillonite clay fraction (Malewar and Randhawa, 1978).

Studies on the distribution of total Mn in soil profiles revealed that its distribution pattern is not regular in a typical profile. Maximum Mn accumulation occurs in the surface layer but, due to leaching, it moves down to the deeper horizons as a result of eluviation of finer particles along with soluble Mn, which is related to clay and variability of total Mn contents in diverse soils (Kanwar and Randhawa, 1974).

4.5.2 Available Manganese Contents in Soils

Available (DTPA-extractable) Mn contents ranged between 2 and 170 mg kg^{-1} in different soils of India. Soils formed under humid and sub-humid climates had 10–15% of the total Mn in available forms compared to 1.3–5.0% in arid and semi-arid region soils (Singh, 2001). Benchmark soils in semi-arid regions had mean available Mn contents of 16 mg kg^{-1} compared to 51 mg Mn kg^{-1} in sub-humid regions (Katyal et al., 1982). Vertisols had lower Mn contents compared to Alfisols. Water soluble and exchangeable-Mn concentrations decrease with increase in soil pH, and with decrease in clay and organic matter contents. The content of available Mn decreases with depth, but increases with altitude (Singh and Rao, 1995).

Manganese deficiency in crops and soils in India occurs sporadically. Only 1–5% of 256,000 surface soil samples representing different parts of India were classed as being Mn-deficient (Singh, 2001b). Deficiency of Mn is not a major problem in most Indian soils except in coarse-textured, leached soils of Punjab (Nayyar et al., 1996), calcareous soils of Bihar (Sakal et al., 1996) and deep swell-shrink clay soils of Madhya Pradesh. Manganese deficiency poses nutritional problems in succeeding crops grown after rice due to excessive leaching of soluble Mn^{2+} below the root zone (Nayyar et al., 1990). Manganese deficiency is commonly reported in soybean grown in poorly drained, deep swell-shrink soils of central India due to poor aeration (Singh, 1998). Over-liming of red lateritic soils in Orissa leads to Mn deficiency which causes low crop yields (Singh, 2006).

Manganese in soils occurs in water soluble, exchangeable, reducible and active forms. Divalent Mn^{2+} gets adsorbed on to clay and organic matter. Oxidation and reduction processes in soil influence the solubility of Mn^{2+} and thereby, its availability to growing plants. Active Mn consists mainly of easily reducible Mn, unless the pH is less than 6, when the exchangeable Mn fraction is considered as a part of the active fraction. Nandi and Mandal (1979) reported that submergence abruptly increased the water soluble and exchangeable Mn concentrations within 2 weeks due to the reduction of higher oxides of Mn. Thereafter, its content decreased due to precipitation of Mn^{2+} as $MnCO_3$ while the reverse was true for easily reducible Mn. Addition of P and Fe decreased the availability of Mn, but green manuring, addition of FYM and gypsum all increased its availability during the submergence of alkaline soils.

4.5.3 Crop Responses to Manganese

The responses of wheat, rice, potato and sorghum to Mn fertilisation in a large number of experiments varied between 20 and 3,400 kg ha^{-1} (2–226%) for wheat, 40 and 1,780 kg ha^{-1} (4–98%) for rice, 290 and 510 kg ha^{-1} (8.5–17%) for sorghum, 1,000 and 3,900 kg ha^{-1} (4–15.6%) for potato, and 30 and 1,030 kg ha^{-1} (2.5–86%) for soybean (Singh and Rao, 1995; Singh, 2001c). Manganese fertilisation also increased the respective economic yields by 110, 550, 430, 600 kg ha^{-1} (9, 45, 55 and 6.5%) of groundnut, sunflower, sesame and tomato (*Lycopersicum esculentum* L.). Soil application of 5, 10 and 20 kg Mn ha^{-1} increased wheat grain yields by 65.6%, 86.5%, 90.8%, respectively, whereas the increase in yield from two foliar sprays of 0.5%, 1.0% and 2% $MnSO_4$ solution was 146%, 180% and 163.4% compared with no Mn in Mn-deficient Ustipsamment soils of Punjab. In another study, two or three sprays of 0.5–1.0% $MnSO_4$ solution on wheat foliage increased the grain yield by 220–3,750 kg ha^{-1} (20–170%) with a mean of 1,400 kg ha^{-1} (63%) (Nayyar et al., 1996). In calcareous soils (Calciorthents) of Bihar, three foliar sprays each of 1.6 kg Mn ha^{-1} in 110 field experiments increased wheat yields by 2–56% (Singh et al., 1979). In 11 fertiliser trials, use of 16 kg Mn ha^{-1} gave a 300 kg ha^{-1} (10.6%) wheat yield response in sandy loam Ustipsamments of Haryana compared to 22.5% in 692 experiments on cultivator's fields in calcareous soils of Bihar (Singh, 2001b).

4.5.4 Correction of Manganese Deficiencies

Among various inorganic and synthetic sources, manganese sulphate ($MnSO_4.4H_2O$) (32% Mn) is the most commonly used source of Mn for correcting its deficiency in crops due to its higher solubility, low cost and easy availability, compared to other Mn sources like $Mn_3(PO_4)$ (20% Mn) and manganese chelates (11–12%). Manganese dioxide (MnO_2) (68% Mn) proved the least efficient among all sources, due to its lower solubility and thereby showed a low response. Manganese deficiency can be corrected by basal soil application through broadcasting, top dressing, foliar sprays or band placement (Singh and Saha, 1992). Basal dressing or placement of Mn fertilisers to wheat leads to their rapid oxidation forming insoluble hydroxide or oxides which have a lower availability to the plants. Soil application requires a higher rate of 10–50 kg Mn ha^{-1} to wheat as a basal dressing, compared to 4.8 kg Mn ha^{-1} required for each spray on a standing crop. Application of 12.5 kg Mn ha^{-1} was found to be optimum for sorghum, which increased the grain yield by 290–640 kg ha^{-1} (9.6–21.3%) (Nayyar et al., 1990). Application of 5 kg Mn ha^{-1} to onions gave an extra bulb yield of 3,960 kg ha^{-1} (21.2%) in Mn-deficient Calcifluvents of Bihar (Sakal et al., 1996).

Foliar sprays of 0.5–1% $MnSO_4$ solution are more efficient than soil application for correcting Mn deficiency in wheat, sorghum, berseem (*Trifolium alexandrinum*), sunflower, green gram and other horticultural crops. Three foliar sprays of 1.0% $MnSO_4$ solution increased the groundnut, green gram, soybean, and pigeon pea yield by 85, 57, 51 and 5%, respectively (Nayyar et al., 1998). Seed coating with 8–10 mL Teprosyn-Mn slurry kg^{-1} seed showed a two to three times higher nutrient-use-efficiency compared to soil application of $MnSO_4$. Thus the Teprosyn-Mn was found to be a good source of Mn for seed coating, especially for crops having large seeds (Singh, 2004). Response of potato crops to foliar sprays of 0.2% solution, or soaking tubers in 0.05% solution of $MnSO_4$ ranged between 1.0 and 3.9 t ha^{-1} (5–14%) (Grewal and Trehan, 1990). Thus, seed coating or soaking required less Mn to ameliorate its deficiency among crops than soil application. The best time for application of Mn is before seeding, or at an early stage of growth, and for foliar sprays, before tillering, prior to first irrigation, and thereafter at 7–10-day intervals before the pre-flowering stage (Zadoks 55) (Nayyar et al., 1990).

4.5.5 Tolerance of Crops and Cultivars to Manganese Deficiency

Fitting the plant to a soil under nutrient stress appears to be an economic solution to control Mn deficiency, because crops differ significantly in their inherent tolerance to Mn stress (Singh and Saha, 1995). A large number of wheat cultivars were shown to respond differentially to Mn fertilization in Mn-deficient coarse textured soil. The magnitude of response to Mn application decreased successively as the rating of tolerance of the cultivars increased (Nayyar et al., 1996). Durum wheat (*Triticum durum* Desf.) genotypes were found to be more susceptible to Mn deficiency compared to inbred wheat (*Triticum aestivum*, L.). Sensitive cultivars of wheat gave 750–1,100 kg ha^{-1} (26.3–55.8%) yield responses to three foliar sprays of 0.5% $MnSO_4$ solution.

Wheat cultivars susceptible to Mn deficiency gave maximum yield responses of 300–850 kg ha^{-1} (12–34%) in Mn-deficient soils (Nayyar et al., 1990). In addition, the tolerant cultivars HD 2329, Raj 3765, WH 542, PBW 343 only needed one foliar spray (1.6 kg Mn ha^{-1}), compared to two to three sprays (3.2–4.8 kg Mn ha^{-1}) which were found to be essential for moderately, or highly susceptible, cultivars like PBW 34, PBW 215, PDW 233. Thus cultivation of tolerant cultivars like HD 2329, PBW 343 is helpful in reducing input costs of Mn fertilisers for achieving higher yields in Mn-deficient soils for resource-poor farmers (Singh, 1998, 2000).

4.6 Molybdenum

4.6.1 Total Molybdenum Contents in Soils

Total contents of Mo in Indian soils ranged between 0.4 and 14.5 mg kg^{-1} soil (Subbarao and Adinarayana, 1995; Singh, 2001b). However, in a majority of soils, the Mo content varied from 1 to 2 mg kg^{-1} compared to values of 0.2–5 mg kg^{-1} reported for soils all over the world (Swaine, 1955). Molybdate anions (MoO_4^{2-}) are strongly adsorbed by soil minerals and colloids (at $pH < 6.0$) and with extensive weathering, the formation of secondary minerals may trap Mo. Hydrous aluminum silicates may also fix Mo strongly. Subbarao and Adinarayana (1995) reported that recent alluvial soils derived from granite and metamorphic crystalline basalt contained 1.5–5.1 mg Mo kg^{-1}. The corresponding Mo values for Vertisols and limestone soils were found to be 1.5 and 1.8 mg Mo kg^{-1}. Soils formed from shale and granite parent materials had high Mo concentrations, whereas those derived from sandstone, basalt and limestone had low Mo contents. Chatterjee and Dakshinamurti (1962) reported a range of total Mo between 2.0 and 5.6 mg kg^{-1} for alluvial soils (Inceptisols), 0.6–11 mg kg^{-1} for swell-shrink (Vertisols) and 1.3–2.00 mg kg^{-1} for lateritic (Alfisol) soils. Alkaline Inceptisol and Vertisol soils contained more Mo than Oxisols and Alfisols and its status was highly correlated with clay content (Balaguru and Mosi, 1973). Deficiency of Mo in Uttar Pradesh was found in the order of Vindhyan > Bhabhar > Alluvial > Bundel khand region. Similarly, soils in Gujarat had 0.5–4.1 mg kg^{-1} total Mo with a mean value of 1.8 mg kg^{-1} (Reddy et al., 1964).

4.6.2 Available Molybdenum Contents in Soils

Available Mo contents, extracted with ammonium oxalate (pH 3.3), in soils in India ranged between 0.07 and 2.67 mg kg^{-1} (Singh, 1999b; Gupta et al., 1994). Its availability in soils depends upon the nature of the parent material and environmental conditions. Thus, soils which are acidic in reaction, whether arising from the parent material, or with a change in climate that leads to higher leaching, had low Mo availability to the growing plants. Water soluble and organically complexed Mo is considered to be relatively more available to the plants. Availability of Mo increases with an increase in soil pH. Availability of Mo is low in acid soil, primarily due to strong adsorption of MoO_4^{2-} onto inorganic soil components. Recent alluvial soils (Typic Haplaquents) derived from granite and metamorphic crystalline basalt had a high available Mo status (Raychaudhary and Datta Biswas, 1964).

Deficiency of Mo is not very common in Indian soils (Reddy, 1964). Singh (2001b) reported that about 11% soils in India are deficient in available Mo. Its deficiency is found to be widespread (<46%) in samples tested from acidic and leached areas in humid zones compared to arid zones. Its deficiency is widely reported in red and lateritic Alfisols of the north and north-eastern Himalaya regions and the Konkan and Malabar regions (Singh, 2006). Nearly 46% of samples from the hill soils (Hapludalfs) of Srikakulam, Vizag and Vijayanagar districts tested deficient in available Mo in Andhra Pradesh (Bhupalraj et al., 2002). Its deficiency is rarely reported in calcareous alkaline soils of arid and semi arid regions as these soils have high available Mo contents.

Continuous submergence, use of organic manure, high soil pH and use of P and S fertilisers increased the availability of Mo. However, regular use of ammonium sulphate in lateritic soils decreased its availability due to the acidifying effect of this fertiliser, leaching and an antagonism between sulphate and molybdate absorption by plants. In soil containing low concentrations of available Mo, its deficiency may become apparent after fertilisation due to increased growth and uptake by crop plants. Use of N fertiliser reduced the response of legume plants to Mo but showed little effect on non-leguminous plants (Gupta et al., 1995). Impeded drainage in the saline alkali soils of the Kandi area of Punjab increased the solubility and availability of MoO_4^{2-} in soils. Forage and other crops grown in such alkali soils with high organic matter contents accumulate excessive Mo which ultimately induces molybdenosis when fed to animals due to the high availability of Mo (Nayyar et al., 1990).

4.6.3 Crop Responses to Molybdenum

Responses of crops to Mo fertilisation have been reported in only a few studies in India. Uptake of Mo by intensive cropping systems ranged between 12.3 and 32 g Mo ha^{-1}. Subbarao and Adinarayana (1995) reported average grain yield responses of 130–880 kg ha^{-1} (5–35.2%) in rice and 20–680 kg ha^{-1} (0.6–22.7%) in wheat to Mo applications in different soils. Crop responses varied widely in different soils depending upon Mo status. The grain yield of Sonalika wheat increased by 38% in sandy soils of West Bengal with the application of 0.5 kg Mo ha^{-1}. Rice showed a response of 340 kg ha^{-1} (13.2%) to Mo sprays in alluvial sandy loam soils (Singh et al., 1979). Application of 1.5 kg sodium molybdate ($Na_2MoO_4.2H_2O$) (39% Mo) increased the potato yield by 2.9 t ha^{-1} in Vertisols, 2.0 t ha^{-1} in mountainous hill soils, 1.3 t ha^{-1} in red and lateritic soils (Alfisols) and by 1.2 t ha^{-1} in alluvial (Entisol) soils (Grewal and Trehan, 1990). Molybdenum responses have been recorded in green gram, black gram, chickpea and lentil in sandy loam soils, groundnut in calcareous soils and mustard in sandy loam soils (Singh, 2001b, 2003). Application of 0.4 kg ha^{-1} Mo significantly increased the yield of maize, soybean, and groundnut pods by 1020 kg ha^{-1} (53.6%), 400 kg ha^{-1} (26%) and 620 kg ha^{-1} (40.3%), respectively, in red loam Inceptisols of Jharkhand (Sakal et al., 1997).

4.6.4 Correction of Molybdenum Deficiency

Ammonium molybdate ($(NH_4)_6Mo_7O_{24}.4H_2O$) (54% Mo), sodium molybdate ($Na_2MoO_4.2H_2O$) (39% Mo) and molybdenum trioxide (MoO_3) (66% Mo) are the most commonly used sources of Mo. However, only ammonium molybdate has been approved for agricultural use in India. Molybdenum exists as a contaminant in various fertilisers like potassium sulphate, urea (5–6 mg Mo kg^{-1}) and phosphatic fertilisers (25–550 mg Mo kg^{-1}). Organic manure also contains Mo (16–21 mg Mo kg^{-1} in FYM, 25–34 mg Mo kg^{-1} in pig manure and 42–65 mg Mo kg^{-1} in poultry manure). Thus, Mo requirements of crops are partially met through the application of these sources.

Deficiency of Mo in crops can be corrected both by soil application and foliar sprays, but seed coating needs lower doses of fertiliser to prevent yield loss (Singh, 2004). Foliar sprays of 0.05–0.1% Na_2MoO_4 solution increased the pod yield of green gram by 300%, or 130 kg ha^{-1}, in a sandy loam swell-shrink clay soil in Tamil Nadu. However, soil application of 2 kg ha^{-1} Na_2MoO_4 showed no effect (Krishnasamy et al., 1992). Seed treatment with 70–140 g ha^{-1} of Na_2MoO_4 significantly increased the seed yield of soybean in red soils of the Bundelkhand region. Basal application of Mo to groundnuts increased the pod yield significantly in calcareous soils in Bihar. Similarly, soil application of 800 g ha^{-1} of Na_2MoO_4 with 120 kg S ha^{-1} also increased the pod yield of groundnuts on Alfisols in Andhra Pradesh (Bhupalraj et al., 2002). Application of 0.50 kg Mo ha^{-1} along with single super phosphate, or spraying of 0.1% Mo solution on the foliage of crops increased the yield of mustard on acid soils. Applications of Mo increased the yield of French bean (*Phaseolus aconitifolius* Jacq.) and also improved the marketable yield of beet root and cauliflower.

Liming of acid soil helps to ameliorate Mo deficiency. Molybdenum fertilisation leaves a residual effect for several succeeding crops. Deficiencies of P, S and Mo have been observed together, therefore, molybdenized single super phosphate is recommended for their correction. Soaking potato tubers in 0.01% $(NH_4)_6Mo_7O_{24}.4H_2O$ solution for 24 h before sowing increased tuber yield by 1.3–2.9 t ha^{-1} (5–8%) in different soils (Grewal and Trehan, 1990). Nayyar et al. (1998) reported that seed treatment of soybean with 3.0 g kg^{-1} of Na_2MoO_4 improved the yield, quality and oil content. Soybean and black gram responded to Mo seed treatment with 1.5 g Na_2MoO_4 kg^{-1} of seed in acidic Mo-deficient Alfisol soils.

Seed treatment with Teprosyn-Mo slurry containing 54% Mo at a rate of 2–3 mL kg^{-1} seed of groundnut and soybean was found to be as efficient in increasing the seed yield and enhancing Mo-use-efficiency as applying 500 g ha^{-1} $(NH_4)_6Mo_7O_{24}$ to soil (Singh, 2003). Molybdenum deficiency is not a common problem in Indian crops except on hilly, acidic red and lateritic soils. Toxicity of Mo is reported only in the Kandi area of Punjab. Legume crops showed a higher response to Mo compared to other crops. Soil and foliar application of Mo fertiliser efficiently corrected the Mo deficiency. Seed treatment also provides better options for the amelioration of Mo deficiency in legume and other crops.

4.6.5 Molybdenum Toxicity in Livestock

Studies carried out in India showed that flood plain and alkali soils in Punjab contain high to excessive levels of Mo. More than 27% of berseem forage samples showed potential toxicity and 66% had moderate contents of Mo which are being fed to cattle (Nayyar et al., 1990). The average content of Mo in these fodders was often more than 10 mg Mo kg^{-1} dry matter. Wheat straw containing high Mo contents in forages proved toxic to cattle. Nayyar et al. (1990), further reported that among various forage crops, maize and sorghum were rated as low accumulators of Mo.

4.7 Zinc

4.7.1 Total Zinc Contents in Soils

Total Zn contents in Indian soils ranged between 2 and 1,205 mg kg^{-1} (Singh, 2001b) compared to 10–300 mg kg^{-1} reported in soils of World (Swaine, 1955). Total Zn in 29 benchmark soil series in India ranged from 20 to 89 mg kg^{-1} with a mean of 59 mg Zn kg^{-1} in arid and semi-arid regions and 52 mg Zn kg^{-1} for soils in humid and sub-humid regions (Katyal et al., 1982). Highly leached, acidic Inceptisols of tropical India contained as little as 2–7 mg Zn kg^{-1}. Its content in Vertisol soils of Maharashtra ranged from 84 to 129 mg kg^{-1} with maximum accumulation in the surface layer (Malewar and Randhawa, 1978). Alkali soils of the Indo-Gangetic alluvial plains had total Zn contents of 62–94 mg Zn kg^{-1} but, due to its low availability, crops suffer with severe Zn deficiency over nearly 2.5 Mha (Singh and Abrol, 1986). Sandy loam to clay loam Inceptisol soils of Punjab had 17–92 mg Zn kg^{-1}. Among the different soils of Rajasthan, alkaline Vertisols had the highest Zn content compared to coarse textured Oxisols (Lal et al., 1960). Similarly, Malewar and Randhawa (1978) found high Zn contents in the Vertisols of Maharashtra but sandy alluvial (Entisols) soils of north Madhya Pradesh generally had low Zn contents (Rathore et al., 1995).

Indian soils developed from limestone are richer in Zn than those formed from sandstone or schists (Raychaudhary and Datta Biswas, 1964). A significant correlation was reported between $CaCO_3$ and the Zn content of soils. Soils developed over sandstone in India differ widely in Zn content regardless of their mineralogical make up. Climatic conditions, parent material and management appeared to be largely responsible for distribution of Zn in soils.

4.7.2 Available Zinc Contents in Soils

Zinc deficiency is a widespread problem in several crops because only a small fraction of the total Zn reserve becomes available during crop growth. Although, there is a good correlation between available (DTPA-extractable) and total Zn in coarse textured Inceptisols (Nair and Mehta, 1959) and swell-shrink Vertisols (Sharma and Motiramani, 1969), the total Zn content is not a good indicator of deficient or sufficient Zn status. Less than 1.75% of the total Zn occurred in bioavailable forms (Singh and Abrol, 1986a). Soils formed from parent materials such as quartz sand contain low total and available Zn and are highly prone to Zn deficiency. The occurrence of Zn deficiency is frequently observed in soils having a high pH, alkalinity, free $CaCO_3$, coarse texture and low organic matter contents (Singh and Abrol, 1985a). Zinc availability is also dependent upon environmental conditions, plant species and their cultivars. Its deficiency is widely reported in flooded rice (Singh and Abrol, 1986b) and in maize and wheat due to the slow release of Zn from soil organic matter complexes, as well as restricted root growth in the winter season leading to lower uptake of Zn by plants.

Mean concentrations of DTPA-extractable Zn varied from 0.1 to 6.92 mg kg^{-1} in most Indian soils, with a mean of 0.87 mg Zn kg^{-1} (Singh, 2001b). The frequency distribution assessed for 14,863 surface soil samples from all over India, indicated that 24%, 23%, and 15% of soils had 0.2–0.4, 0.4–0.6 and 0.6–0.8 mg kg^{-1} available Zn, respectively. Mean available Zn status declined in the order of: Alfisols > Mollisols > Inceptisols > Entisols. In Central India, Inceptisols and Vertisols had lower DTPA-Zn concentrations of 0.65 mg kg^{-1} than the 1.50 mg Zn kg^{-1} found in brown forest soils (Alfisols) of hilly regions due to the high organic matter contents in these latter soils (Singh, 1996).

Zinc deficiency is the most common micronutrient problem limiting crop yields in several soils of India. Analysis of 256,000 surface soils and 25,000 plant samples has indicated that 48.5% of soil and 44% of plant samples are potentially Zn-deficient. The extent of Zn deficiency in different agro-ecological zones, which ranged from 20% to 77%, is shown in Fig. 4.2 (in Colour Section). The magnitude of deficiency is higher in the red loam (Entisols) and black clayey (swell-shrink Vertisol) soils of the central Deccan plateau and lowest in brown forest hill soils of the Himalayas (Singh, 2001c). Soils of the semi-arid tropical parts of India showed more Zn deficiency than those found in the humid and sub-humid tropics. Several workers have established relationships between available Zn and soil properties.

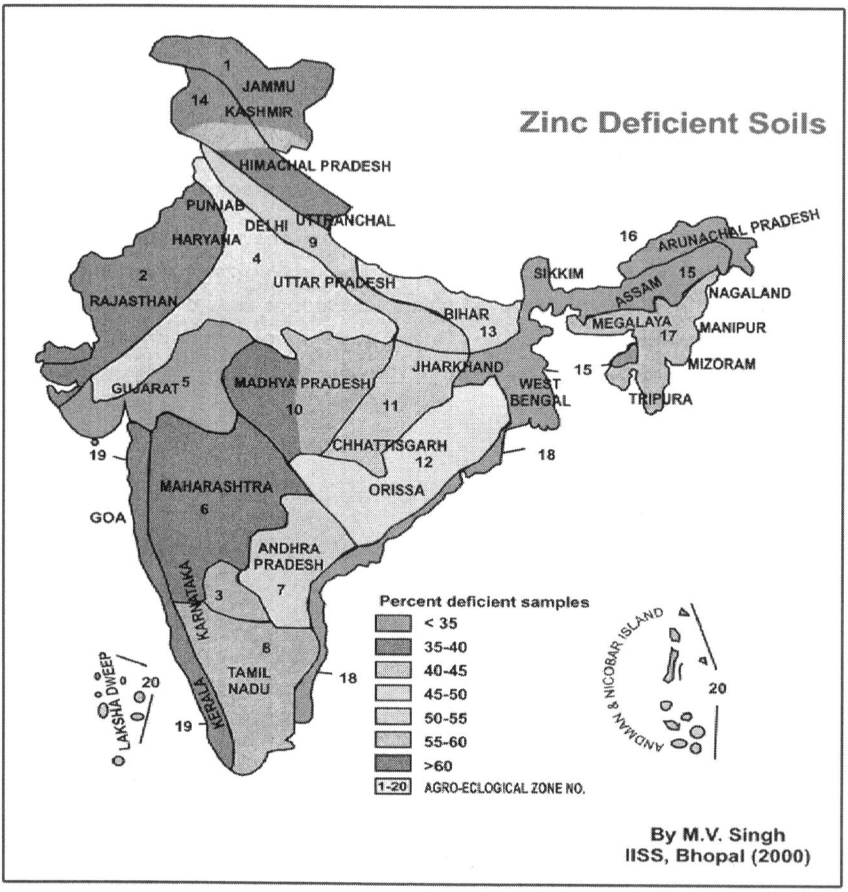

Agro-ecological Zones

1. Western Himalayas (cold arid), 2. Western Plain (hot arid), 3. Deccan Plateau (hot arid), 4. Northern Plain and Central Highlands (hot semi-arid), 5. Central (Malwa) Highlands, Gujarat plains and Kathiawar Peninsula, 6. Deccan Plateau (hot semi-arid), 7. Deccan Plateau (Telangana) and Eastern Ghat (hot semi-arid), 8. Eastern Ghats, Tamil Nadu Uplands and Deccan (Karnataka) Plateau (hot semi-arid), 9. Northern Plan (hot subhumid/dry), 10. Central Highlands (Malwa and Bundelkhand) (hot subhumid/dry), 11. Chattisgarh/Mahanadi Basin, 12. Eastern Plateau (Chhotanagpur) and Eastern Ghats (hot subhumid), 13. Eastern Plain (hot subhumid/moist), 14. Western Himlayas (warm subhumid to humid/perhumid), 15. Assam and Bengal Plain (hot suhumid to humid/perhumid), 16. Eastern Himalayas (warm perhumid), 17. North-Eastern Hills (Purvachal) (warm perhumid), 18. Eastern Coastal Plain (hot subhumid to semi-arid), 19. Western Ghats and Coastal Plain (hot humid/perhumid), 20. Islands of Andaman-Nicobar and Lakshadweep (hot humid/perhumid).

Fig. 4.2 Percentage of zinc-deficient soils in the different agro-ecological zones of India (Singh, 2001b) (*See Colour Plates*)

Soil pH and $CaCO_3$ content showed negative relationships, but organic matter and clay content had a positive influence on the availability of Zn in most soils (Nair and Mehta, 1959; Singh and Abrol, 1986b; Gupta et al., 1994).

Besides parent materials and climate, the type of cropping system and management practices also play important roles in influencing the build-up or depletion of native soil Zn. Available Zn was higher (1.06 mg kg^{-1}) in rice fields, due to the regular addition of organic manure, other inputs and Zn fertilisers, than 0.65 mg Zn kg^{-1} in cotton and mustard fields (Singh, 1991b). Liming of acid soils, land leveling, terracing, exposure of subsoil, high rates of P fertilisation and failure to recycle crop residues may lead to Zn deficiency in plants. Calcareous soils having concretions of $CaCO_3$ or a hardpan in the subsurface layer also induce Zn deficiency in several horticultural crops (Subbaiah et al., 1998; Bhupal Raj et al., 2002).

During the last two decades, the status of available Zn in Indian soils has shown an improvement due to regular fertilisation of crops with Zn (Singh and Abrol, 1986a). Zinc deficiency has decreased by 15–45% in certain parts of India compared to Zn deficiency recorded in the early 1980s (Singh and Saha, 1994). However, intensive cropping and enhanced productivity in marginal soils have caused greater depletion of native soil micronutrients, which, in turn, has resulted in multi-nutrient deficiencies and higher responses to micronutrient fertilisation (Singh et al., 2003).

4.7.3 *Responses of Crops to Zinc*

Zinc uptake by plants is poorly related to its total content in soil because several soil properties such as pH, $CaCO_3$ and organic matter contents, crop and cultivars, nutrient interactions and environmental factors can all influence the emergence of micronutrient deficiency or toxicity in plants. Occurrence of Zn deficiency has been confirmed through the biological responses achieved in a large number of experiments on cultivator's fields conducted throughout India. A soil is classified as being 'responsive to Zn' only when it gave more than a 200 kg ha^{-1} increase in grain yield (Singh et al., 1979; Kanwar and Randhawa, 1974; Singh et al., 2003). Further increases in economic yield of <200, 200–500, 500–1,000 and >1,000 kg ha^{-1} representing <6%, 6–15%, 15–30% and >30% responses with a basal application of 5 kg Zn ha^{1} were considered as indicative of 'high', 'marginal', 'low' and 'very low' fertility status of the soils (Singh, 2001c). Based on the above criteria, 37%, 37%, 19% and 7% of the 5,807 trials conducted on farmers fields showed high, marginal, low, very low fertility status of Indian soils, respectively. Therefore, Zn fertilisation proved a highly profitable option in 63% of cultivated soils in India (Singh, 1999b, 2001).

The data in Table 4.3 show that Zn deficiency assessed through soil testing as being 53%, 46%, 24%, 61%, 63%, 47%, 64% and 64% of samples in the states of Andhra Pradesh, Bihar, Gujarat, Haryana, Madhya Pradesh, Punjab, Tamil Nadu and Uttar Pradesh, respectively. This has been confirmed through biological responses in 66%, 84%, 51%, 52%, 44%, 64%, 67% and 67%, respectively, of the 5,807 field trials

Table 4.3 Matching deficiency of zinc assessed by soil analysis and through biological response assessed in field experiments on cultivator's fields (ECF) conducted in different states (Compiled by Singh, 2001c; Singh et al., 2003)

State	No. of ECF trials	Percentage distribution of ECF based on response to Zn (kg ha^{-1}) <200 (>6%)[a]	200–500 (6–15%)[a]	500–1,000 (15–30%)[a]	>1,000 (>30%)[a]	Above >200[aa] (>6%)	% Zinc deficiency based on soil analysis
Haryana	850	48	39	9	4	52	61
Punjab	1,014	48	32	16	4	64	47
Gujarat	704	49	21	15	15	51	24
Maharashtra	60	37	38	25	0	63	73
Andhra Pradesh	696	34	30	23	13	66	53
Tamil Nadu	446	36	50	9	5	67	64
Uttar Pradesh	321	33	36	25	6	67	64
Madhya Pradesh	249	56	18	20	6	44	63
Orissa	76	18	78	4	0	82	66
Bihar	1,391	16	48	29	7	84	46
All states	5,807	38	37	19	7	63	48

[a] Expected percentage increase in average grain yield over NPK control
[aa] Total percentage of experiments showing response of above 200 kg ha^{-1} to Zn over controls

conducted in these states (Singh, 1999a). Thus, a crop response to Zn was found in 63% of field trials as compared to 48% deficiency assessed through soil analysis. Thus, about 15% soils are marginally deficient. Crops grown in such soils may be likely to be affected by hidden Zn deficiency (Singh, 2001b, 2002).

Among all the micronutrients, Zn assumes the greatest significance in exploiting the high yield potentials of modern crop varieties. The maximum crop responses were recorded on those sites where application, or no application of Zn, decided the success or failure of crops (Singh et al., 1987). Cereals, millet, oilseed, fodder, pulses, vegetables, fruit, plantation and medicinal crop plants have all shown marked responses to Zn application in a divergent range of soils (Table 4.4). The crop responses varied widely from soil to soil, ecological situations, among plant species, genotypes and the degree of deficiency existing in different soils.

Information compiled to date for 15,000 on-station field experiments all over the country, indicated that crop responses to Zn fertilisation ranged from 0.42 to 0.55 t ha^{-1} (15.7–23%) in cereals, 0.19–0.36 t ha^{-1} (13.4–41.2%) in pearl millet (*Pennisetum typhodium* Stapf.), 0.17 to 0.46 t ha^{-1} (7.3–28.2%) in pulses, 0.11–0.36 t ha^{-1} (11.4–40%) in oilseeds, 0.09–4.62 t ha^{-1} (5–34%) in fodder crops and 0.02–17.7 t ha^{-1} (24.4–53.8%) in cash crops such as sugar cane and cotton (Table 4.4) (Singh et al., 1979; Nayyar et al., 1990; Singh, 1999c, 2003). Cereal crops were found to be the most susceptible and showed a high response to Zn fertilisation. Most often Zn deficiency became so crucial that it decided either a complete failure or success of the crop in several soils. In the case of acute deficiency, crop

Table 4.4 Crop responses to zinc in field experiments in India (Compiled by Singh, 1999c; Singh et al., 2003)

Crop	No. of field experiments	Response range ($t\ ha^{-1}$) Individual experiments	Mean of experiments	Mean response $t\ ha^{-1}$	%
Wheat	2,447	0.01–4.70	0.01–1.47	0.42	18
Rice	1,652	0.02–5.47	0.14–1.27	0.54	23
Maize	280	0.01–3.09	0.11–1.37	0.47	16
Pearl millet	236	0.00–1.17	0.17–0.46	0.19	13
Sorghum	83	0.07–1.35	0.21–0.65	0.36	41
Finger millet	47	0.00–1.25	0.08–0.42	0.35	14
Lentil	16	0.03–0.58	0.18–0.39	0.22	28
Chickpea	15	0.10–1.01	0.23–0.56	0.36	24
Black gram	10	0.07–1.12	0.11–1.12	0.24	28
Green gram	9	0.05–0.45	0.06–0.30	0.17	7
Groundnut	83	0.04–0.60	0.21–0.47	0.32	18
Soybean	12	0.08–0.69	0.16–0.39	0.36	11
Mustard	11	0.02–0.34	0.14–0.26	0.27	18
Linseed	5	0.12–0.21	0.15–0.20	0.16	20
Sunflower	8	0.01–0.67	0.15–0.20	0.24	40
Sesamum	6	0.08–0.15	0.08–0.15	0.11	18
Potato	45	1.10–7.60	2.40–3.90	2.96	17
Sugarcane	6	8.00–4.30	1.72–2.40	17.70	24
Cotton	27	0.01–0.78	0.06–0.24	0.22	54

responses to Zn application were quite striking and recorded as 5.5, 4.7, 3.1, 1.2 and 0.8 t ha^{-1}, with respective increases of 92%, 78%, 79%, 48%, 88% for rice, wheat, maize, barley and oat crops over application of NPK in controls (Singh, 2001c).

Little information is available on responses to Zn in horticultural crops in India, except for citrus, grapes and mango. Citrus decline, also referred to citrus 'Die back disease', is a serious problem in major citrus growing areas in India (Singh and Saha, 1995). Responses of potatoes ranged from 1.0 to 7.6 t ha^{-1} (mean 12%) in alluvial soils. Studies conducted in Gujarat suggest that deficiencies of N, Zn and B were mainly responsible for the 'Fatio disease' of guava (*Psidium* spp) (Dangarwala et al., 1994).

4.7.4 Correction of Zinc Deficiency in Crops

Zinc deficiency can be corrected through addition of inorganic fertilizers, synthetic chelates, natural organic manures, recycling crop residues and, to some extent, by cultivation of tolerant crops and their cultivars. The effectiveness of different sources varies considerably with mode and rate of application, nutrient content, residual fertility build up and market price.

Zinc sulphate heptahydrate ($ZnSO_4.7H_2O$) (21–22% Zn), Zn sulphate monohydrate ($ZnSO_4.H_2O$) (33% Zn), sparingly soluble Zn oxide (ZnO) (67–80% Zn), Zn carbonate ($ZnCO_3$) (56% Zn), Zn phosphate ($Zn_3(PO_4)_2$ (50% Zn), Zn frits (4–16% Zn), Zn chelates (12–14% Zn) and Teprosyn-Zn slurry (55% Zn) are the important sources for ameliorating Zn deficiency. Zinc sulphate is the most common source of Zn in India due to its high water solubility, easy availability and relatively low price compared with other sources and it is being widely used to correct Zn deficiency in different crops and soils (Singh, 1991c). Application of Zn-EDTA was found comparable to $ZnSO_4$ in the calcareous soils of Pusa and Inceptisols of Hisar (Gupta et al., 1994). It proved better than $ZnSO_4$ in combating Zn deficiency in rice on loamy sand soils of Punjab, but the higher cost of chelates made them less popular than $ZnSO_4$ (Nayyar et al., 1990).

The efficiency of sparingly soluble Zn sources such as ZnO, $ZnCO_3$ and Zn frits in fine textured, high Zn-fixing soils was found to be comparable to that of highly soluble $ZnSO_4$. However, when sparingly soluble and soluble Zn sources were compared in coarse textured soils, it was the soluble sources that gave the best performance (Patel et al., 2003; Gupta et al., 1994). Zinc phosphate is found to be a good source, but less effective than $ZnSO_4$ in Zn-deficient soils. Zinc sulphate monohydrate and hepta hydrate were found equally efficient in calcareous (Calciorthents, Calcifluvents) soils in correcting Zn deficiency in the rice–wheat sequence (Sakal et al., 1997). Blended sources of micro and macronutrients, like zincated urea, when added on a Zn equivalent basis to rice in calcareous soils produced comparable yields to those with $ZnSO_4$ and left significant residual effects to succeeding wheat crops compared with controls. Zincated super phosphate (2.5% Zn) was either comparable, or less efficient, than $ZnSO_4$. A mismatch in Zn supply in meeting nutritional requirements of the crop made these sources less efficient than $ZnSO_4$. Promotion of the regular use of such blended sources may, however, be helpful in preventing hidden hunger in crops. Application of green manure, farmyard manure and city compost were found beneficial for ameliorating nutritional deficiencies. Seed coating with a concentrated slurry of Teprosyn-Zn (55% Zn) proved good for correcting deficiency in crops having large seeds compared to sesame, small millets and mustard (Singh, 2004).

Zinc deficiency can be corrected by applying Zn through seed coating, foliar sprays or top dressing the crop, or to soil through broadcasting and band placement methods. Zinc applied to soil through broadcasting and mixing into the topsoil proved superior to top dressing, side-dressing or band placement and foliar application of 0.5–2% $ZnSO_4$ solution. Pre-soaking, or coating of seed, or seedlings in concentrated Zn slurry effectively corrected its deficiency in maize, wheat, gram, groundnut, soybean and potato crops (Singh, 2004). Dipping rice seedlings in 2–4% ZnO before transplanting proved equally efficient as the broadcasting of 11 kg ha^{-1} Zn as $ZnSO_4$ in combating Zn deficiency. Dipping vegetable plant seedlings or sugarcane sets in 3% ZnO suspension for 24 h could not meet the full Zn requirement (Nayyar et al., 1990). However, the pre-soaking treatment of seed potatoes with $ZnSO_4$ solution proved equally effective as that of foliar sprays or soil application in zinc-deficient alluvial soils (Grewal and Trehan, 1990). Foliar feeding is often considered an emergency treatment to save field crops from unexpected Zn deficiency, but it is the method of preference in horticultural and plantation crops.

Seed coating may be a better option for several crops with large seeds like maize, soybean, wheat and gram.

Plants absorb the maximum amount of Zn up to the tillering stage (Zadoks 21), or pre-flowering stage (Zadoks 55). Time of application is also influenced by severity of deficiency and nutrient reserves of the seed. Thus, the best time to apply Zn to most field crops is prior to seeding or transplanting (Zadoks 10). Wheat grain yield was lower when Zn was top-dressed compared to basal application (Gupta et al., 1994; Sakal et al., 1996). Zinc application to rice by splitting half the dose as a basal dressing prior to seeding, and half at the tillering stage (Zadoks 21) was found equally efficient as that of the full basal application. However, it was superior to top-dressing half of the quantity of Zn at tillering stage (Zadoks 21) and the remaining half at the panicle initiation stage (Zadoks 30) (Singh, 1995, 2000; Nayyar et al., 1990). If a soil application is missed, top dressing of $ZnSO_4$ up to flowering stage (Zadoks 55) has been found effective in correcting Zn deficiency with some sacrifice in yield.

The quantity of Zn required to alleviate deficiency varies with the severity of the deficiency, nature of crop and soil type. Data from a number of field experiments have indicated that Zn deficiency in the majority of instances can best be alleviated by application of 11 kg Zn ha^{-1} to wheat and rice, 5.5 kg Zn ha^{-1} annually to maize, soybean and sugarcane and 2.5 kg Zn ha^{-1} to groundnut, soybean, gram, raya and millet crops. In addition, it has been found that the optimum rate of Zn application not only varied among crop species, but their cultivars too (Singh and Saha, 1995).

In Alfisol soils of Chhatishgarh, wheat responded to 5.0 kg Zn ha^{-1}, but in alluvial (Entisols) and swell-shrink clayey (Vertisol) soils of Madhya Pradesh, crops responded to 10 kg Zn ha^{-1}. Double the rate is required for less soluble Zn sources compared with the rate of soluble $ZnSO_4$ (Rathore et al., 1995; Subbaiah et al., 1998). Application of 10 kg Zn ha^{-1} left enough residual effect to meet the Zn requirements of four succeeding crops for improving and sustaining the high productivity of the rice–wheat sequence (Singh and Abrol, 1985b; Singh, 1991c).

Alkali soils (Natrustalfs, Ustorthents) occupying nearly 2.5 million ha in the Indo-Gangetic plains are generally deficient in Zn, therefore crops grown in these soils showed high responses to Zn fertilisation. Higher yields of rice–wheat, rice–berseem and other crops cannot be achieved unless deficiencies of Zn and Ca, and toxicity of Na are corrected simultaneously (Takkar and Nayyar, 1981; Singh et al., 1985b). The fertiliser Zn requirement of rice, wheat, maize and other crops during reclamation of alkali soils can be reduced substantially by 25–75%, depending upon the rate of soil amendment applied or level of soil sodicity (Singh et al., 1987).

Balanced nutrition is the key for efficient utilization of various inputs and Zn fertilisation plays an important role in it. Application of 5.5 kg Zn ha^{-1} enhanced the input use efficiency of gypsum, N, P and K in sequential rice–wheat crops to 200% compared to controls without Zn in alkali soils (Singh, 1988). Rice is found to be more susceptible to Zn deficiency than wheat. In coarse textured sandy soils in Punjab, wheat needs Zn at twice the rate applied in fine textured soils (Nayyar et al., 1990).

Zinc fertilisers leave a residual effect and therefore it is not necessary to apply Zn to every crop. On the Fatehpur sandy loam alkaline alluvial soil (Typic Ustochrepts) of Punjab, Nayyar et al. (1990) found that 5.5 kg Zn ha^{-1} for the first four crops and for next 8 and 12 crops, repeat applications of 2.75 and 5.5 kg Zn ha^{-1} respectively, gave the largest yield response for different crops. Under brackish water irrigated alkali soil (Typic Cambiorthents), the residual effect of 22 kg Zn ha^{-1} could last for only two rice–wheat sequences and the fifth rice crop required a repeat application of Zn (Takkar and Nayyar, 1981; Nayyar et al., 1990). In contrast, with rice grown in alkali soil irrigated with normal quality water, the yield achieved with continuous application of 2.25 kg Zn ha^{-1} was no different from a single initial application of 18 kg Zn ha^{-1}, even after four cropping cycles, thus indicating that its effectiveness had not diminished (Singh and Abrol, 1985b). Savithri et al. (1996) reported that the use of Zn-enriched organic manure reduced the fertiliser Zn needs of rice in alkali soils. Thus use of brackish water, or a high level of alkalinity shortened the residual effectiveness of Zn in the former case compared with the use of normal quality water.

Studies to evaluate the optimum rate and frequency of Zn applications to rice–wheat systems revealed that Zn application of 5.5 kg Zn ha^{-1} to every third crop of rice in Ustochrepts in Haryana, 10 kg Zn ha^{-1} to every fifth crop in Haplaquents in Bihar and 11 kgZn ha^{-1} to every seventh crop in Ustipsamments of Punjab would be sufficient for maximizing yield response and benefits (Singh, 1991c; Gupta et al., 1994; Nayyar et al., 1995; Singh and Saha, 1995). Application of 11 kg Zn ha^{-1} to first rice followed by 5.5 kg Zn ha^{-1} to every fifth successive crop gave the maximum yield in a rice–rice system in a coarse textured Vertisol in Hyderabad. In a rice–groundnut sequence in red soil (Typic Haplustalfs) in Coimbatore, application of 9 kg Zn ha^{-1} with 1.0 t ha^{-1} coconut coir pith, or FYM, increased crop productivity and response to Zn in this cropping system (Savithri et al., 1996).

The nutrient use efficiency (NUE) of inorganic micronutrient fertilisers seldom exceeds 5%. This NUE increases when fertilisers are applied with organic manures or synthetic materials due to a higher degree of chelation and subsequent slow release (Singh, 2006). In view of this, Zn-enriched manure (fresh cow dung) was prepared using very low doses of manure and Zn. Zinc at the rate of 1.25–2.5 kg ha^{-1} was thoroughly mixed with about 200–500 kg of fresh cow dung or well-pulverized FYM, incubated for 25–30 days under optimum moisture conditions in the shade to facilitate greater chelation of Zn. The material was intermittently mixed and the moisture content was maintained at 60–70%. This material was then dried and finely ground to facilitate its uniform application in the fields. The efficiency of these treatments is shown in Table 4.5. Results indicated that manure enriched with 2.5 kg Zn ha^{-1} gave similar yields of groundnut pod, mustard, gram, pea and sesame to those achieved by direct application of 5 kg Zn ha^{-1} alone in soil. Zinc application at 5 kg ha^{-1} increased the seed yield of various crops significantly over NPK controls without Zn. Thus, enrichment of manures with small doses of Zn not only enhanced the fertiliser Zn use efficiency by almost 200%, but was also helpful in economizing on the cost of 2.5 kg ha^{-1} Zn without any loss in crop productivity in acid soils (Singh, 2006).

Table 4.5 Effect of zinc alone or zinc enrichment to organic materials on the pod[a] or seed yield and agronomic efficiency (AE) of crops (Singh, 1998, 2000, 2006)

		Seed yield (kg ha^{-1})						
		Levels of Zn added, kg ha^{-1}					AE kg seed kg^{-1} Zn	
Crops	No. of experiments	0	2.5	5.0	2.5 enriched	LSD >0.05	5 kg ha^{-1}	2.5 kg ha^{-1} enriched
Groundnut[a]	2	1,027	1,231	1,258	1,340	82	46	125
Mustard	2	941	1,161	1,036	1,198	112	19	103
Pea	2	1,016	1,853	1,833	2,067	230	163	420
Gram	2	900	1,070	1,033	1,207	121	27	123
Sesame	1	660	670	680	717	25	4	23

[a] pod yield in the case of groundnuts

Application of 10 kg Zn ha^{-1} to alternate cotton crops gave the highest cotton seed yield in Zn-deficient Aridisols of Haryana, but application of 2.8, 5.6 and 11.2 kg Zn ha^{-1} only once to the first cotton crop in a loamy sand (Typic Ustochrept) soil in Punjab was sufficient for 2, 4 and 6 crops, respectively, in a cotton–wheat rotation. Application of 5.6 kg Zn ha^{-1} once to the first green gram crop was found to be optimum to meet the Zn requirement of three cycles of green gram–wheat rotation on a sandy loam (Typic Ustochrept) soil in Ludhiana (Nayyar et al., 1998). The sorghum–cotton sequence in Coimbatore gave significantly higher yields with the application of 7.5 kg Zn ha^{-1} to every sorghum, or 5 kg Zn ha^{-1} to sorghum plus 2.5 kg Zn ha^{-1} to the following cotton crop. In medium Vertisols, an application of 15 kg Zn ha to the first soybean crop left a significant residual effect in improving productivity of three soybean–wheat sequences with a benefit:cost ratio of Rs. 13.9:1 for each rupee spent on Zn (Savithri et al., 1996; Singh, 1998).

In Vertisol soils of Madhya Pradesh, the Zn fertiliser requirement of the soybean–wheat sequence can be met by using either 8–16 t ha^{-1} FYM, or 6–12 kg Zn ha^{-1}, or by applying 4 t ha^{-1} FYM with 3 kg Zn ha^{-1} (Singh, 1994). Application of organic manure at 10 t ha^{-1} FYM, 5 t ha^{-1} poultry manure and 2.5 t ha^{-1} of piggery manure was found as efficient as that of 11.2 kg Zn ha^{-1} in meeting the Zn requirements of the maize–wheat rotation. About 50%, or even lower, rates of manures proved equally efficient or better when amended with 5.6 kg Zn ha^{-1} for the maize–wheat rotation (Nayyar et al., 1990). Thus, the integrated use of Zn and FYM resulted in better Zn fertiliser utilization efficiency by crops.

In alkali soils (Vertic Haplustalfs) maximum yields of rice and cowpea (*Vigna sativa* Lens), Zn uptake and the highest fertilizer use efficiency were achieved when the alkali soil was amended with gypsum at 50% of its requirement followed by mulching with dhaincha (*Sesbania bispinosa*) at 5 t ha^{-1} along with 5.5 kg Zn ha^{-1}. Enrichment of organic matter with an inorganic Zn source improved the utilization efficiency of fertiliser Zn by increasing the yield of cropping system (Singh, 1998, 2006). In a Haplaquent soil in Bihar, application of 2.5 t ha^{-1} of biogas slurry with 2.5 kg Zn ha^{-1} to rice enhanced fertiliser Zn use efficiency by 200% in rice–wheat sequence crops (Singh et al., 2006). In Haplustalf soils of Tamil Nadu, FYM and composted coir pith enriched with, or without, 2.5 kg Zn ha^{-1} by spraying 0.25%

$ZnSO_4$ solution before its decomposition, improved the rice productivity by 620–775 kg ha^{-1} (15–18%) and that of groundnut by 165–250 kg ha^{-1} (11–17%) compared with their individual application (Savithri et al., 1996). In Zn-deficient light textured Vertisols in Hyderabad, application of 2.8 kg Zn ha^{-1} with 2 t ha^{-1} FYM or 11 kg Zn ha^{-1} to the first crop and later 5.5 kg Zn ha^{-1} at 2-year intervals, successfully satisfied the Zn requirements of a rice–rice cropping sequence (Subbaiah et al., 1998).

4.8 Summary and Recommendations

India has achieved self-sufficiency in food production but intensive cropping has resulted in the marked depletion of the native fertility of the soils. Emerging deficiencies of micronutrients in soils are causing either low yields, or complete failure of crops. Rice, wheat, maize and other crops suffer with severe deficiency of Zn unless adequate amounts of Zn are supplied. The introduction of the rice–wheat cropping system in sandy, alkaline soils caused severe Fe deficiency in rice initially, and later it accentuated a deficiency of Mn in succeeding crops. This has posed a severe threat to the sustainability of the rice–wheat cropping system. However, such deficiencies are not observed in maize–wheat, cotton–wheat or cereal–oilseed based sequences, despite long term cropping.

Analysis of a large number of soil and plant samples collected from various parts of country indicated 49%, 12%, 5%, 3% and 33% deficiencies of Zn, Fe, Mn, Cu and B, respectively, and micronutrient fertility maps have been prepared. Zinc fertilisation has become the common practice in important crops. As a result of this, in many areas the Zn status of soils is improving and its deficiency is declining. Multi-micronutrient deficiencies are, however, emerging in certain areas. Existence of such deficiencies, assessed by soil analysis, has been confirmed through biological responses. Yield increases of <6%, 6–15%, 15–30% and >30% (equivalent to around <200, 200–500, 500–1,000 and >1,000 kg ha^{-1}) were considered to represent Zn fertility of soils as: adequate, marginal, deficient, and highly deficient. The magnitude of response accrued by the application of 5 kg Zn ha^{-1} to cereals like wheat, rice, maize and barley was found to be 360, 540, 460 and 550 kg ha^{-1}, respectively, 170–470 kg ha^{-1} of pulses, and 110–340 kg ha^{-1} of oilseeds, with benefit:cost ratios ranging between 5 and 70:1.

Efficient sources of micronutrients, including inorganic, synthetic chelates, natural organic complexes and mixtures, together with on-site and off-site farm wastes and organics have been suggested. Also, the mode, rate and frequency of application, residual effects on succeeding crops and tolerance of genotypes were compared and the most cost-efficient management technologies which were suited to specific location, soil type, crop, cropping and management options have been recommended. Zinc deficiency can be corrected efficiently by applying 5–10 kg Zn ha^{-1} which leaves a residual effect of Zn for the succeeding three to six crops. Deficiencies of Zn, Fe and Mn can also be corrected efficiently with two to four sprays of 0.5–1.0% solutions of sulphate salts on standing crops at 7–10-day intervals, or top dressing in the early stages of crop growth. In the case of B, basal soil application of 0.5–1.5 kg B ha^{-1} to alternate crops was found optimum and better than foliar sprays. Borax and Granubor-II (borax pentahydrate) were found equally efficient sources in correcting B deficiencies in various crops. Regular use of B at rates of more than 2.0 kg ha^{-1} caused adverse effects on crop growth. Sulphate salts of Zn, Fe, Cu and Mn were found to be the most efficient, cheap and easily available sources. Chelated compounds were found initially to be more efficient, but are more costly and uneconomical for regular use. Inclusion of citric acid in foliar sprays of $FeSO_4$ solution enhances its use-efficiency in correcting Fe chlorosis. Soil application of Fe salts results in low fertiliser use efficiency.

In order to ameliorate micronutrient deficiencies, it is estimated that by the year 2025, India will need to apply 324,000 t Zn, 130,000 t Fe, 11,000 t Cu, 3,900 t B and 22,000 t Mn as fertiliser annually (Takkar et al., 1997).

A balanced and integrated supply of organic manure and micronutrient has been found to be more effective than the use of micronutrients alone in correcting micronutrient deficiencies and increasing crop yields. Regular application of 8–10 t ha^{-1} FYM annually corrected Zn and other deficiencies in most crops and soils. Use of 4–5 t ha^{-1} FYM or recycling of crop residues improves yields by 50–75% in rice–wheat, soybean–wheat or maize–wheat cropping systems. Studies in Andhra Pradesh have shown that men and women living in areas with Zn-deficient soils had lower Zn contents in their blood serum than those consuming farm produce from Zn-sufficient soils. Toxicity of Zn is uncommon in Indian crops and soils, but Fe toxicity in rice is common in lateritic soils

4.9 Future Research Needs in Micronutrients Nutrition in Crops of India

There is a need to plan strategies for meeting future demands of micronutrients in soil, plant, animal and human continuum. Nutrient indexing for forecasting emerging micronutrient deficiencies and delineation of deficient areas using GIS should be a priority. Assessment of 'hot spots' of multi-micronutrient deficiencies and their impact on crop yields needs urgent attention. The development and evaluation of alternate techniques, customized fertiliser products, complex fertiliser formulations, and organic matter enrichment are needed to ensure balanced fertilisation of crops and cropping systems. There is a need for developing a total micronutrient ameliorative package for different crops, areas and cropping systems. Research to evaluate interactions of micronutrients with other nutrients, physical and environmental factors, plant stress and disease needs to be focused on enhancing NUE and reducing input costs. Policies to create awareness amongst the farmers for diagnosing micronutrient disorders precisely, and ensuring balanced fertilisation, are very much needed.

References

Ali, M.H. (1992) Micronutrients in soils of Assam. Proceedings of workshop meetings of Indo-British fertilizer educational project. HFC Ltd., Calcutta, 1: 204–206.

Balaguru, T. and Mosi, A.D. (1973) Studies on soil molybdenum in Tamil Nadu soils. Madras Agric. J., 70: 147–151.

Bhupalraj, G., Patnaik, M.C. and Khadke, K.M. (2002) Molybdenum status in soils of Andhra Pradesh. Progress report of AICRP micro and secondary nutrients in soils and plants of Andhra Pradesh. ANGRAU, Hyderabad, 36: 1–87.

Chatterjee, R.K. and Dakshinamurti, C. (1962) Available molybdenum status of some Indian soils. J. Sci. Indus. Res., 21 (8): 597–598.

Datta Biswas, N.R. and Dakshnamurty, C. (1958) A note on the micronutrients distribution in the farm profile. Indian J. Agron., 3: 48–60.

Dangarwala, R.T., Patel, K.P., George, V., Patel, K.C., Ramani, V.P., and Patel, M.S. (1994) Micronutrient and sulphur research in Gujarat. Res. Bull., GAU, Anand, 2: 1–92.

Deore, S.K., Dhake, A.V., Tupatka, P.N. and Sonar, K.R. (1994) Micronutrients in agriculture. Proceedings of seminar on recent trends in micronutrient research in soils and plants in Maharashtra. MAU, Parbhani, p. 29.

Dutta, H., Rahman, S., Kanjilal, A. and Ali, A.H. (1992) Micronutrients. Workshop of Indo-British fertilizer education project. HFC Ltd., Calcutta, 1: 180–186.

FAI (2002) Fertilizer use and crop production. Fertilizer Statistics, Fertilizer Association of India. New Delhi, pp. 1–100.

Grewal, J.S. and Trehan, S.P. (1990) Micronutrient deficiency and amelioration in potato. Shimla Tech. Bull., CPRI, 23: 1–26.

Gupta, V.K., Gupta, S.P., Kala, R., Potalia, B.S. and Kaushik, R.D. (1994) Response of crops to micronutrients and amelioration. Twenty five years of micronutrient research in soils and crops of Haryana. Res. Bull., CCSHAU, Hisar, 1: 1–99.

Kanwar, J.S. and Randhawa, N.S. (1974) Micronutrient research in soils and plants in India: A review. ICAR Tech. Bull. (Agric.), 50: 1–60.

Katyal, J.C. and Vlek, P.G.L. (1985) Micronutrient problems in tropical Asia. Fert. Res., 7: 69–74.

Katyal, J.C., Sharma, B.D. and Gupta, S.K. (1982) Annual report of the AICRP micronutrients in soils and plants for the year 1982–83. ICAR, New Delhi, 16: 1–66.

Krishnasamy, R., Shanmugasundarum, M., Poongothai, S., Valliappan, K., Vanangamudi, M., Mani, S., Savithri, P. and Devarajan, R. (1992) Twenty five years of micronutrient research in soils and crops of Tamil Nadu. Res. Bull., TNAU, Coimbatore, 1: 1–192.

Lal, B.M., Sahu, D. and Das, N.B. (1960) Available zinc status of Indian soils. Curr. Sci., 29: 316–322.

Lal, F. and Biswas, T.C. (1973) Factors affecting the distribution and availability of micronutrient elements in major soil groups of Rajasthan. I. Surface soils. J. Indian Soc. Soil Sci., 21: 455–466.

Lal, F. and Biswas, T.C. (1974) Factors affecting the distribution and availability of micronutrient elements in major soil groups of Rajasthan. II. Soil profiles. J. Indian Soc. Soil Sci., 22: 333–346.

Lindsay, W.L. and Norvell, W.A. (1978) DTPA soil test method for determining available micronutrient cations. Soil Sci. Soc. Am. J., 42: 421–428.

Malewar, G.U. and Randhawa, N.S. (1978) Clay mineralogical makeup of Marathwada soils and their relationship with content of zinc, iron, manganese and copper. J. Maharashtra Agric. Univ., 3 (1): 1–4.

Mathur, C.M., Moghe, V.B. and Talati, M.R. (1964) Distribution of boron in soils of western Rajasthan irrigated with high boron waters. J. Indian Soc. Soil Sci., 12: 319–324.

Nair, G.G.K. and Mehta, B.V. (1959) Status of zinc in soils of western India. Soil Sci., 87: 155–159.

Nandi, S. and Mandal, L.N. (1979) Effect of application of P, Fe and Mn under different moisture regimes on the yield and nutrition of rice. J. Indian Soc. Soil Sci., 27: 150–157.

Nandi, T. (1992) Boron deficiency in soils of Orissa. Proceedings of IBFEPO-HFC workshop on micronutrients, Bhubaneswar, 1: 157–160.
Nene, Y.L. (1965) Symptoms, causes and control of Khaira disease of paddy. Bull. Indian Phytopath. Soc., 3: 97–101.
Nayyar, V.K., Takkar, P.N., Bansal, R.L., Singh, S.P., Kaur, N.P. and Sadana, U.S. (1990) Micronutrients in soils and crops of Punjab. Res. Bull., PAU, Ludhiana, 1: 1–148.
Nayyar, V.K., Takkar, P.N., Bansal, R.L. and Singh, S.P. (1996) Annual progress reports of AICRP micro and secondary nutrients and pollutant elements in soils and plants. Department of Soils, PAU, Ludhiana, 32: 1–90.
Nayyar, V.K., Bansal, R.L. and Khurana, V.K. (1998) Amelioration of micronutrient deficiencies. Annual progress reports of AICRP micro and secondary nutrients and pollutant elements in soils and plants. PAU, Ludhiana, 34: 1–112.
Nayyar,V.K., Bansal, R.L. and Khurana, M.P.S. (2005) Response of copper to soybean, Progress report AICRP micro and secondary nutrients and pollutant elements in soils and plants, PAU, Ludhiana, 39: 1–96.
Patel, K.P. and Singh, M.V. (1995) Copper research and agricultural production. Micronutrient Research and Agricultural Production (Ed. Tandon, H.L.S.), FDCO, New Delhi, 1: 15–33.
Patel, K.P., Patel, K.C., George, V., Singh, M.V. and Ramani, K.C. (2005) Effect of farm yard manure and gypsum on yield of sweet corn and Cu fractions of sewage irrigated soil. Pollution Res., 24 (1): 127–134.
Rathore, G.S., Khamparia, R.S., Gupta, G.P., Dubey, S.B., Sharma, B.L. and Tomar, V.S. (1995) Twenty five years of micronutrient research in soils and crops of Madhya Pradesh. Res. Bull., JNKV, Jabalpur, 2: 1–101.
Raychaudhari, S.P. and Datta Biswas, N.R. (1964) Trace elements status of Indian soils. J. Indian Soc. Soil Sci., 12: 207–214.
Reddy, G.R. (1964) Molybdenum status of western India soils. J. Agric. Sci., 34: 219–233.
Sakal, R. and Singh, A.P. (1995) Boron research and agricultural production. Micro nutrient Research and Agricultural Production (Ed. Tandon, H.L.S.), FDCO, New Delhi, 1: 5–21.
Sakal, R., Singh, A.P., Sinha, R.B. and Bhogal, N.S. (1996) Twenty five years of research on micro and secondary nutrients in soils and crops of Bihar. Faculty of Agric. RAU, Pusa, 1: 1–135.
Sakal, R., Nayyar, V.K. and Singh, M.V. (1997) Micronutrient management for rice-wheat system. Bull. Indian Soc. Soil Sci., 18: 39–47.
Saha, J.K. and Singh, M.V. (1997) Effect of CaCO3 on the adsorption and desorption of boron in swell-shrink soils in Madhya Pradesh. J. Indian Soc. Soil Sci., 45: 57–61.
Saha, J.K., Singh, M.V. and Sharma, B.L. (1998) Available boron status in major soils of Madhya Pradesh. J. Indian Soc. Soil Sci., 46: 478–479.
Satyanarayan, Y. (1958) Trace element status of cotton soils of Gujarat. Indian Cotton Cong. Rev., 12: 138–141.
Savithri, P. and Poongothai, S., Shanmugasundarum, R. (1996) Annual report of AICRP micro and secondary nutrients and pollutant elements in soils and plants. TNAU, Coimbatore, 32: 1–98.
Sharma, R.B. and Motiramani, D.P. (1969) Zinc status of the soils of Madhya Pradesh. J. Indian Soc. Soil Sci., 17: 19–26.
Singh, A.L. and Chaudhari, V. (1991) Screening of groundnut cultivars tolerant to iron chlorosis. Indian J. Agric. Sci., 61: 12–17.
Singh, A.P., Singh, M.V., Sakal, R. and Chaudhary, V.C. (2006) Boron nutrition of crops and soils of Bihar. Tech. Bull., 6: 1–80.
Singh, D., Leelawathi, C.R., and Surup, S. (1979) Responses of crops to micronutrients fertilization. Indian Agricultural Statistical Research Institute, New Delhi. Res. Bull., 1: 1–36.
Singh, M.V. (1988) Response of rice and wheat to phosphorus and zinc fertilization in sodic soils. Indian J. Agric. Sci., 58 (11): 823–827.
Singh, M.V. (1991a) Micronutrient management in rice–wheat cropping system. Proceedings of rice–wheat, PDCSR, Modipuram, 1: 133–144.
Singh, M.V. (1991b) Micronutrient deficiency delineation and mapping. Proceedings of national seminar on zinc sulphate industry – whither too. Uttar Pradesh zinc sulphate manufacturer association, Lucknow, II: 1–20.

Singh, M.V. (1994) Fertilizer zinc requirement of soybean and wheat crops in swell-shrink soils, Extended summary. Indian Soc. Soil Sci., 446–448.
Singh, M.V. (1986a) Relevance of micronutrients in long term sustainable crop production. Proceedings of NRI workshop on integrated nutrient management on farmers fields: Approaches that work (Eds. Gregory, P.J., Philbeam, C.J. and Walker, S.H.), University of Reading, Reading, UK, 418–424.
Singh, M.V. (1998) Micronutrients in Indian soils and plants. Progress report of AICRP of micro- and secondary-nutrients and pollutant elements in soils and plants, 28: 1–122.
Singh, M.V. (1999a) Micronutrient status in Indian soils. Micronutr. News, 46: 25–42.
Singh, M.V. (1999b) Current status of micronutrients in Indo-Gangetic alluvial plains. J. Sustai. Use of Chemicals in Agr., 1: 49–59.
Singh, M.V. (2000) Micronutrients in Indian soils and plants. Progress report of AICRP of micro- and secondary-nutrients and pollutant elements in soils and plants. 29: 1–122.
Singh, M.V. (2001a) Importance of sulphur in balanced fertilizer use in India. Fert. News, 46 (10): 13–35.
Singh, M.V. (2001b) Evaluation of micronutrient status in different agroecological zones of India. Fert. News, 46 (2): 25–42.
Singh, M.V. (2001c) Response of micronutrient to crops in different agroecological regions – experiences of AICRP Micronutrients. Fert. News, 42 (10): 43–49.
Singh, M.V. (2002) Micro and secondary nutrient management for doubling productivity of oilseed-based cropping systems (Ed. Gangwar et al.), Proceedings of oilseed based cropping system – issues and technologies, PDCSR, Modipuram, 1: 51–67.
Singh, M.V. (2004) Micronutrients seed treatment to nourish the crop plants at critical stages of growth. Res. Bull., IISS, Bhopal, 2: 1–87.
Singh, M.V. (2005) Efficiency of Granubor II for enhancing crop productivity in boron deficient soils. AICRP Micronutr., IISS, Bhopal, 3: 1–82.
Singh, M.V. (2006) Micro and secondary nutrients and pollutant elements research in India. Progress report, IISS, Bhopal, pp. 1–110.
Singh, M.V. and Abrol, I.P. (1985a) Solubility and adsorption of zinc in sodic soils. Soil Sci., 140: 406–411.
Singh, M.V. and Abrol, I.P. (1985b) Direct and residual effect of fertilizer zinc application on the yields and chemical composition of rice–wheat system in alkali soils. Fert. Res., 8: 179–191.
Singh, M.V. and Abrol, I.P. (1986a) Transformation and movement of zinc in alkali soil and their influence on zinc uptake by rice. Plant and Soil, 94: 445–449.
Singh, M.V. and Abrol, I.P. (1986b) Transformation and availability of zinc in alkali soils. Fert. News, 31 (7): 17–27.
Singh, M.V. and Saha, J.K. (1992) Annual report of AICRP of micro- and secondary nutrients in soils and plants. IISS, Bhopal, 24: 1–86.
Singh, M.V. and Saha, J.K. (1994) Annual report of AICRP of micro- and secondary nutrients in soils and plants. IISS, Bhopal, 26: 1–86.
Singh, M.V. and Saha, J.K. (1995) Progress report of AICRP of micro and secondary- nutrients and pollutant elements in soils and plants. IISS, Bhopal, 26: 1–118.
Singh, M.V. and Saha, J.K. (1995) Micronutrients in Indian soils and plants. Progress report of AICRP of micro- and secondary-nutrients and pollutant elements in soils and plants. IISS, Bhopal, 27: 1–100.
Singh, M.V. and Subba Rao, A. (1995) Manganese research and agricultural production. Micronutrient Research and Agricultural Production (Ed. Tandon, H.L.S.), FDCO, New Delhi, 1: 33–56.
Singh, M.V., Chhabra, R. and Abrol, I.P. (1987) Interaction between application of gypsum and zinc on yield and chemical composition of rice in alkali soil. J. Agric.Sci. Camb., 108: 275–279.
Singh, M.V., Patel, K.P. and Raman, V.P. (2003) Crop responses in swell-shrink soils. Fert. News, 48 (4): 38–44.

Singh, M.V., Wanjari, R.H. and Adhikari, T. (2004) Nutrient dynamics, crop productivity and sustain-ability under long term fertilizer experiments in India. AICRP LTFE, IISS, Bhopal, 1: 1–120.

Subba Rao, I.V. and Adinarayana, V. (1995) Molybdenum research and agricultural production. Micronutrient Research and Agricultural Production (Ed. Tandon, H.L.S.), FDCO, New Delhi, 1: 60–77.

Subbaiah, V.V., Patnaik, M.C., Raj, B. and Subbarao, I.V. (1998) Two decades of research on micro and secondary nutrients in soils and plants of Andhra Pradesh, Res. Bull., ANGRAU, Hyderabad, 1: 1–263.

Swaine, D.J. (1955) The trace element content of soil. Commonw Bur Soil Sci. Tech. Comm., Farnham Royal, England, 48.

Takkar, P.N. and Nayyar, V.K. (1981) Gypsum and zinc requirement of rice–wheat under sodic water irrigation. Expt. Agric., 17: 49–55.

Takkar, P.N., Singh, M.V. and Ganeshmurthy, A.N. (1997) Micronutrient deficiency and fertilizer requirement by 2025. Proceedings of national symposium on plant nutrient needs, supply, efficiency and policy issues for 2000–2025 (Eds. Kanwar, J.S. and Katyal, J.C.), NAAS, New Delhi, 1: 218–224

Takkar, P.N., Mehta, S.K. and Chhibba, I.M. (1990) A decade of micronutrient research. Res. Bull., AICRP Micronutr., IISS, Bhopal, 1: 1–136.

Chapter 5
Micronutrient Deficiencies in Crop Production in China

Chunqin Zou, Xiaopeng Gao, Rongli Shi, Xiaoyun Fan, and Fusuo Zhang

Abstract The essential micronutrients for field crops are Fe, Zn, B, Mo, Cu and Mn. The incidence of micronutrient deficiency has increased in recent years. Iron and Zn deficiency are paid more attention because they negatively affect both food production and human health in a major part of the world. In this paper, the Fe, Zn, B, Mo, Mn and Cu deficiency status in soils and crops in China are reviewed. Iron and Zn deficiencies cause some serious problems in crop production in China, and B, Mn and Mo deficiency are second in importance. The corrections of these micronutrient deficiencies by fertilization, agronomic strategies and genotypic exploitation are also discussed. In China, a scarcity of water has caused a shift from flooded to aerobic conditions for rice production. The consequence for micronutrient availability is a function of the changes in both soil and plant factors.

5.1 Introduction

China is situated in eastern Asia on the western shore of the Pacific Ocean, with an area of 9.6 M km^2. It is located in north latitude 53°34′, south latitude 30°51′, east longitude 135°05′ and west longitude 73°. The greater part of the Chinese territory is situated in the Temperate Zone, its southern part in the tropical and subtropical zones, and its northern part near the Polar Zone. China is divided into 23 provinces, 5 autonomous regions, 4 municipalities under the direct jurisdiction of the Central Government, and one special administrative region (Fig. 5.1).

Multiple soil micronutrient deficiencies are one of many factors limiting crop yields, crop product quality and human health in many areas of China. Iron (Fe) and zinc (Zn) deficiencies cause serious problems in crop production in China. As fertiliser and other agronomic strategies are shown to be inadequate for overcoming the limitations of micronutrient deficiencies, the large variation in micronutrient efficiency among plant genotypes offers opportunities for breeding as a tool to resolve micronutrient deficiency problems. New challenges of micronutrient deficiency in crop production are now becoming apparent because of water constraints.

B.J. Alloway (ed.), *Micronutrient Deficiencies in Global Crop Production.*

Fig. 5.1 Map of People's Republic of China: Locations of provinces, autonomous regions and municipalities

5.2 Micronutrient Status of Soils in China

5.2.1 Distribution of Micronutrient-Deficient Soils

Micronutrient deficiencies are widespread in China because of the generally low micronutrient availability in soils and also because of increasing nutrient demands from increasingly intensive cropping practices (Table 5.1).

There are more than 48.6 and 20.3 Mha of soils deficient in zinc (Zn) and manganese (Mn), respectively, which are mainly distributed on calcareous soils in the northern part of China. There are 44.5 and 32.8 Mha of soils deficient in molybdenum (Mo) and boron (B), respectively, which are mainly distributed in the eastern part of China (Lin and Li, 1997). The copper (Cu) content in the majority of soils is adequate with the exception of organic soils and paddy soils with permanent waterlogging. Iron deficiency is widespread on calcareous soils and alkaline soils (DTPA-Fe < 4.5 mg kg^{-1}) and affects around 40% of farmland depending on crop genotypes and agronomic strategies.

Table 5.1 Micronutrient-deficient soil areas and the critical level used in China (Lin and Li, 1997)

		Below critical level area	
Element	Critical level (mg kg^{-1})	Million hectares	% of total farmland
Zn	0.5a	48.6	51.1
B	0.5b	32.8	34.5
Mo	0.15c	44.5	46.8
Mn	5.0a	20.3	21.3
Cu	0.2a	6.5	6.9
Fe	4.5a	4.7	5.0

Extractants: a = DTPA, b = hot water, c = NH_4Ac

5.2.2 *Zinc*

The total Zn content in China soils varies from 3 to 790 mg kg^{-1}, with an average of 100 mg kg^{-1} (Liu, 1996). There are large variations in Zn content among the soils of different areas in China which are mainly due to the different Zn contents in diverse types of soils and parent materials. Geographically, soil Zn contents tend to decrease from the south to the north in China. The average Zn content in soils of the south is 163 mg kg^{-1}, while only 78 mg kg^{-1} in the north. Two markedly different regions of Zn-rich and Zn-limited soils are identified. The former mainly includes the red soils (FAO-UNESCO: Ferralsols, Luvisols and Cambisols) of southern China and the latter mainly includes the calcareous soils (Calcisols) of northern China.

Different forms of Zn in soils vary in their availability to plants. The total Zn content is not a reliable index to reflect the soil's ability to supply Zn for plants. For characterisation of the available fraction of Zn in soils, DTPA and HCl are frequently used as extractants for soils with high and low pHs, respectively. The critical levels for DTPA-Zn and HCl-Zn are 0.5 mg kg^{-1} and 1.5 mg kg^{-1}, respectively. Based on the analytical results and the responses of crop plants to Zn fertilisation, soils in China can be divided into five classes with different available-Zn contents (Fig. 5.2). It could be concluded that the available-Zn contents in acid soils of southern China are much higher than those in calcareous soils of northern China. This is in accordance with the distribution of Zn deficiency problems in crops on the calcareous soils.

5.2.3 *Iron*

The total Fe contents in soils are relatively high and vary from 1.05% to 4.84%, with an average of 2.94%. There are large variations in Fe content among the soils of different areas of China because of different soil types and parent materials. However, the total soil content is not a good indication of the soil's ability to supply Fe for plants.

Extraction with DTPA is generally used to represent the availability of Fe in soils with high and low pH. The critical level for DTPA-Fe is 4.5 mg kg^{-1}. Based on the analytical results and the response of crop plants to Fe fertilisers, soils in China can

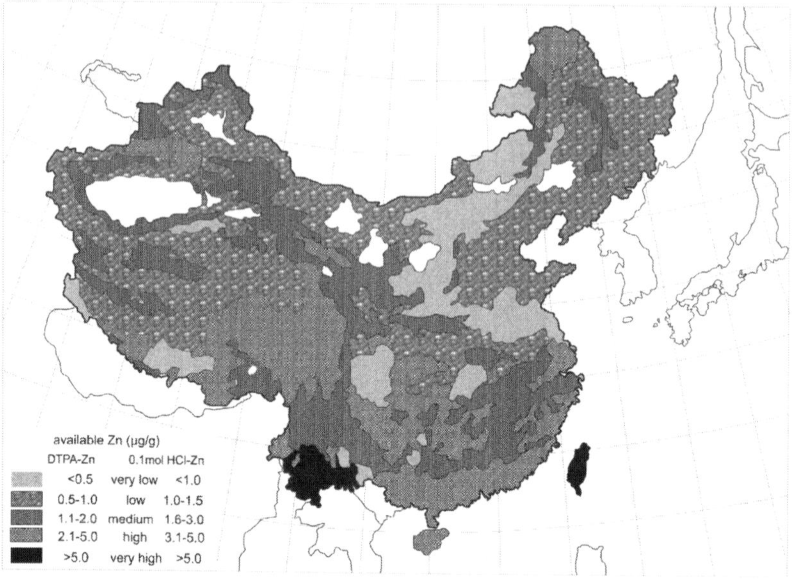

Fig. 5.2 Map of the distribution of available zinc in soils in China (Liu, 1996)

be divided into five classes with different available-Fe contents (Fig. 5.3). It could be concluded that the available-Fe contents in acid soils of southern China are much higher than those in the calcareous soils and alkaline soils of northern China. This is in accordance with the distribution of Fe deficiency problems in crops on the calcareous soils with higher soil pH. Iron deficiency occurs from the Sichuan basin in the south to the Inner Mongolia highlands in north and from the North China Plain (NCP) in the east to the loess highlands in the west (Zou and He, 1985).

Iron deficiency in crops is strongly induced by high pH and a high $CaCO_3$ content, which occur widely in the north of China. So, the actual area of Fe-deficient soil in crop production is more than the statistical data from the soil test for DTPA extractable Fe. About 40% of the soil area is deficient in Fe, but this depends on the crops, fertilisation management and soil conditions.

5.2.4 *Manganese*

Soil Mn contents are relatively higher than for other micronutrients. Manganese deficiency in plants is normally caused by low availability instead of low total content. Total Mn contents in China soils vary from 10 to 5,532 mg kg^{-1} with an average of 710 mg kg^{-1}. It differs markedly among different soil types. In the same way as Zn, Mn contents are much higher in the acid soils of southern China than those in the calcareous soils of northern China.

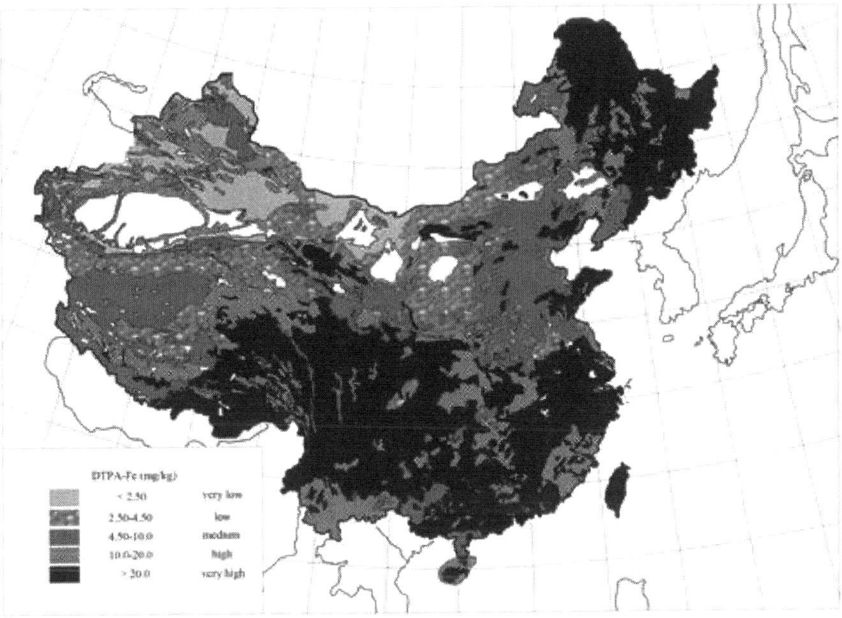

Fig. 5.3 Map of the distribution of DTPA-extractable iron in soils in China (Liu, 1996)

The soluble and exchangeable Mn in soils is thought to be directly available to plants. Active-Mn is thought to be a suitable index to reflect the ability of soil to supply Mn to plants, which is determined by extraction with 1 mol L^{-1} NH_4OAc + 2 g L^{-1} hydroquinone. Manganese availability is influenced by many soil factors, including soil type, parent materials, pH, redox potential, soil organic matter, and so on. It differs significantly among different areas in China (Fig. 5.4). As found with total Mn contents, acid soils in southern China have much higher available-Mn (active-Mn) contents than calcareous soils in northern China. The critical concentration for Mn deficiency is 100 mg kg^{-1} as active-Mn. This critical level is especially for calcareous soils. Manganese deficiency is seldom found on acid soils. The critical concentration is identified as the level for Mn-sensitive plants to grow normally.

5.2.5 *Boron*

Soil B contents in China range from 2 to 500 mg kg^{-1}, with an average of 64 mg kg^{-1}. Boron contents are mainly determined by the distributions of soil types and parent materials. In brief, soil B contents in the north and west of China are higher than those in the south and east of China.

Boron availability in soils is negatively related to the soil pH. Soil B availability is highest with a pH between 4.7 and 6.7. Boron deficiency of plants mainly occurs

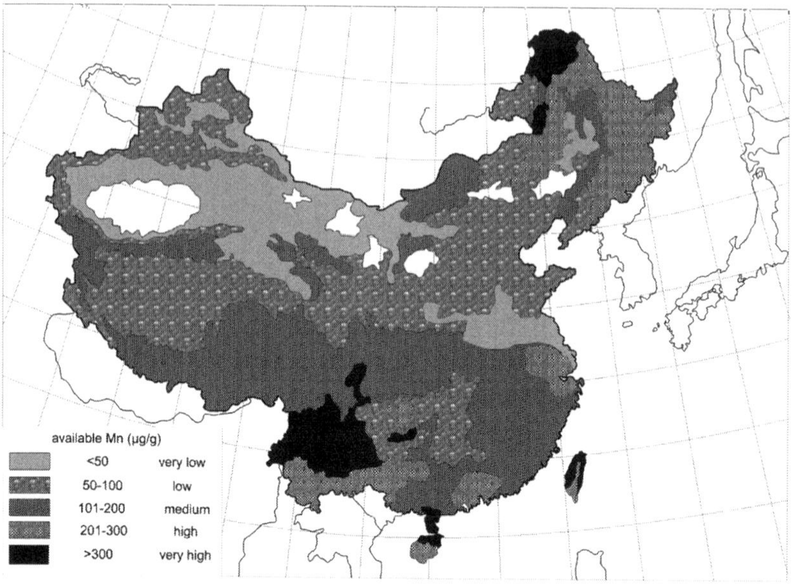

Fig. 5.4 Map of the distribution of available manganese in soils in China (Liu, 1996)

on soils with pH values higher than 7. Hot water-soluble-B is a reliable index to evaluate the soil B availability. In China, the critical concentration for normal crop plants is 0.5 mg kg^{-1}. When the water-soluble-B content is lower than the critical value, the yields of some sensitive plant species may decrease. Figure 5.5 is a map of the water-soluble-B distribution in China. It could be concluded that arid, inland regions of the western part are B-rich areas and the moist regions of the southeast seaboard are low in B, or B-deficient areas. A transition area with a moderate B content can be found between the two regions.

5.2.6 *Molybdenum*

Soil Mo contents are much lower than for other micronutrients. It is higher in acid soils of southern China than calcareous soils of northern China. The total Mo content in China ranged from 0.1 to 6 mg kg^{-1} and is averaged to 1.7 mg kg^{-1}. In brief, the Mo content in southern China is higher than that in northern China. The distribution is mainly determined by the soil types, which can reflect the differences in geography and climate between the north and south of China.

Ammonium acetate (NH_4Ac) is used as an extractant to evaluate the Mo availability. It may account for 5–20% of the total Mo content, depending on soil types. The critical concentration is 0.15 mg kg^{-1} for legume plants, which are more sensitive to B deficiency than other plant species. When soil NH_4Ac-Mo is lower than 0.15 mg kg^{-1}, Mo fertiliser application may increase the yields of legume plants. Around 47% of the agricultural area in China is Mo-deficient, which is

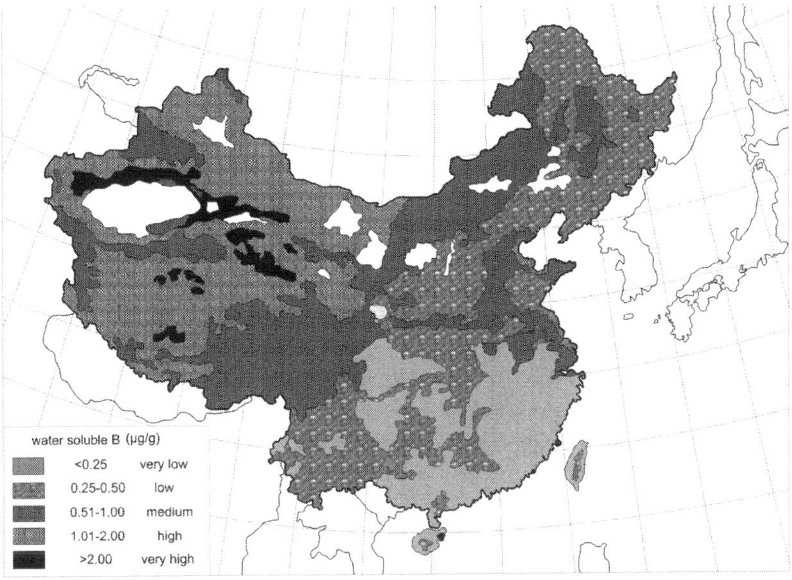

Fig. 5.5 Map of the distribution of water-soluble boron in soils in China (Liu, 1996)

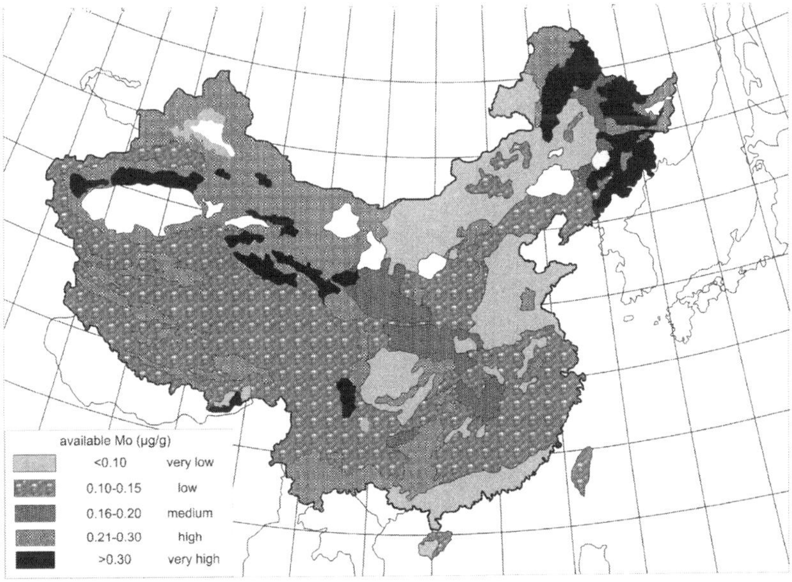

Fig. 5.6 Map of the distribution of available molybdenum in soils in China (Liu, 1996)

mainly distributed in the east of China (Fig. 5.6; Lin and Li, 1997). It may include red soils and laterite soils (equivalent to Ferralsols, Luvisols, Cambisols) in south and yellow soils (equivalent to Luvisols, Acrisols and Cambisols) in the north.

5.2.7 *Copper*

Total soil Cu contents ranged from 2 to 500 mg kg^{-1} with an average of 22 mg kg^{-1}. For most soils, the Cu content is between 20 and 40 mg kg^{-1}. Some special soils, such as red soils have relatively high Cu contents of about 100 mg kg^{-1}. The soil Cu content is mainly determined by the parent materials, which are influenced by many geographical and climatic factors. The effect of soil types on Cu content is weaker than for other micronutrients in soils.

As with Zn, DTPA and HCl are frequently used as extractants to evaluate Cu availability for soils with high and low pH values, respectively. The critical concentrations for DTPA-Cu and HCl-Cu are 0.2 mg kg^{-1} and 2.0 mg kg^{-1}, respectively. It could be concluded from Fig. 5.7 that for most of the agricultural area of China, the available-Cu content is high enough for agricultural production. In some areas of north-east and south-west China, Cu deficiency may occur because of the low Cu availability.

5.3 Micronutrient Deficiencies in Crops and Their Correction

5.3.1 *Zinc*

Zinc deficiency is widely reported in crop production in China and has been reported on most soil types. The calcareous soils (Calcisols) and neutral paddy

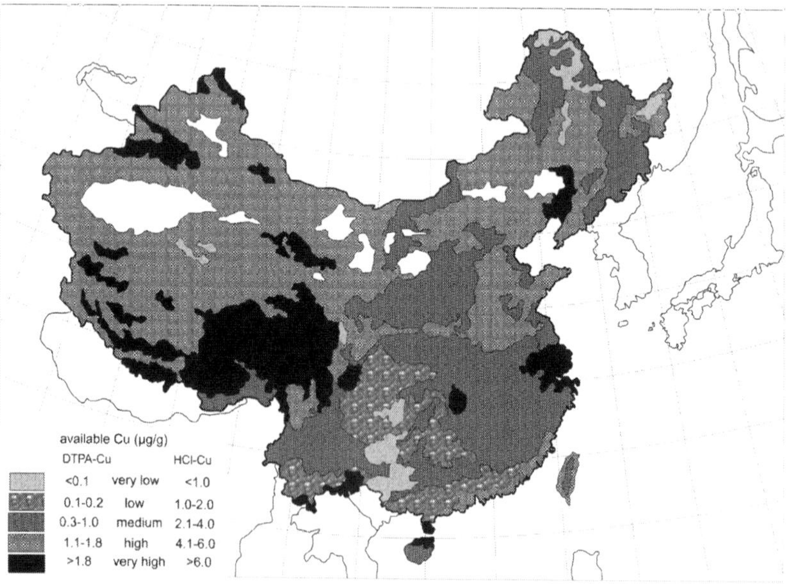

Fig. 5.7 Map of the distribution of available copper in soils in China (Liu, 1996)

Fig. 5.8 Zinc deficiency in maize grown in the north of China (*left*) and in lowland rice grown on paddy soil in the south of China (*right*) (F.S. Zhang and S.H. Lv, 1999, personal communication) (*See Colour Plates*)

soils in northern China are more deficient than the acid soils in southern China (Liu, 1996). Crop species exhibit differential tolerance to Zn deficiency. Maize (*Zea mays*) is the most sensitive crop to Zn deficiency (Gao et al., 2001). Lentil (*Lens culinaris*), pea (*Pisum sativum*), sorghum (*Sorghum bicolor*), cotton (*Gossypium linn*), rice (*Oryza sativa*), apple tree (*Malus sylvestris*) and peach (*Prunus persica*) are also very sensitive to Zn deficiency and are used as indicator plants in many areas of China. Field surveys have shown that maize production has the largest area of Zn fertiliser application (Fig. 5.8 in Colour Section). Rice is in the second place. In Sichuan, Shandong and Hebei provinces of China, Zn fertiliser has been widely used in maize production. Because of the evident effect on yield improving in these areas, Zn fertilisation has been thought to be one of the effective ways to increase crop yields. Zinc deficiency in rice also occurs widely on alkaline paddy soils (Fig. 5.8 in Colour Section).

Zinc fertilisers are commonly applied to many crops in China. The most commonly used sources are $ZnSO_4$ and ZnO. Some other inorganic products such as $ZnCl_2$ and chelates are also used. Zinc fertilisers can be applied as soil applications, foliar sprays, seed coatings, and so on. The different types of application are mainly determined by the crop species. Soil applications are mainly used on food crops including maize, wheat and rice. Foliar sprays of Zn are used for certain crops, especially fruits and vegetables. These practices are widely used by farmers.

In recent years, the correction of Zn deficiency has turned its focus from obtaining high yields to improving the Zn density in edible parts of crop plants. This shift is caused by the fact that Zn deficiency among humans is now becoming a serious problem. Foods derived from plants are major contributors to the Zn requirements of most people in developing countries, including China (Welch, 1993). However, some important crop plants, such as rice, wheat (*Triticum aestivum*) and maize, have relatively low Zn concentrations in their edible parts. So, many researchers are now focusing on improving Zn concentrations in the grains of crop plants.

Some researchers reported that, for wheat, the most simple and effective method of applying Zn is soil application (Zhou, 1995; Table 5.2). Compared with the control treatment, the yield of wheat increased 18.5% with a 16.5 kg ha^{-1} soil

Table 5.2 Effects of Zn fertiliser on the yield of wheat from field experiment (Zhou, 1995)

Treatment ($ZnSO_4$ as fertiliser)	Wheat yield	
	Average (kg ha^{-1})	Increased rate (%)
Soil application (16.5 kg ha^{-1})	12,090	18.5
Soil application (6 kg ha^{-1})	11,325	11.8
Seed dressing (1.5 kg ha^{-1})	10,800	5.8
Spray application (0.1% water solution)	10,680	4.4
Control	10,200	–

Table 5.3 Mean content and range of Fe, Zn and Se in brown rice (Liu et al., 1995)

Micronutrient	Mean content (mg kg^{-1})	Range (mg kg^{-1})
Fe	21.31 ± 12.17	2.99–75.71
Zn	42.12 ± 24.11	13.45–145.78
Se	0.0621 ± 0.0635	0.0187–0.3392

Zn application. This is based on the analysis of the contents of Cl, N, Zn and the 17 different amino acids as well as the yield of wheat. Despite the fact that the effect of Zn fertilisers on crop yield is considered to be generally good, its effect on Zn concentration in grain may be absent or only minor. Zinc application to soil didn't increase grain Zn concentrations in lowland rice and aerobic rice grown on a calcareous soil in Beijing, China (Gao et al., 2006). Similar results have been reported for wheat and maize.

As fertiliser and other agronomic strategies are shown to be inadequate for overcoming the limitations of Zn deficiency, genotypic differences in Zn efficiency offer a sustainable solution to Zn malnutrition in humans. By exploiting plant genetic potential for enhanced ability to accumulate more micronutrients in their edible part, micronutrient concentrations in cereal grains can be increased and better human nutrition may be achieved. Significantly different Zn contents were found in the grains of 115 rice varieties (including wild rice and cultivated rice) collected from various areas of China. The variations in Fe and Zn contents were about 25- and 11-fold, respectively (Liu et al., 1995; Table 5.3). Moreover, the contents of Fe, Zn and Se in the tested rice varieties were closely correlated to the availability of nutrients in soil, geographical conditions and cultivation measures. These studies show that it is possible to increase seed Zn concentration by genetic improvement. Breeding for Zn-dense crop varieties with agronomic advantages is just as important for plant nutrition as for human nutrition.

5.3.2 *Iron*

Generally, Fe deficiency occurs in arid and semiarid regions in China. These areas are mainly located in the north of China. The fact that most of the soils in north China are calcareous and alkaline soils aggravates the Fe deficiency problem. Some high-order peat land, sandy soils, and soils with poor aeration or excessive P fertiliser are also Fe-deficient. Even some low organic matter acid soils can also be Fe-deficient (He, 2002).

Iron deficiency results in decreased dry matter production, reduced chlorophyll concentration in leaves, and reduced activity of enzymes involved in sugar metabolism. Iron chlorosis is one of the limiting factors controlling crop production in areas of alkaline calcareous soils. Many fruit, horticultural and agronomic crops are sensitive to Fe deficiency, such as apple tree, gooseberry (*Ribes* spp.), grape (*Vitis vinifera*), peach, orange (*Citrus sinensis*), peanut (*Arachis hypogea*), soybean (*Glycine max*) and sorghum (*Sorghum bicolor*) (He, 2002). Some data showed that in many soils, chlorosis in rice is a kind of physiologic disease which is induced by high soil Mn content and lower effective Fe/Mn ratios. This may increase Mn accumulation in rice, but decrease the active Fe contents to the plant. Iron-Mn nutrition imbalance impeded the development of rice and significantly decreased the yield (Li et al., 1995).

As with the correction of Zn deficiency, Fe fertiliser is recommended to be applied to the soil. The use of acid fertilisers (e.g., ammonium sulfate instead of urea) on high pH soils is an effective method of overcoming Fe deficiency. Seed soaking is also a good approach to germination and stress-resistance of soybean to Fe deficiency (Yang et al., 2004). Iron application by spraying 2,500 mg L^{-1} $FeSO_4$ solution two or three times during the flowering period could increase the yield of soybean by 17.5–22.9%. Ferrous sulphate mixed with acetic acid was more effective to increase yields than any other types of Fe fertilisers (Han et al., 1994).

Iron deficiency is a serious nutritional problem in fruit production in China, especially in the north of China (Liu et al., 2002). It is one of main factors affecting fruit yield and quality. Apple, peach and orange are the main Fe-deficient fruit trees. Traditional practices are foliar sprays and soil fertilisation. But the effect of both practices is not very good. So other simple and effective practices have been studied, proposed and widely used by farmers, such as normal trunk injection (injecting Fe^{2+}), higher pressure trunk injection, root suction (putting cut roots in Fe fertiliser solution), and bag fertilisation (putting cut branches in a bag with Fe fertiliser solution) (Fig. 5.9 in Colour Section).

Although there are several Fe fertilisation practices used in crop production in China, Fe deficiency in crops is still a major nutritional problem, limiting yield and

Fig. 5.9 Special practices to ameliorate iron deficiency in fruit trees. *Left*: Bag fertilisation; *Right*: Root suction (S.H. Lv and J.J. Xue, 2001, personal communication) (*See Colour Plates*)

Fig. 5.10 Fe chlorosis of peanut in a monocropping system on a calcareous soil (*left*) and its improvement by intercropping with maize (*right*) (From Zuo et al., 2000) (*See Colour Plates*)

quality in China. It is widespread in crop production on calcareous and alkaline soils. Iron bioavailability is very low and ineffective remedial practices are used. Therefore, some biological practices have been put forward and perfected, for example, the intercropping system.

In China, the peanut (*Arachis hypogea*) is the major oilseed crop. It accounts for 30% of the cropped area and 30% of the total oilseeds production in the country. However, Fe chlorosis is one of the most common yield-limiting nutrient problems in peanuts grown in monocropping systems on calcareous soils in northern China (Fig. 5.10 in Colour Section). Zuo et al. (2000) reported that there was a marked improvement in the Fe nutrition of the peanut when it was intercropped with maize in the field. The chlorophyll and HCl-extractable Fe concentrations in young leaves of peanut in the intercropping system with unrestricted interactions of the roots of both plant species were also much higher than those of peanut in monoculture. The results suggested that the improvement in the Fe nutrition of peanut intercropped with maize was mainly caused by rhizosphere interactions between the two species. Further studies showed that the effect of intercropping peanuts with five gramineous species with different phytosiderophore release rates could improve the Fe nutrition of the peanut (Zuo and Zhang, 2003).

Nitrogen forms had a significant influence on Fe uptake, distribution and remobilization in maize plants (Table 5.4). Compared with NO_3-fed plants, a higher proportion of ^{59}Fe was observed in young leaves of the Fe-deficient plants fed with NH_4^+-N. The ammonium supply greatly improved ^{59}Fe re-translocation from primary leaves and stem to young leaves. Under Fe deficiency, about 25% of Fe in primary leaves of the NH_4-fed plants was mobilized and re-translocated to young leaves. Exogenous Fe supply decreased the efficiency of such ^{59}Fe re-translocation. The results suggest that Fe can be remobilized from old to young tissues in maize plants but the remobilization depends on the form of N supply as well as supply of exogenous Fe (Zou et al., 2001). Similar results were found in aerobic rice (unpublished data). On the basis of this study, a special Fe-NH_4 complex fertiliser was produced and used to correct Fe deficiency in fruit crops in China.

Table 5.4 Effect of N form and Fe supply on distribution of ^{59}Fe in maize plants. Plants were pre-cultured in NO_3-N solution with ^{59}FeEDTA for 5 days and were then grown in treatment solutions for 8 additional days (Zou et al., 2001)

	^{59}Fe proportion (%)			
	−Fe		+Fe	
Organ	NO_3-N	NH_4-N	NO_3-N	NH_4-N
Young leaves	14b	22a	15b	19a
Primary leaves	32ab	29c	33a	30bc
Roots	54a	49c	52ab	51bc

5.3.3 *Boron*

Boron deficiency is widely found in oil seed-rape in the south of China. Boron deficiency is one of the most important factors that limited oil seed-rape (*Brassica napus*) production. Hubei Province has the largest oil seed-rape cropping area. Until 2004, the growth area of oil seed-rape in Hubei province was almost 1.4×10^7 ha and the total yield in 2004 was around 2.5×10^8 kg. The soils for oil seed-rape production are mainly red soils, yellow soils, and yellow-brown soil (equivalent to Luvisols and Cambisols). The hot water-soluble-B content is lower than 0.5 mg kg^{-1}. For some typical deficient areas, it may be lower than 0.25 mg kg^{-1}. Therefore, the low soil B availability is one of the most important causes of B deficiency in rape plants. Boron fertiliser application can significantly and sustainably increase the yield and quality of oil seed-rape, using methods including seed fertilisation, soil fertilisation and foliar application. In general, the yield of oil seed-rape could increase 10–20% by B application. When the soil is seriously deficient in B, the yield increase of oil seed-rape is around 30–50%, and even 100%, as a result of B application (Ren, 2004).

Boron deficiency is also found in other crops, such as cotton (*Gossypium hirsutum* L.) and sugar beet (*Beta vulgaris* L.). The cropping area of cotton is about 5.1×10^7 ha, and most of this is in the B-deficient-soil region. As with oil seed-rape, B fertilisation could increase the yield of cotton to a large extent. Seed treatment, soil fertilisation and foliar application are effective in improving the B nutrition of cotton.

5.3.4 *Manganese*

Rice–wheat and rice–oil seed rape cropping systems are long-established major crop production systems in China. Manganese deficiency is common in wheat, oil seed rape and other upland crops in rice–wheat, rice–rape rotation systems in China (Lv and Zhang, 1997). For example, Hu (1981) found that Mn deficiency in wheat occurred on flood plain soils (equivalent to Fluvisols) along the sides of rivers of the Chengdu Plain. The increasing occurrence of Mn deficiency and decline in wheat grain yield at many sites without Mn application stimulated studies to develop techniques for efficient management of this problem.

At five field sites surveyed, total Mn and active Mn concentrations in the topsoil layers under rice–wheat rotations were only 42% and 11%, respectively, of those under systems without paddy rice. Both total and available Mn increased with depth in soils with rice–wheat rotations, showing significant spatial variability of Mn in the soil profile. Manganese leaching was the main pathway for Mn loss in coarse-textured soil with high pH, while excessive Mn uptake was the main pathway for Mn loss in clay-textured and acid soils. When Mn was deficient in the topsoil, sufficient Mn in the subsoil contributed to better growth and Mn nutrition of wheat but insufficient Mn in the subsoil resulted in Mn deficiency in this crop. The main mechanism of Mn deficiency in soils with rice–wheat rotation can be explained as follows: (1) spatial Mn variation induced by rice–wheat rotation leads to the low availability of Mn in soils and (2) cultivars with high-yielding potential are unable to utilize sufficient Mn in subsoil due to the hard plough-layer. Therefore this nutrient deficiency problem can be solved by rhizosphere management such as deep ploughing, Mn application to the root zone, or selection of wheat genotypes with deep root systems (Lv et al., 2002; Fig. 5.11 in Colour Section).

Manganese deficiency can be corrected by foliar application of Mn or by banding Mn with an acid starter fertiliser. Various Mn fertilisers (Mn salts, chelates, such as Mn-EDTA, and preparations consisting of mixtures of micronutrients) and various

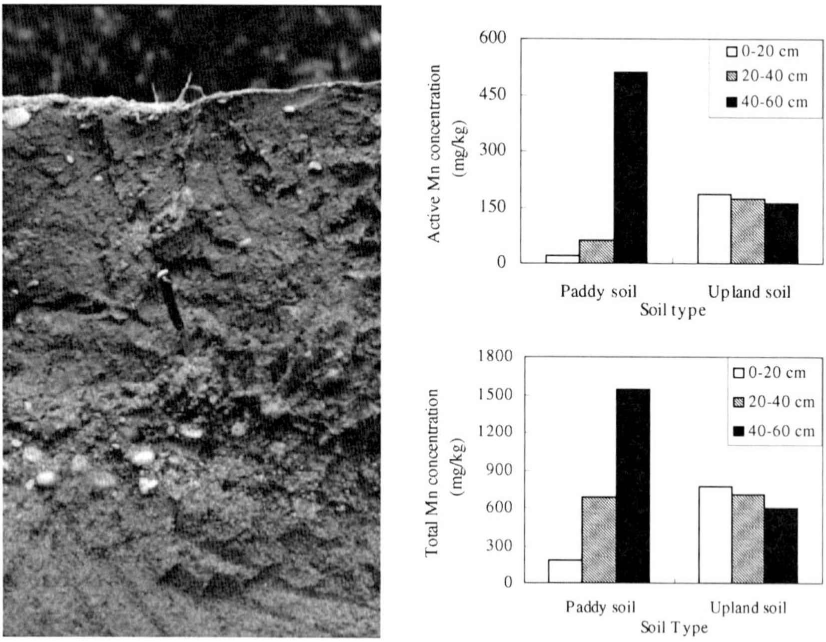

Fig. 5.11 Spatial distribution of Mn in a paddy soil with rice–wheat rotation compared with an upland soil (Lv et al., 2002) (*See Colour Plates*)

application methods (soil, foliar and seed application) have been evaluated for their efficiency in correcting Mn deficiency in wheat (Liu et al., 2001). Most studies have sought to overcome the deficiency of this micronutrient through the application of Mn fertilisers. Research has shown that with wheat, the most effective method of applying Mn fertiliser is seed treatment with Mn, followed by foliar spraying.

Breeding and selecting some Mn-efficient genotypes is also a good approach to overcome Mn deficiency. Fang et al. (1998) reported that there were considerable differences in tolerance to Mn-deficiency among 15 wheat varieties. Most of wheat varieties were sensitive to Mn deficiency but a few varieties had a high tolerance. The tolerant varieties were very useful for growing on Mn-deficient soils. The effect of Mn deficiency on the metabolism of carbohydrate, amino acid and protein differs among crop species (Tai et al., 2004).

5.3.5 Molybdenum

Most Mo deficiency occurs in leguminous plants and some *Cruciferae* species. In the northwest of China, Mo deficiency caused by continuous cropping is one of the important reasons for the yield decrease in soybean (Han et al., 1998). Molybdenum-deficient plants show a light green colour and interveinal yellow patches on old leaves. Molybdenum deficiency results in reduced root nodulation and stunted plants (Liu et al., 2001). Where soybean plants suffered from Mo deficiency, the root weight was 74–84% lower compared with plants under normal Mo conditions. The shoot length was decreased by 83–90%. The numbers of root nodules were decreased by 70–98% and the yield was decreased by 20% (Wu and Xiao, 1994).

More and more studies have shown that Mo deficiency has become a limiting factor in winter wheat production in southern China (Fig. 5.12 in Colour Section), such as in Hubei, Chongqin, Sichuan, Henan, Anhui and Shandong provinces (Wang et al., 1995; Li et al., 2001). Wang et al. (1995) but was first found in the yellow brown soils in Hubei province. The results of some field and pot experiments

Fig. 5.12 Molybdenum deficiency in winter wheat cultivated in the field (*left*) and special symptom of the leaf (*right*) (Y.H. Wang, 1997, personal communication) (*See Colour Plates*)

showed that the death of winter wheat seedlings from Mo deficiency was closely related to low temperatures and a high level of N supply. Molybdenum deficiency occurs in winter wheat where the soil pH is higher than 7.0–7.5, the available soil Mo concentration is lower than 0.20–0.25 mg kg^{-1}, the temperature is lower than 5 °C and the N fertilizer application rate is higher than 300 kg ha^{-1}. Thus Mo fertilisation could increase wheat yields by about 700–750 kg ha^{-1}.

Correction of Mo deficiency can be achieved by applying Mo fertiliser. Commonly used Mo fertilisers are ammonium molybdate and sodium molybdate. The application methods may include soil application, seed dressing and foliar spraying. It has been reported that soybean yields could be increased by around 50% through Mo fertilisation (Dong and Sun, 1990).

Genotypic differences have been identified in soybean and wheat with tolerance to Mo deficiency. Under nutrient solution culture, eight genotypes of soybean were selected for their tolerance to Mo deficiency from 37 varieties (Liu et al., 2001). The Mo efficient and inefficient wheat genotypes were selected from 34 winter wheat genotypes grown on an acid yellow brown soil (equivalent to Luvisols and Cambisols). The differences may be caused by different capacities for Mo uptake and phloem transportation (Yu et al., 2003).

5.4 New Challenges Caused by Water Constraints

Rice production in China is now undergoing an important change brought about by water constraints. Traditional high water-consuming flooded rice cultivation is gradually being replaced by the new upland rice cultivations, including an aerobic rice production system and the Ground Cover Rice Production System (GCRPS, Liang et al., 2000). Both of these rice production systems could save 30–50% of water used. It is very important to improve water resource scarcity in China. However, micronutrient deficiencies, especially Fe, Zn and Mn, in these new rice production systems are serious nutritional problems, which limit the quality and yield increase of rice (Fig. 5.13 in Colour Section).

Fig. 5.13 Comparison of three rice production systems. *Left*: Paddy rice production system, flooding; *Middle*: Ground cover rice production system, no flooding; *Right*: Aerobic rice production system (H.Q. Wang, 2002, personal communication) (*See Colour Plates*)

5.4.1 Aerobic Rice

In aerobic rice cultivation systems, rice is grown as a dry field crop in irrigated, but non-flooded and non-puddled fertile soils (Bouman et al., 2002). Because of water limitations, the area of aerobic rice has increased gradually, accounting for 2% of the total rice area at present. Moreover, newly bred, high-yielding rice varieties are currently being developed by crossing lowland with upland varieties. In China, breeders have produced aerobic rice varieties with an estimated yield potential of 6–7 t ha^{-1} which are now being pioneered by farmers on some 190,000 ha in northern China where water is getting increasingly scarce and where water scarcity makes lowland rice uneconomic (Wang et al., 2002).

The shift from flooded to aerobic conditions sets the problem of micronutrient deficiency in rice in a new perspective. In order to establish this point, we need to understand how factors that influence micronutrient availability change when a crop is moved from a lowland paddy field to an aerobic field. Aerobic soils will differ greatly from lowland soils with respect to factors that control micronutrient availability of plants. Plant acquisition of micronutrients is affected by numerous soil, plant, microbial, and environmental factors.

Many factors that determine micronutrient bioavailability are expected to change after a shift to aerobic cultivation. Bulk soil pH may either increase or decrease depending on the original soil pH (Liu, 1996), and redox potential will increase (Gao et al., 2002). Increased nitrification may cause plants to take up NO_3^- instead of NH_4^+ (Voesenek and Veen, 1994), which causes the rhizosphere pH to decrease. Organic matter will be oxidised. Furthermore, the reduction in the soil water content may restrict micronutrient transport towards the plant root (Yoshida, 1981). The consequences of all these changes for micronutrient bioavailability are hard to predict.

5.4.2 Consequences for Zinc Availability

Zinc deficiency is the most common micronutrient problem in flooded rice, especially on calcareous and/or alkaline soils. These soils under flooded rice cultivation in north China are usually Zn-deficient soils with DTPA-extractable Zn levels of <1.0 mg kg^{-1}. The major causes are high pH, high carbonate content and low redox potential. Iron oxidation by root-released oxygen causes a reduction of the rhizosphere pH and limits release of Zn from highly insoluble fractions (Kirk and Bajita, 1995). Many factors that determine Zn bioavailability are expected to change after a shift to aerobic cultivation, such as soil pH, Eh, precipitation of $Fe(OH)_3$, NO_3^- increase instead of NH_4^+, organic matter, etc. The consequence of all these changes is that it is hard to predict their effects on Zn bioavailability.

Results from a field experiment on a calcareous soil showed that the crop in aerobic fields took up less Zn than in flooded fields. The Zn harvest index significantly decreased in these fields and also large genotypic differences were found (Table 5.5; Gao et al., 2006). These results demonstrate that the introduction of

Table 5.5 Effects of cultivation and genotype on Zn and Fe harvest index (field experiment, Zn (Fe) harvest index = grain Zn (Fe) content/(shoot + grain Zn (Fe) content) (Gao et al., 2006)

	Zn harvest index (%)		Fe harvest index (%)	
Genotypes	Flooded	Aerobic	Flooded	Aerobic
Qiuguang	46.9	24.6	30.0	14.5
K150	48.3	37.2	36.0	18.5
Han72	56.4	23.5	27.5	18.0
89B-271–17(Hun)	53.1	46.1	40.0	10.5
Han277	41.7	34.5	45.5	26.5
Han297	42.6	44.3	27.0	13.0
Average	48.2	35.0	34.3	16.8

aerobic rice systems on calcareous soils may increase Zn deficiency problems. It is not clear whether the decrease in Zn availability and Zn uptake by plants in aerobic fields is caused by changes in chemical properties of the soils or some plant factors, such as root morphology and mycorrhizae.

Substantial variation (50–98%) in Zn efficiency among aerobic rice cultivars was found under both field and pot experimental conditions (Gao et al., 2005). Zinc efficiency correlated significantly ($P< 0.05$) with Zn uptake ($R^2 = 0.34$) and Zn translocation from root to shoot ($R^2 = 0.19$). Furthermore, variation in Zn uptake could be explained only for 32% by root surface area. These results indicate that Zn uptake may be an important determinant of Zn efficiency and that mechanisms other than root surface area are of major importance in determining Zn uptake by rice.

5.4.3 *Consequences for Iron Availability*

Under the anaerobic conditions in flooded soils, Fe occurs in the Fe^{2+} state, which is readily soluble and available to plants. In general, solution Fe^{2+} increases sharply after flooding, especially in acid soils. Consequently, Fe toxicity can be a problem for lowland rice even though it is adapted to anaerobic conditions and has developed mechanisms to tolerate certain Fe levels. However, Fe deficiency under flooded conditions can also be caused by insufficient organic matter and thus delayed reduction of Fe^{3+} to Fe^{2+}. Iron deficiency also occurs if the redox potential remains high (i.e., >200 mV) at pH 7 after flooding, due to the Fe concentration in the soil solution remaining small. In low-Fe soils, the oxidation of Fe^{2+} in the rhizosphere may lead to poor Fe absorption capacity, making rice more susceptible to Fe chlorosis than other crops.

Under aerobic conditions, N is applied mainly in the form of NO_3^-. Increasing NO_3^- uptake may cause an imbalance in the cation/anion ratio, resulting in exudation of OH^- into the rhizosphere with a subsequent reduction in Fe uptake. Moreover, NO_3^- application usually leads to a rhizosphere pH increase compared with NH_4^+ supply. Thus, under aerobic cultivation, Fe bioavailability is decreased because of NO_3^- application.

The results from field experiments showed that Fe application did not improve the Fe nutrition of aerobic rice. It also did not increase the yield harvest index and Fe harvest index. Significant genotypic differences were observed in both Fe harvest index and yield harvest index among tested genotypes. This suggests that development of aerobic rice genotypes with a high Fe harvest index seems a more promising strategy to increase Fe availability to a human consumer than Fe fertilisation of the soil (Table 5.5).

5.5 The Ground Cover Rice Production System

The Ground Cover Rice Production System (GCRPS) was introduced because of water shortage. In the GCRPS, lowland rice is cultivated without a standing water layer during the entire growth period and is irrigated when the soil water tension drops below a certain threshold value (e.g., 80% of the water holding capacity). For reduction of evaporation, the soil surface is covered by covering material, such as plastic film, paper or plant mulch. In recent years, the micronutrient problems caused by the introduction of this system are now being paid more and more attention.

Soil chemical and physical characters, including the available Fe, Mn and Zn content in the soils, change dramatically with the introduction of GCRPS. The Fe and Mn availability in soil is much lower compared with the traditional lowland rice cultivation system (Lv and Zhang, 1997). Compared with conventional flooded soil, non-flooded film mulching slightly decreased the available Fe concentration in the 0–10 cm layers of the soil. The GCRPS increased Zn availability in the 0–15 cm soil layers and reduced Mn availability in the 0–15 cm soil layers (see Fig. 5.14 in Colour Section). Compared with dryland treatment, GCRPS increased Fe and Zn availability in the 0–20 cm layers (Table 5.6; Liu et al., 2004).

5.6 Concluding Remarks

Micronutrient deficiencies are a problem that adversely affects world food production and human health. The latter specifically holds for developing countries, in which diets mainly consist of cereals such as rice. Knowledge of micronutrient acquisition by crop plants is necessary to develop appropriate strategies to prevent its deficiency in crops and the human body.

Micronutrient deficiencies are widespread in China because of soils with generally low levels of available micronutrients and also because of increased nutrient demands from more intensive cropping practices. As fertiliser and other agronomic strategies are shown to be inadequate for overcoming the limitations of micronutrient deficiencies, genotypic differences in micronutrient efficiency offer a sustainable solution to micronutrient malnutrition. By exploiting the genetic potential of

Fig. 5.14 Manganese deficiency of rice plants cultivated with plastic (*left*) and wheat straw mulch (*right*) (S.H. Lv, 2001, personal communication) (*See Colour Plates*)

Table 5.6 Changes of available Fe and Mn content in 0–60 cm soil layers of paddy soils under GCRPS, dryland and conventional flooded condition

	Fe (mg kg^{-1})			Mn (mg kg^{-1})		
Soil layer (cm)	Traditional rice	Dryland rice	Film mulching rice	Traditional rice	Dryland rice	Film mulching rice
0–5	39.7a	38.8a	39.3a	22.7a	21.6a	19.8b
5–10	39.4a	38.3a	39.2a	22.3a	20.1b	19.6b
10–15	35.00b	36.6b	38.8a	21.0a	18.9b	18.9b
15–20	31.5b	32.2b	34.9a	18.6a	18.3a	19.7a
23–30	27.3a	27.8a	28.6a	17.1a	17.9a	16.8a
30–60	23.3a	23.1a	24.1a	17.1a	16.7a	16.0a

Note: The small letter and the capital letter on the same line respectively show the significant difference of P = 0.05 and 0.01(LSD) (From Liu et al., 2004)

plants for enhanced ability to accumulate more micronutrients in their edible parts, micronutrient concentrations in cereal grains can be increased and better human nutrition may be achieved.

At present, rice production in north China is undergoing important changes from traditional high water-consuming lowland (paddy) rice cultivation to a promising new cultivation method of "aerobic rice" because of water constraints. This shift puts the problem of micronutrient deficiency in rice in a new perspective, especially

Fe and Zn deficiency in aerobic cultivation. The effects of the cultivation shift from flooded to aerobic on the availability of micronutrients are determined by many factors, including the soil and plant factors. The consequence for micronutrient availability is the function of the changes of all these factors.

References

Bouman, B.A.M., Hengsdijk, H., Hardy, B., Bindraban, P.S., Tuong, T.P., Ladha, J.K. (2002) Water-wise rice production. Proceedings of the International Workshop on Water-wise Rice Production, 8–11 April, Los Baños, Philippines. International Rice Research Institute, Los Baños, Philippines, p. 356.

Dong, Y., Sun, Y. (1990) Research on Mo nutrition of soybean, Soybean Science, 9 (4): 332–336 (in Chinese).

Fang, Z., Lv, S., Zhang, F. (1998) Study on tolerance of different wheat cultivars or lines to manganese deficiency, Plant Nutrition and Fertilizer Science, 4 (3): 277–283 (in Chinese).

Gao, S., Tanji, K.K., Scardaci, S.C., Chow, A.T. (2002) Comparison of redox indicators in a paddy soil during rice-growing season, Soil Science Society of America Journal, 66: 805–817.

Gao, X.P., Zou, C.Q., Fan, X.Y., Zhang, F.S., Hoffland, E. (2006) From flooded to aerobic conditions in rice cultivation: Consequences for zinc uptake, Plant Soil, 280: 41–47.

Gao, X.P., Zou, C.Q., Zhang, F.S., van der Zee, S.E.A.T.M., Hoffland, E. (2005) Tolerance to zinc deficiency in rice correlates with zinc uptake and translocation, Plant Soil, 278: 253–261.

Gao, Z., Lin, B., Zhou, W. (2001) Effect of zinc nutrition on endogenous hormone and free radicals in spring maize, Plant Nutrition and Fertilizer Science, 7 (4): 424–428 (in Chinese).

Han, C., Li, C., Li, R., Wang, Q., Wu, M., Xiao, C. (1994) Technical approach to application of iron fertilization on soybean, Oil Crop of China 16 (2): 43–45 (in Chinese).

Han, L., Ju, H., Zou, Y., Fu, H., Yan, F., Liu, J. (1998) The study on trace nutrient in soybean continuous cropping: The effect of continuous cropping on Mo nutrition, Soybean Science, 17 (2): 135–141 (in Chinese).

He, X. (2002) Ferrous fertilizer and its requirements of use, Phosphorus Fertilizer and Complex Fertilizer, 17 (4): 69–71 (in Chinese).

Hu, S. (1981) Study of Mn fertilizer application on wheat in Wenjiang, Chinese Journal of Soil Science, 3: 100–103 (in Chinese).

Kirk, G.J.D., Bajita, J.B. (1995) Root-induced iron oxidation, pH changes and zinc solubilization in the rhizosphere of lowland rice, New Phytologist,131: 129–137.

Li, W., Wang, Z., Mi, G., Han, X., Zhang, F. (2001) Molybdenum deficiency in winter wheat seedlings as enhanced by freezing temperature, Journal of Plant Nutrition, 24 (8): 1195–1203.

Li, X., Lv, S., Chen, H., Gu, M., (1995) Iron-Manganese nutrient balance and rice development, Journal of Guangxi Agricultural University, 14 (3): 217–222 (in Chinese).

Liang, Y., Hu, F., Shen, Q., Lu, S., Wu, L., Zhang, F. (2000) An overview of rice cultivation on plastic film mulched dryland. In: An Overview and Outlook of Plant Nutrition. Feng, F., Zhang, F., Yang, X. (Eds.), China Agricultural University Publisher, Beijing, China, pp. 114–127 (in Chinese).

Lin, B., Li, J.K. (1997) Problems and strategies of fertilizer in China, Phosphorus Fertilizer and Complex Fertilizer, 23 (2): 1–5 (in Chinese).

Liu, M., Wu, L., Lu, X., Zhang, F. (2004) Changes of available Fe, Mn, Zn and Cu contents among different layers of non-flooded paddy soils with ground mulching, Journal of Zhejiang University, 30 (6): 646–649 (in Chinese).

Liu, P., Yang, Y., Zhao, Y. (2001) The screening of soybean genotypes with tolerance to Mo and B deficiency, Chinese Journal of Oil Crop Sciences, 23 (4): 65–71 (in Chinese).

Liu, W., Du, L., Liu, D. (2002) A review on diagnosis for iron nutrition status of fruit trees and fertilization correction for Fe-deficiency chlorosis, Chinese Agricultural Science Bulletin 18 (4): 67–69 (in Chinese).

Liu, X.H., Sun, C.Q., Wang, X.H. (1995) Studies on the content of four elements Fe, Zn, Ca, and Se in rice varieties from various area of China, Acta Agriculture Universitatis Pekinensis, 21 (2): 139–142 (in Chinese).

Liu, X., Zeng, X., Lv, S., Zhang, F. (2001) Effects of Mn application periods on Mn nutrition and grain yield of different wheat genotypes, Southwest China Journal of Agricultural Sciences, 14 (4): 39–43 (in Chinese).

Liu, Z. (1996) Microelements in Soils of China, Jiangsu Science and Technology Publishing House, Nanjing, China.

Lv, S., Zhang, F. (1997) Study on iron and manganese deficiency in dry nursery cultured rice plants and measures of prevention, Journal of China Agricultural University, 2 (3): 90–100 (in Chinese).

Lv, S., Zeng, X., Liu, X., Zhang, F. (2002) Effects of root penetration restriction on growth and Mn nutrition of winter wheat with different genotypes in paddy soils, Scientia Agricultural Sinica, 35 (7): 809–814 (in Chinese).

Ren, Q. (2004) Effects on crop yields and quality of B, Zn and Mn, Journal of Jining Teachers College, 26 (3): 44–46 (in Chinese).

Tai, C., Wu, T., Wang, J. (2004) Studies of Mn deficiency on some physiological effect of wheat genotypes, North-China Agronomic Bulletin, 19 (2): 53–56.

Voesenek, L.A.C.J., van der Veen, R. (1994) The role of phytohormones in plant stress: Too much or too little water, Acta Botanica Neerlandica, 43: 91–127.

Wang, H., Bouman, B.A.M., Zhao, D., Wang, C., Moya, F. (2002) Aerobic rice in northern China-opportunities and challenges. In: Water-wise Rice Production. Proceedings of the International Workshop on Water-wise Rice Production. Bouman, B.A.M., Hengsdijk, H., Hardy, B., Bindraban, P.S., Tuong, T.P., Ladha, J.K. (Eds.), International Rice Research Institute, Los Baños, Philippines, pp. 143–154.

Wang, Y., Wei, W., Tan, Q. (1995) A study on molybdenum deficiency and molybdenum application of winter wheat in yellow-brown soil of Hubei province, Soil Fertilization, 3: 24–28 (in Chinese).

Welch, R.M. (1993) Zinc concentration and forms in plants for humans and animals. In: Zinc in Soils and Plants. Robson A.D. (Ed.), Kluwer, Dordrecht, The Netherlands, pp. 183–195.

Wu, M., Xiao, C. (1994) Study on molybdenum in soybean, Soybean Science, 13 (3): 245–251 (in Chinese).

Yang, W., Xu, Z., Qian, B., Liu, P. (2004) The effect of Fe^{3+} on seed germination of soybean, Seed, 23 (4): 32–34 (in Chinese).

Yoshida, S. (1981) Fundamentals of Rice Crop Science, IRRI, Los Banos, Philippines, p. 269.

Yu, M., Hu, C., Wang, Y. (2003) Response of different winter wheat cultivars to Mo deficiency, Journal of Huazhong Agricultural University, 22 (4): 360–364 (in Chinese).

Zhou, W. (1995) Impacts of applying zinc with different methods on zinc content and yield of wheat, Chinese Journal of Eco-Agriculture, 3 (1): 34–38 (in Chinese).

Zou, B., He, X. (1985) Nutrients in Plants, Agriculture Press, Beijing, China.

Zou, C., Shen, J., Zhang, F. (2001) Impact of nitrogen form on iron uptake and distribution in maize seedlings in solution culture. Plant Soil, 235 (2): 143–149.

Zuo, Y., Zhang, F., Li, X., Cao, Y. (2000) Studies on the improvement in iron nutrition of peanut by intercropping with maize on a calcareous soil, Plant Soil, 220: 13–25.

Zuo, Y., Zhang, F. (2003) The effects of peanut intercropping with different gramineous species and their intercropping model on iron nutrition of peanut, Agricultural Sciences in China, 2 (3): 289–296 (in Chinese).

Chapter 6
Micronutrient Constraints to Crop Production in the Near East

Potential Significance and Management Strategies

Abdul Rashid and John Ryan

Abstract This review focuses on the Near East, a region of the world where there is a dearth of information on micronutrients in soils and plants. The long history of cultivated agriculture in the region and the peculiar characteristics of the soils and the climate predispose towards problems with micronutrient deficiencies. Over 2 decades ago, a global study on micronutrients indicated deficiencies of iron (Fe) and zinc (Zn), and the likelihood of excess levels of boron (B), in some countries of the Near East. This review primarily addresses two focal points in the region, Pakistan on one side and the Syria/Lebanon/Turkey region on the other, reflecting the zones of activity of the two authors. It brings together and summarises published work in the areas of crop and soil micronutrient availability, their behaviour in soils in relation to crop growth, and strategies to deal with either deficiency or toxicity, including crop selection for tolerance and subsequent genetic manipulation. The review highlights the implications of micronutrient constraints in the soil–plant–human–animal continuum. While intensification of agricultural production is likely to accentuate micronutrient deficiencies in the region, other developments may counteract this trend. Nevertheless, as the trend for land use intensification increases, and as other nutrient constraints are eliminated, micronutrients will inevitably assume greater significance in future agriculture of the Near East.

6.1 Introduction

The urgency of addressing the issue of an expanding world population, in particular by eliminating hunger and malnutrition in less-developed countries, underlines the need for policies that ensure sustainable agricultural productivity while preserving the environment and natural resource base. In a compelling lecture, Nobel Laureate, Norman Borlaug, highlighted the central role of soil fertility and mineral nutrition, along with improved crop varieties and water availability, in ensuring the nutritional welfare of mankind (Borlaug, 2003). Much of the world's food supply today is attributed to the use of chemical fertilisers (Stewart et al., 2005); future increases will even be more dependent on fertiliser inputs, particularly nitrogen

B.J. Alloway (ed.), *Micronutrient Deficiencies in Global Crop Production.*

(N), phosphorus (P) and, to a lesser extent, potassium (K). Soil deficiencies of these major nutrients are well understood and largely eliminated in modern commercial agriculture through the routine use of fertilisers. However, in many developing countries, chemical nutrient infertility still poses a major limit on crop productivity (Loneragan, 1997).

Diagnosis of deficiency and toxicity of nutrients, especially micronutrients, can contribute to better nutrition of crops and greater productivity. While the state of knowledge on micronutrients was brought together by Mortvedt et al. (1972) in the first-ever monograph *Micronutrients in Agriculture* published by the Soil Science Society of America and later updated (Mortvedt et al., 1991) to include new developments in micronutrient research, most of the contributions came from the developed world. Extensive reviews are available on the geographic distribution of micronutrient problems and their correction in many parts of the world, e.g., USA (Mortvedt et al., 1991, Fageria et al., 2003), Australia (Robson, 1993), and tropical countries (Vlek, 1985; Katyal and Vlek, 1985); however, the state of knowledge of micronutrients in less-developed areas of the world such as the Near East, in contrast, is poorly described, and sketchy at best.

Despite being the centre of origin of settled agriculture and western civilization, as well as the location where many of the world's major crops, especially cereals, pulses, and nuts, evolved (Damania et al., 1998), the Near East is still largely a food-deficit region. The vast swathe of the globe, from Morocco to Pakistan, is characterized by a Mediterranean-type climate merging into a continental one (Kassam, 1981). As the dominant climatic feature is a cool, relatively moist season followed by a long dry one, and as annual rainfall is low, 200–600 mm, the dominant system of rainfed cropping, involving cereals and legumes and associated livestock production, is invariably restricted by drought (Cooper et al., 1987). Due to increased land use pressure, in the past few decades, there has been increasing emphasis on irrigation where water is available, either from rivers or groundwater, mechanization, and the use of chemical fertilisers, mainly N and P (Ryan, 2002). In essence, the developments that have occurred earlier in the West are now emerging in the Near East.

However, the state of awareness on micronutrients in agriculture of the Near East region has lagged behind that of the major nutrients. For example, in two workshops of the soil test-calibration program involving soil fertility and crop scientists from the West Asia–North Africa region (Ryan and Matar, 1990, 1992), no mention was made of micronutrients as the general perception was that the only nutrients of importance were N and P. In a subsequent, more comprehensive international workshop on soil fertility, there were only a few reports on micronutrients, i.e., Turkey, Iraq and Syria (Ryan, 1997). Similarly, a recent major review of dryland agriculture (Rao and Ryan, 2004), including several contributions from the Near East, contained only one paper that referred to micronutrients. However, some earlier internal publications at the American University of Beirut (Ryan et al., 1981a) and Pakistan (Anonymous, 1998) did bring together various in-country respective publications dealing with micronutrients, which were not typical of the Near East as a whole.

The first publication ever that indicated any potential problems with micronutrients, especially iron (Fe), zinc (Zn) and boron (B), emanated from a Food and Agriculture Organization (FAO)-sponsored study on micronutrient status in selected countries around the world, led by Sillanpää (1982), which formed the basis of subsequent reports (Sillanpää, 1990; Katyal and Vlek, 1985). Subsequently, a recent review by Rashid and Ryan (2004) was the first attempt to develop a coherent picture of micronutrient research in the soils and crops of the Mediterranean climatic region.

This chapter elaborates on micronutrient research, highlighting recent developments, particularly in relation to Zn and B, and stressing the role of plant adaptation to micronutrient deficiencies and toxicities, as well as the implications of these problems for crop productivity and human health. As a background to the review, some pertinent generalisations are presented for the Near East region as well as broad aspects of micronutrients that set the scene for subsequent studies on micronutrients in the region.

6.2 The Near East: Background Considerations

The area of the world focussed on in this review, the Near East, is in some ways synonymous with terms such as the "Middle East" and more recently, "West Asia–North Africa". To be specific, according to the FAO of the United Nations, the Near East region comprises of 31 countries, i.e., Afghanistan, Algeria, Azerbaijan, Bahrain, Cyprus, Djibouti, Egypt, Iran, Iraq, Israel, Jordan, Kuwait, Kyrgyz Republic, Lebanon, Libya, Malta, Mauritania, Morocco, Oman, Pakistan, Qatar, Saudi Arabia, Somalia, Sudan, Syria, Tajikistan, Tunisia, Turkey, Turkmenistan, United Arab Emirates, and Yemen (Fig. 6.1). As in other areas of the world, the soils of the Near East are varied, being the outcome of the factors of soil formation, with the arid to semi-arid climate having a disproportionate influence on soil properties (Yaalon, 1997). As the limited rainfall allows for only minimal leaching of cations, most soils in the Near East are calcareous (Calcisols) with a high pH. While the environment promotes limited biomass accumulation, and at the same time favours mineralisation, low levels of soil organic matter ($<1\%$) are common features in the Near East (Matar et al., 1992).

The soil and environmental conditions of the region dictate that N and P are deficient in unfertilised soils and are required to be added for cropping, but K is generally adequate due to the weak weathering environment and K-rich parent materials from which the soils were derived. However, given the fundamental relationship between soil properties, especially the free calcium carbonate ($CaCO_3$), high pH, and low reserves of organic matter, and micronutrient chemistry, one could anticipate deficiencies of some micronutrients, especially Zn and Fe, as the global study of Sillanpää (1982) had shown.

The observed micronutrient deficiencies in countries of the Near East are listed in Table 6.1. In contrast to major nutrients, however, micronutrient problems are more

Fig. 6.1 Political map of the Near East (labelled as Asia – Middle East)

location-specific (Rashid et al., 2002d; Welch et al., 1991), as well as genotype-specific (Rashid and Din, 1992; Rashid et al., 2000b, 2002a, c; Cakmak, 1998). Also, some micronutrient deficiencies, such as those of Fe and Zn, not only reduce crop productivity, but also low micronutrient concentrations in plant parts (and hence in plant-derived feeds and foods) can adversely affect human health and well-being (Graham and Welch, 2000). Given these considerations, it is germane to this review to give a brief outline of the approaches used in the Near East for the diagnosis of micronutrient disorders.

6.3 Diagnosis of Micronutrient Disorders

Diagnosis refers to identification of a disorder at the time of sampling by determining the micronutrient status in the plant, or in specific plant part(s), whereas prognosis is the prediction of the possibility of a micronutrient malnutrition impairing

Table 6.1 The observed micronutrient deficiencies, based on crop responses, in countries of the Near East (Cakmak, 1998, 2005; Anonymous, 1998; Cakmak et al., 1995; Saxena et al., 1990; Erskine et al., 1993; Rafique et al., 2006; Rashid, 2006a; Rashid and Ryan, 2004; Rashid et al., 1981a, 2002d, 2006b; Ryan et al., 1981a; Hamze et al., 1986, 1987; Shorrocks, 1997; Sillanpää, 1982; Zahid et al., 2000)

	Countries of the Near East[a]		
Crop species	Zinc	Iron	Boron
Barley (*Hordeum vulgare*)	Tk		
Chickpea (*Cicer arietinum*)	Pk	Ln, Pk, Sy	Pk
Citrus (*Citrus spp.*)	Ln, Pk, Tk	Ln, Pk	Pk
Cotton (*Gossypium hirsutum*)	Pk	Pk	Pk
Deciduous fruits (*apple, apricot, peach, plum*)	Pk	Pk	Eg
Forage legumes (*Trifolium spp.*)	Sy		
Lentil (*Lens culinaris*)		Sy	
Maize (*Zea mays*)	Lb, Pk		Pk
Mustard (*Brassica juncea*)	Pk		Pk
Medic (*Medicago spp.*)	Sy		
Millet (*Panicum miliaceum*)	Lb		
Peanut (*Arachis hypogaea*)		Pk	Pk
Potato (*Solanum tuberosum*)	Pk		Eg, Pk
Rapeseed (*Brassica napus*)	Pk		Pk
Rice (*Oryza sativa*)	Pk	Pk	Pk
Sorghum (*Sorghum bicolor*)	Lb, Pk	Lb	Pk
Sugarbeet (*Beta vulgaris*)	Pk	In	Eg, Ir, Mo, Pk
Wheat (*Triticum spp.*)	Ir, Lb, Pk, Tk	Lb, Pk, Sy	Pk

[a] Countries: Eg, Egypt; Ir, Iran; Iq, Iraq; Lb, Libya; Ln, Lebanon; Mo, Morocco; Pk, Pakistan; Sy, Syria; Tk, Turkey

growth at some later crop growth stage, after the sample is taken. Macronutrient deficiency prognosis can be made from either soil or plant analysis. By contrast, usually only plant analysis can reliably diagnose micronutrient deficiencies. The approaches employed for ascertaining micronutrient deficiencies in the Near East soils and crops are: (1) consideration of crop sensitivity to a micronutrient deficiency; (2) deficiency symptoms in plants; (3) soil testing and plant analysis; and (4) crop responses to micronutrient application.

6.3.1 Crop Sensitivity to Micronutrient Deficiencies

Crop genotypes vary in their micronutrient requirement and/or ability to utilise soil micronutrient supplies; various species have been observed to differ in their susceptibility to micronutrient deficiencies. For example, dicotyledons, such as cotton (*Gossypium hirsutum*), and legumes have higher B requirements than monocotyledons, e.g., cereals (Shorrocks, 1997); hence, they are more susceptible to B deficiency. Despite the small amount of B uptake by cereals, however, these crops have a relatively high B requirement for flower fertilization and seed-set (Shorrocks, 1997) and, thus, suffer yield losses in low B soils.

On the other hand, because of their low B requirements or more efficient soil B utilization, some crop species, like maize (*Zea mays*), pea (*Pisum sativum*), and rye (*Secale cereale*), are considered to be tolerant to B deficiency. Similarly, maize and onion (*Allium cepa*) are highly susceptible to Zn deficiency, whereas sugarcane (*Saccarum officinarum*) and wheat (*Triticum aestivum*) are quite tolerant (Rashid and Fox, 1992). Marked variations in susceptibility to micronutrient deficiencies have been observed within varieties of the same species (Table 6.2).

For example, Chaudhry et al. (1977) in Pakistan observed that the cultivar (cv.) IR-6 of rice (*Oryza sativa*) was highly susceptible to the incidence of Zn deficiency and cv. Basmati-370 was quite tolerant. Thus, it is well recognised that micronutrient fertiliser needs depend on crop genotypes (Rashid and Ryan, 2004) rather than soil micronutrient status per se. The predominant field crops, vegetables and fruits grown in the region and considered susceptible to micronutrient deficiencies are listed in Table 6.3.

6.3.2 *Deficiency Symptoms in Plants*

As micronutrient deficiency symptoms in plants can prove useful in some situations, this approach is invariably relied upon in situations such as Fe chlorosis in most fruit trees as well as in some field crops, such as peanut (*Arachis hypogaea*) and chickpea (*Cicer arietinum*), and ornamentals. However, it is also well realized that micronutrient

Table 6.2 Differential susceptibility of crop genotypes to micronutrient deficiency: Some examples from the Near East

	Genotype/Cultivar		
Species	Efficient	Inefficient	Reference
Zinc			
Wheat (*Triticum spp.*)	AK-702, Kirac-66, Sertak-52	BDME-10, Kunduru-1149, Kiziltan-91, Cakmak-79	Cakmak (1998)
Rice (*Oryza sativa*)	Basmati-370	IR-6	Chaudhry et al. (1977)
Iron			
Chickpea (*Cicer arietinum*)	CM-72	C-44, CM-88	Rashid and Din (1992)
	ILC-263	ICCL-81131, ICCL-81192	Saxena et al. (1990)
Lentil (*Lens culinaris*)			Erskine et al. (1993)
Boron			
Rapeseed (*Brassica napus*)	Sheerali, Westar	CON-II, CON-III	Rashid et al. (2002c)
Lentil (*Lens culinaris*)	Simal, Simrik	ILL-1744, ILL-6459, ILL-6465	Srivastava et al. (2000)
Wheat (*Triticum aestivum*)	Sindh-81, Rahtas-90	Inqlab-91, Bakhtawar	Rashid et al. (2002a)

Table 6.3 Crop species observed sensitive to micronutrient deficiencies in the Near East countries[a] (Rashid and Ryan, 2004; Rashid, 1996, 2006a; Shorrocks, 1997)

ZINC	
High	Rice (*Oryza sativa*)
Apple (*Malus domestica*)	Soybean (*Gycine max*)
Beans, field (*Vicia faba*)	**Medium**
Citrus (*Citrus spp.*)	Alfalfa (*Medicago sativa*)
Cowpea (*Vigna unguiculata*)	Barley (*Hordeum vulgare*)
Grape (*Vitis vinifera*)	Clover (*Trifolium spp.*)
Maize (*Zea mays*)	Cotton (*Gossypium hirsutum*)
Millet (*Panicum miliaceum*)	Potato (*Solanum tuberosum*)
Onion (*Allium cepa*)	Sorghum (*Sorghum bicolor*)
Peach (*Prunus persica*)	Sugarcane (*Saccharum officinarum*)
Pear (*Pyrus communis*)	Sugar beet (*Beta vulgaris*)
Pines (*Pinus spp.*)	Sunflower (*Helianthus annuus*)
Plum (*Prunus domestica*)	Tomato (*Lycopersicon esculentum*)
BORON	
High	
Alfalfa (*Medicago sativa*)	Cherry (*Prunus spp.*)
Apple (*Malus domestica*)	Grape (*Vitis vinifera*)
Brassica (*Brassica spp.*)	Lettuce (*Lactuca sativa*)
Cauliflower (*Brassica oleracea*)	Mustard (*Brassica juncea*)
Clovers (*Trifolium spp.*)	Olive (*Olea europaea*)
Cotton (*Gossypium hirsutum*)	Peach (*Prunus persica*)
Eucalypt (*Eucalypus spp.*)	Peanut (*Arachis hypogaea*)
Rapeseed (*Brassica napus*)	Pear (*Pyrus communis*)
Sugar beet (*Beta vulgaris*)	Radish (*Raphanus sativus*)
Sunflower (*Helianthus annuus*)	Rice (*Oryza sativa*)
Turnip (*Brassica rapa*)	Spinach (*Spinacia oleracea*)
Medium	Sweet Potato (*Ipomoea batatas*)
Cabbage (*Brassica oleracea var. capitata*)	Tobacco (*Nicotiana tabaccum*)
Carrot (*Daucus carota*)	Tomato (*Lycopersicon esculentum*)
IRON	
High	
Bean, field (*Vicia faba*)	Barley (*Hordeum vulgare*)
Berries (*Morus spp.*)	Bean, field (*Vicia faba*)
Chickpea (*Cicer arietinum*)	Cotton (*Gossypium hirsutum*)
Citrus (Citrus spp.)	*Maize (Zea mays)*
Grape (*Vitis vinifera*)	Oats (*Avena sativa*)
Peanut (*Arachis hypogaea*)	Pea (*Pisum sativum*)
Soybean (*Glycine max*)	Rice (*Oryza sativa*)
Walnut (*Juglans spp.*)	Sorghum (*Sorghum bicolor*)
Medium	Soybean (*Glycine max*)
Alfalfa (*Medicago sativa*)	Wheat (*Triticum spp.*)

[a] Species listed under both categories exhibit differential response of their varieties'

deficiency diagnosis based on crop symptoms has certain serious disadvantages, e.g., apparent visual symptoms may be caused by many factors other than a specific micronutrient stress, or a peculiar visual symptom may be caused by more than one nutrient or even by more than one factor. Also, deficiency symptoms are difficult to be distinguished in the field, and very frequently deficiency symptoms appear too late during the growth period for taking an effective remedial measure. Moreover, all crops do not exhibit deficiency symptoms and may suffer yield losses without showing any symptoms (Rashid et al., 2004). Thus, in countries of the Near East region, micronutrient deficiency symptoms are used only as a supplement to other diagnostic techniques such as soil testing and plant analysis.

6.3.3 Soil Testing

Soil testing, being practically more feasible, is the most widely used technique for the diagnosis or prognosis of micronutrient disorders. The conventional soil tests are the DTPA (diethyelene triamine pentaacetic acid) test of Lindsay and Norvell (1978) for Cu, Fe, Zn, and Mn, and hot-water extraction (HWE) for B (Berger and Truog, 1944). Realizing that increased reliability can be achieved when soil tests are calibrated for local soil types and crop genotypes, soil test calibration has been carried out in some countries of the region (Ryan et al., 2001; Rashid and Ryan, 2004).

In Pakistan, for example, it has been established that the multi-element soil test for alkaline soils, ammonium bicarbonate-DTPA (AB-DTPA) (Soltanpour and Workman, 1979), is comparable in effectiveness to the conventional soil test for micronutrients (i.e., Cu, Fe, Zn and Mn), the DTPA test (Rashid et al., 1997d). Because the macronutrient as well as micronutrient fertility status of soils can be determined in an economical and more efficient manner by using the "universal" soil test for alkaline soils, AB-DTPA, this test is also being used in some countries in the region, like Pakistan (Anonymous, 1998). It has also been established that a simpler and less prone-to-error test for B, i.e., dilute HCI method, is comparable in effectiveness to the HWE method. However, dilute HCI extracts slightly less soil B, and the relationship is: HCI B = 0.030 + 0.841 (HWE-B) (Rashid et al., 1994b; Rafique et al., 2002). Thus, the latter test, being less tedious and less prone to error, is also being used in some countries.

Generalised interpretation criteria for micronutrient soil tests and plant analysis, applicable for the Near East countries, are available in an ICARDA-published *Soil and Plant Analysis Laboratory Manual* in English, Russian and Arabic languages (Ryan et al., 1996a, 2001). Generalized guidelines for interpreting soil test results are presented in Table 6.4.

6.3.4 Plant Analysis

Like in many other parts of the world, much less plant analysis information is available for countries of the Near East compared with soil test data in the developed world. However, systematic nutrient indexing and field experimentation have

Table 6.4 Generalized criteria for interpreting micronutrient soil test data in countries of the Near East (mg kg^{-1}) (Ryan et al., 2001; Rafique et al., 2002)

Micronutrient	Soil test	Low	Marginal	Adequate
B	Hot Water	<0.5	–	>0.5
	HCl	<0.45	–	>0.45
Zn	[a]DTPA	<0.5	0.5–1.0	>1.0
	[b]AB-DTPA	<1.0	1.0–1.5	>1.5
Cu	DTPA	<0.2	0.2–0.5	>0.5
	AB-DTPA	<0.2	–	> 0.5
Fe	DTPA	<4.5	–	>4.5
	AB-DTPA	<2.0	2.1–4.0	>4.0
Mn	DTPA	<1.0	1.0–2.0	>2.0
	AB-DTPA	<1.8	–	>1.8

[a] DTPA = Diethylene triamine pentaacetic acetic acid
[b] AB-DTPA = Ammonium bicarbonate-diethylene triamine pentaacetic acetic acid

Table 6.5 Generalized interpretation of plant analysis for micronutrients (Mortvedt et al., 1991)

Micronutrient	Deficient	Sufficient or normal	Excessive or toxic
		(mg kg^{-1})	
B	5–30	10–200	50–200
Cl	<100	100–500	500–1000
Cu	2–5	5–30	20–100
Fe	<50	100–500	>500
Mn	15–25	20–300	300–500
Mo	0.03–0.15	0.1–2.0	>100
Zn	10–20	27–150	100–400

provided comprehensive information about the micronutrient status of some farmer-grown crops and fruits in many countries of the region. For example, in Pakistan nutrient indexing information is available for cotton, wheat, sorghum, rapeseed-mustard (*Brassica napus, B juncea*), peanut (Anonymous, 1998; Rashid and Rafique, 2002; Rashid et al., 1997a, b, c, d), citrus and apple (*Malus domestica*) (Rashid, 2006a). Generalized guidelines for interpreting plant analysis laboratory data are given in Table 6.5.

Though the use and utility of plant analysis for micronutrients is much less than for soil testing, plant analysis is used routinely in research (e.g., diagnosing micronutrient problems by nutrient indexing of farmer-grown crops, and determining the micronutrient status of crops in management studies). In fact, in some countries of the region, internal micronutrient (especially B, Zn and Fe^{2+}) requirements of locally grown crop genotypes have been determined (e.g., Table 6.6) to arrive at more effective diagnostic criteria.

Although most of the locally developed criteria are not very far from values in the literature, some of the determined critical levels are considerably different. For example, the critical level of B for cotton leaves in Pakistan, 53 mg B kg^{-1} (Rashid

Table 6.6 Locally developed micronutrient plant analysis diagnostic criteria for selected crops (Rashid, 2006a; Rashid and Fox, 1992; Rashid et al., 1994a, b 1997a, b, c, d, 2002a, b, c, d; Rafique et al., 2005; Rashid and Rafique, 2005)

	Critical concentration/range (mg kg^{-1})				
	Zn			B	
Crop species	[a]WS	[b]Leaves	Seeds	WS	Leaves
Wheat	16–20	12–16	20–24	4–6	5–7
Rice	20	19	15		6
Cotton					53
Maize	18	24	18		
Sorghum	27–33	20–22	10–14	17–18	25–31
Chickpea					49
Soybean		22	43		
Cowpea		21	36		
Rapeseed	29	33	29	32	38
Mustard	35	41	33	41	49
Peanut			[c]29		
Sunflower					
4-week				46–63	
8-week				36	

[a]Whole shoots = Young whole shoots (≤30 cm tall)
[b]Leaves = Youngest fully expanded leaf blades, at flowering/heading (unless stated otherwise)
[c]~4 cm shoot terminals

and Rafique, 2002), is about three times greater than the listed valued in the literature, i.e., 15–20 mg B kg^{-1} (Shorrocks, 1992; Reuter and Robinson, 1997). Critical levels of Fe^{2+} have also been developed for a few legume species, like chickpea and peanut (Rashid and Din, 1992; Rashid et al., 1997a). However, in most instances, plant tissue sampling guidelines and procedures and data interpretation largely rely upon international literature (e.g., Jones, 1991; Reuter and Robinson, 1986, 1997; Tandon, 1995). Frequently, the task of data interpretation remains complex; thus, more elaborate locally relevant interpretation criteria are warranted.

Critical levels of some micronutrients (especially Zn and B) have also been developed for the seeds of some grain crops (Rashid and Fox, 1992; Rashid et al., 1994a, 1997d; Rafique et al., 2006). The use of seeds as diagnostic tissue has advantages in terms of ease of sampling, processing, and/or analyzing. Though seed analysis diagnosis poses problems, it can help identify field sites where future crops are likely to respond to micronutrient applications and also regional trends in micronutrient status of crops.

6.3.5 *Crop Responses to Micronutrients*

The most convincing diagnosis of a deficiency is attained by alleviation of the constraint as a result of micronutrient application in greenhouse or field situations. In fact, this is an essential requirement for convincing the stakeholders (i.e., agricultural extension staff and growers) to adopt micronutrient fertiliser use in a crop. The approach is rather costly and time-consuming, and, in fact, impractical to perform at each and every field site. Following identification of a micronutrient deficiency, on the basis of a soil test and/or plant analysis, the deficiency is initially verified in greenhouse situations and in the major crop-growing regions.

Only upon obtaining consistent positive crop responses to the micronutrient's use in farmers' fields, are the agricultural extension service and the fertiliser industry involved in field demonstrations of the technology. Two salient examples, both from Pakistan, are the establishment of B and Zn deficiency in cotton and B deficiency in rice. Where crops are deficient in a micronutrient, and the deficiency is severe and growth-limiting, the use of that micronutrient is almost always highly cost-effective, especially as foliar feeding of high-value crops. Common micronutrient fertilisers, and general application rates and methods for micronutrients are presented in Table 6.7.

Table 6.7 Common micronutrient fertilisers, general application rates and methods (Rashid, 1996, 2006a, b; Rashid and Ryan, 2004)

Micronutrient source	Formula	Concentration
Zinc sulphate	$ZnSO_4\ H_2O$	35% Zn
	$ZnSO_4\ 7H_2O$	22% Zn
Zinc oxide	ZnO	78% Zn
Ferrous sulphate	$FeSO_4\ 7H_2O$	20% Fe
Sequestrene	NaFe-EDDHA	6% Fe
Borax (*Fertiliser borate*)	$Na_2B_4\ O_7\ H_2O$	11% B
Boric acid	H_3BO_3	17% B
Solubor	$Na_2B_8O_{13}\ 4H_2O$	20% B
Copper sulphate	$CuSO_4\ 5H_2O$	25% Cu
Copper chelates	Na_2Cu EDTA	13% Cu
	NaCu EDTA	9% Cu
Manganese sulphate	$MnSO_4 \times H_2O$	24–30% Mn
	General application rate and method	
Micronutrient	Soil application	Foliar feeding[a]
Zinc	2–5 kg Zn ha^{-1}, as zinc sulphate; one application lasts 3–4 crop seasons	0.1% Zn solution; 2–3 sprays
Iron	–	0.5% ferrous sulphate solution or 1% *Sequestrene* solution; 3–4 sprays
Boron	0.75–1.0 kg B ha^{-1}, as borax	0.05–1.0% B solution, as boric acid or *Solubor*; 3–4 sprays
Copper	5 kg Cu ha^{-1}, as copper sulphate	0.5% copper sulphate solution; 2–3 sprays
Manganese	5–10 kg Mn ha^{-1}, as manganese sulphate	0.2–0.5% manganese sulphate; 2–3 sprays

[a] Add detergent powder (0.05%) to spray solutions as surfactant; where necessary, neutralize spray solution with lime or urea

6.4 Micronutrients in the Near East: Significance, Soil Behaviour, Responses

In this section, we focus on studies of individual micronutrients in terms of what has been published regarding the incidence of deficiency, chemical behaviour that impinges on their effectiveness as nutrients, and the implications for crop growth and quality with inferences to the end-user, the human being. The review is not exhaustive, nor intended to cover all studies on micronutrients in the Near East region, but rather broadly reflects the evolution of micronutrient research, from problem assessment and identification to approaches to solution of the problem, with particular emphasis on current strategies using plant adaptation as an alternative to soil or foliar fertilisation.

6.4.1 Iron

Of the soil micronutrients, Zn and B are the most important in terms of the areal extent of deficiency and are serious constraints to crop yields at the global level. However, because of the negative influence of $CaCO_3$, and consequently pH, on Fe solubility, deficiency of Fe is ubiquitous in areas of the world that are dominated by calcareous soils. Many other factors influence Fe availability, including bicarbonate (HCO_3), high soil moisture and cold temperatures. In addition, there is a high degree of plant specificity to Fe deficiency in some crops, in particular, citrus, deciduous fruits, legumes and some acid-loving (calcifuge) ornamental plants. Soil fertilisation with soluble Fe sources is problematic and rarely effective because Fe is readily precipitated as insoluble Fe compounds in a high pH/$CaCO_3$-dominated soil environment.

An alternative approach is to use chelated forms of Fe (EDDHA, EDTA, etc.), but these are costly and only justified with high-value cash crops. In practice, the only alternative is foliar feeding or the longer-term strategy of breeding for tolerance to Fe deficiency. Symptoms of Fe deficiency are expressed in typical interveinal chlorosis. While these symptoms may disappear with time, or with increasing temperature; severe deficiency generally reduces the ultimate crop yield. Although the test for available Fe in soil is reliable, i.e., using DTPA (Lindsay and Norvell, 1978), the concentration of Fe in the plant issue bears little or no relationship with the extent of Fe deficiency or sufficiency. Given the ubiquitous nature of Fe deficiency, and the many factors impinging upon it, it was hardly surprising that many of these aspects related to Fe were addressed in the Near East, where all soil and environmental conditions conducive to Fe deficiency occur.

Despite the generalizations about Fe deficiency at the eco-regional level, the extent of the problem varies from one country to another. Clearly, despite the low levels of soluble soil Fe, many indigenous crops that evolved in the region are adapted to such conditions. For instance, in a review of various field and greenhouse studies and observations, Ryan et al. (1981a) indicated that there was little evidence of any widespread problem with Fe nutrition in Lebanon, especially in the country's typical "terra rosa" (Calcic Luvisol) soils which are high in total and citrate-dithionate-extractable Fe (Arshad et al., 1980). Lebanon has a relatively high rainfall and cold winter–spring conditions. Only one case of Fe deficiency was noted, i.e., a citrus orchard in which a spray of chelated Fe (Na Fe-EDTA) eliminated the symptoms. However, the occurrence of Fe chlorosis in certain crops led to individual studies of Fe in plant nutrition.

One report based on field and greenhouse studies (Hamze et al., 1985) dealt with the Fe nutrition of an ornamental plant, *Hydrangea macrophylla*, which is prone to chlorosis. Of the various Fe sources tested, Fe-EDDHA was the most effective when soil applied, but its effect on alleviating chlorosis symptoms decreased with time and was not as effective as $FeSO_4$ foliar sprays. Decreases in chlorosis were directly related to phenanthroline extractable-Fe (Fe^{2+} form) content (Rashid and Din, 1992; Rashid et al., 1997a). Another study (Ryan et al., 1985a) assessed the stability of a range of Fe-containing materials (EDDHA, DTPA, "Metalosate", $FeSO_4$, "Iron Sul", "Micronized Iron") and H_2SO_4 acidification in incubation for 5 months at field capacity. Only the EDDHA chelate maintained Fe in an available form throughout the incubation period.

A novel approach to the problem of lime-induced chlorosis in citrus involved the grafting of citrus scions on sour orange rootstocks (Hamze et al., 1986). By growing a range of rootstocks in highly calcareous soil, it was possible to identify Fe-tolerant rootstocks and thus, following grafting, bypass the perennial problem of Fe chlorosis for susceptible crops in calcareous soils. In another study with a Fe-inefficient variety of soybeans, Ryan et al. (1988) tested the hypothesis that an acid material, urea phosphate might alleviate chlorosis. However, the material accentuated chlorosis instead, an effect that was attributed to an increase in pH following hydrolysis of the urea moiety of the compound.

As Fe deficiency tends to occur in food legumes, where a wide variation exists with respect to chlorosis susceptibility (Hamze et al., 1987), but rarely in cereals, a few reports emanated from ICARDA, which has a mandate for legume crops. Of the 3267 lines of Kabuli-type chickpea tested in calcareous soil, most were tolerant of low soil Fe, with only 25 lines considered susceptible (Saxena et al., 1990); in the latter case, foliar spraying with 0.5% $FeSO_4$ eliminated the chlorosis symptoms but there was no effect on yield. Studies on inheritance of resistance to Fe-deficiency revealed that resistance was dominant and is governed by a single gene. The breeding program involved negative selection for chlorosis in the segregating populations in the field. While Fe deficiency is most pronounced in the winter (cool-season period), 99% of the 6,224 lines tested under such conditions were tolerant to low Fe (Bejiga et al., 1996). A similar type of study evaluated accessions of lentil (*Lens culinaris*). A sizeable proportion of the lines that originated from India (37.5%) and Ethiopia (30%) were susceptible to Fe chlorosis, whereas chlorosis was rare in germplasm from Syria and Turkey (Erskine et al., 1993). These authors noted that symptoms were transient, and disappeared with increasing ambient temperatures; the yield was negatively affected only where symptoms were severe.

Research results from Pakistan broadly coincided with these from West Asia. The same discrepancy between chlorosis intensity and total plant Fe was noted by Rashid and Din (1992); indeed chlorotic leaves were shown to contain much more total Fe than green healthy leaves. As with Hamze et al. (1985), the physiologically active ferrous Fe was shown to be directly related to chlorosis. Where chlorosis occurred, symptoms were characteristically uneven in affected fields. Examples of Fe chlorosis in crops in Pakistan are chickpea (Rashid and Din, 1992), peanut (Rashid et al., 1997a), and deciduous fruits in other countries (Anonymous, 1998; Tandon, 1995). A listing of crops that are commonly susceptible to Fe deficiency is presented in Table 6.3. As genotypes within species vary in susceptibility to Fe chlorosis (Table 6.2), the problem can be effectively dealt with by screening, as has been demonstrated by the previously cited authors (Saxena et al., 1990; Bejiga et al., 1996; Erskine et al., 1993).

Various other studies from the Near East dealt with Fe in soils and its indirect influence on crop nutrition. For instance, Fe oxides, especially the relatively soluble or acid-oxalate Fe, were the dominant soil component in the precipitation/adsorption of soluble P following initial reactions with soil (Ryan et al., 1985b) and also in dictating reversion to less soluble forms in the longer term of several months (Ryan et al., 1985c). These studies indicated the extent to which Fe influences P fertilizer use efficiency for crops. Another study of Fe oxides revealed implications for N-fertilizer efficiency. The laboratory work of Ryan et al. (1981b) showed that Fe oxide coating in soil can reduce the influence of $CaCO_3$ on promoting volatile loss of ammonia following the application of urea to calcareous soils.

As a major agricultural country in the Near East, especially with respect to cereals and food legumes, and one encompassing several contrasting zones, Turkey produced a number of Fe-related studies. For instance, a survey of central Anatolia (Eyupoglu et al., 1998), involving 1,511 soil samples indicated that a quarter of the samples had less than the critical value of DTPA-extractable Fe of 4.5 mg kg^{-1} (Lindsay and Norvell, 1978). This indicated that with Fe-sensitive crops, deficiency of Fe may be a constraint. Another study in the laboratory showed that Fe application can accentuate Zn deficiency and therefore indirectly influence crop nutrition (Uygur, 1998).

6.4.2 Zinc

Of the micronutrients required by crops, Zn deficiency is probably the most important in terms of its negative influence on crops worldwide (Welch et al., 1991), especially in calcareous soils of arid and semi-arid areas. In Sillanpää's (1982) study, more than half the samples collected from 30 countries were Zn-deficient. Deficiency is especially common in cereal-growing areas of the world (Graham and Welch, 2000). Zinc availability in soils and plant uptake is influenced by many factors, e.g., alkaline pH, P, $CaCO_3$, organic matter, redox potential in flooded soils, and nutrient interactions (Mortvedt et al., 1991). Soil Zn is also influenced by the parent material from which the soil was derived and by erosion, natural or man-made, that causes removal of organic matter-rich topsoil and exposed subsoils. As superphosphate, formerly the most common form of P fertilizer, contains various amounts of Zn, the trend towards more concentrated P forms (Triple superphosphate, diammonium phosphate) is likely to exacerbate problems with Zn deficiency. As is evident from the commentary, conditions are such in the Mediterranean agro-ecosystems as to make Zn deficiency the most common micronutrient crop growth constraint. While the major work on Zn has been by Cakmak (1998) and colleagues in Turkey, brief mention will be made of other studies prior to focusing on the significant contributions from Turkey.

One of the earlier studies of Zn in soils and field crops in Lebanon did not indicate any problem with Zn deficiency to any significant extent (Bhatti et al., 1982), although extractable soil Zn varied regionally within the country. However, citrus and other orchard crops were not included in this survey-type study. The study assessed various common tests for available Zn ($NH_4H_2PO_4$, DTPA, $MgCl_2$) and concentration with soil properties. Although this study was preliminary, it did reflect the beginning of an awareness of micronutrients, especially Zn, in soils of thc Near East.

While much soil fertility research had been reported from Morocco (Ryan and Matar, 1990, 1992), studies related to Zn were few. The only crop-related work involved P and Zn nutrition of maize in the greenhouse (Ryan et al., 1995), as Zn deficiency is commonly noted in this crop and the soil had low available concentrations of both elements; there were significant responses in growth to both elements, with differences between the Mollisol and Vertisol soils. Added P caused a growth response that induced Zn deficiency by a dilution effect. Other Zn-related studies in Morocco involved detailed assessment of DTPA-extractable Zn in experimental plots in a range of agricultural experiment stations (Ryan et al., 1990) and in farmers' fields in the cereal-growing dryland agriculture zone (Abdel Monem et al., 1990).

In Syria, where much research emphasis was on cereals and food and forage legumes, the first recorded study of Zn-addition, and *Rhizobia* inoculation in relation to annual self-regenerating medic (*Medicago* spp.) was conducted in the greenhouse. While all species responded to P application, there was also a significant response to applied Zn (at 5 mg kg^{-1}), but only when there was adequate P and N from fixation as a result of the inoculation (Materon and Ryan, 1995). A subsequent field trial examined the effects of the same three factors on the same medic cultivars in addition to vetch *(Vica sativa)* and grasspea *(Lathyrus sativus)*. Again, as with the greenhouse study, P fertilizer evoked a significant response as the level of available P was 3.4 mg kg^{-1}, or less than the critical range of 5–7 mg kg^{-1} (Materon and Ryan, 1996). While the effect of inoculation or mineral N was available, there was a pronounced effect of Zn given that the level of DTPA-extractable soil Zn was low (0.6 mg kg^{-1}). When the trial was repeated for a second year on an adjacent site with a higher level of available Zn (1.2 mg kg^{-1}), there was no response to added Zn for any crop species, thus indicating that the soil test for available Zn was a good indicator of crop response and growth in the field.

Several studies in the Indian subcontinent also highlighted the role of Zn in crop nutrition. Among the crops that have shown widespread Zn deficiency in Pakistan are: rice (Chaudhry et al., 1977), maize (Anonymous, 1998), cotton (Rafique et al., 2002; Rashid and Rafique, 2002), citrus and deciduous fruits (Anonymous, 1998; Rashid, 2006a). Research in Pakistan showed that not only does Zn deficiency reduce crop yields, but it also delays maturity (Rashid and Fox, 1992; Chaudhry et al., 1977; Anonymous, 1998). As Zn deficiency is common in flooded rice, much effort has been focused on finding a solution to the problem. Having assessed various corrective measures, Rashid et al. (2000a) found that addition of 20 kg Zn ha^{-1} as $ZnSO_4$ to the rice nursery area was most effective. The crop species responses to micronutrients in Near East countries are shown in Table 6.3, as well as genotypic/cultivar interactions for Zn deficiency are shown in Table 6.2.

Few research programs anywhere have been as successful and as comprehensive as the NATO project headed by Ismail Cakmak in Turkey (Cakmak, 1998; see also Chap. 7, this volume). The research program built on the author's expertise in plant physiology and nutrition, especially with respect to Zn nutrition of wheat (Cakmak et al., 1995; Cakmaks, 1998), and other physiological aspects such as the interaction of P and Zn in cotton, Zn and cell membrane permeability, enzymatic behaviour in beans in relation to Zn nutrition, and Zn and Fe in relation to phytosiderophore release in wheat genotypes has added to the status of knowledge.

These contributions to the scientific knowledge of Zn in plant nutrition have to be seen in the context of the observations of Sillanpää (1982) that many soils in Turkey were Zn-deficient, as well as the survey of Eyüpoglu et al. (1998) in Central Anatolia which showed that 50% of the samples in the survey were low in Zn. Despite the various studies involving Zn, it was not until the early 1990s that commonly observed necrotic symptoms in wheat were attributed to Zn deficiency (Cakmak, 1998). Field trials with Zn, Fe, Mn, and Cu, individually and in combination, definitively identified Zn as the causative deficiency factor (see also Chap. 7).

The subsequent range of field, greenhouse and laboratory studies, summarized by Cakmak (1998), revealed the extent and significance of Zn deficiency for wheat production in Anatolia (Cakmak et al., 1996a), differences between bread and durum wheat (*T. durum* Desf.) in Zn deficiency (Cakmak et al., 1996b), and the role of phytosiderophores in explaining the sensitivity of durum wheat to Zn deficiency (Cakmak et al., 1996c) as well as with grasses (Cakmak et al., 1996d)

Other papers dealt with Zn application methods (Yilmaz et al., 1996); rye chromosome in Zn deficiency (Cakmak et al., 1997); morphological and physiological differences in cereal responses to Zn deficiency (Cakmak et al., 1998) and uptake of Zn and efficiency in cereal cultivars (Erenoglu et al., 1999). Related studies addressed Zn in goat grass (*Aegilops tauschii*) and einkorn wheat (*Triticium monoccum*) with respect to synthetic hexaploid wheats (Cakmak et al., 1999) and the influence of Zn-efficient wild grasses on phytosiderophore release under Zn deficiency (Cakmak et al., 1996d). Subsequently, Cakmak (2005) in Turkey, and Malakouti et al. (2005) in Iran showed that Zn-fortified seed improved crop yield and quality.

Among the major conclusions emerging from this Project were:

1. A large proportion of the soils and crops in central Anatolia are Zn-deficient.
2. Durum wheat is more sensitive to Zn deficiency than bread wheat.
3. Soils with less than 0.2 mg kg^{-1} DTPA-extractable Zn are deficient and highly likely to respond to Zn fertilisation.
4. Resistance to Zn deficiency is in the decreasing order: rye > triticale > barley > oat > bread wheat > durum wheat.
5. Within any species, there was considerable variation among genotypes.
6. Zn efficiency was related to the activity of Zn-containing superoxide dismutase and the rate of root uptake and translocation of Zn.
7. The phytic acid-to-Zn molar ratios are high for both Zn-efficient and Zn-inefficient wheat's grown on Zn-deficient soils, causing poor bioavailability in cereal grain; this ratio could be reduced by soil and/or foliar application of Zn.
8. Zn deficiency in soil and crops had a negative effect on the health of children.

A major outcome at the practical level from the Project was the practice of incorporating a trace amount of Zn oxide in the compound fertilisers in use in Turkey. This has had a major impact in reducing the incidence of Zn deficiency and increasing cereal yields with minimal extra costs. The linkage between Zn deficiency and differential responses of crop species and genotypes gave impetus to the concept of breeding for Zn efficiency and increasing the content of Zn in cereal grain in the interest of improved human nutrition (Graham et al., 1993). This realization led to various international agricultural research centres and advanced institutions collaborating in the formation of a Global Challenge Program on "Biofortification". The target for the Zn-enhanced cereals is poorer communities in the developing world who depend heavily on bread and have limited protein sources. The program represents a major paradigm shift in dealing with a problem of micronutrient deficiency, from feeding the soil to manipulating the plant, and going beyond crop yields to consider the health of the end-user.

6.4.3 *Boron*

As with other micronutrients, problems with B nutrition, whether deficiency or toxicity, are widespread (Gupta, 1979). Indeed, B is unique in that the range between deficient and toxic levels is relatively narrow. As with most micronutrients, the emphasis has been on deficiency rather than toxicity as crop growth constraints (Berger, 1962). The Near East is no exception, as reflected in work from two focal points: Syria/Lebanon/Turkey, and Pakistan. Thus, some of the earliest research on micronutrients was initiated by an iconic figure in B research, Kermit Berger (Berger, 1962). As with initial studies of other micronutrient deficiencies, Berger used a countrywide survey of soils and plants in Lebanon according to ecological zones (Khan et al., 1979). The survey showed that available B in the soil was closely related to B in the plant tissue, while available B was negatively correlated with soil pH. However, based on plant observations, and considering norms in both soil and the 33 crops sampled, there was no evidence of B deficiency, nor any indication of excess B either.

The global micronutrient study of Sillanpää (1982), which was subsequently expanded by Sillanpää and Vlek (1985), threw further light on the Near East countries, with specific reports on countries like Egypt, Lebanon, Iraq and Turkey. There was little evidence from the soil analyses presented that B deficiency was a crop constraint to any extent. In the only other country from the Near East included in the survey, Pakistan, there were indications of B deficiency based on soil test values, especially in un-irrigated soils. Observing relatively high soil B values, Sillanpää (1982) also suspected B toxicity in salt-prone irrigated soils of this country. Subsequent extensive research, however, revealed the incidence of widespread B deficiency in many field and horticultural crops throughout the country, rather than any evidence of toxicity (Rashid et al., 2002d; Rashid, 2006b). In consideration of its high cost-effectiveness, B use is now being rapidly adopted in many crops, including rice and cotton, in Pakistan (Rashid 2006a, b).

Boron deficiency has not been observed widely in most countries of the Near East; however, research in the recent past has revealed this deficiency to be a widespread problem in many crops in the Indian subcontinent (Anonymous, 1998; Srivastava et al., 2000; Rashid et al., 2002d; Rashid, 2006b, See also Chap. 4). Boron deficiency is highly genotype-specific (Table 6.2), occurring most frequently in sugar beet (*Beta vulgaris*), alfalfa (*Medicago sativa*), cauliflower (*Brassica oleracea* var capitata), rapeseed, turnip (*B. rapa*), and apple. Boron is immobile in the plant, with the result that deficiency symptoms first appear in the new growth. Salient examples are B deficiency in cotton in Pakistan (Anonymous, 1998; Rafique et al., 2002; Rashid and Rafique, 2002), rapeseed (Rashid et al., 1994b, 2002c; Wei et al., 1998), peanut (Rashid et al., 1997b) and wheat (Rashid et al., 1997b) and potato in Pakistan (Anonymous, 1998; Rashid et al., 2002d).

Considerable research by Rashid and co-workers (Rashid et al., 1994b, 1997b, c, 2002d) confirmed the critical threshold level of 0.5 mg kg^{-1} for B deficiency in various crops in Pakistan. As with Khan et al. (1979) in Lebanon, the research in Pakistan showed that available B was negatively correlated with pH and was related to low soil organic matter. One of the few countries of the Near East where B toxicity is a problem is Iraq. Following the early indications of Sillanpää (1982), considerable research emphasis was placed on the element in soils and irrigation water of Iraq (Al-Khafagi, 1997). While some soils were low in available B, most were considerably in excess of the 0.5 mg kg^{-1} critical level. High soil B increased from north to south in the country and was associated with soil derived from marine sediments and with irrigated areas. An inverse relationship was shown with mean annual rainfall. While most freshwaters had good quality in terms of soluble B, levels were high in many wells and in drainage water. A recent study examined B species in relation to salinity (Al-Falahi et al., 2001).

While deficiency, generally indicated by less than 0.5 mg kg^{-1} hot-water soluble B in the soil, is the more common problem globally, the limited soil and plant analysis data suggest that deficiency may not be common in the Near East region (Khan et al.,

1979; Sillanpää and Vlek, 1985). However, in semi-arid and arid areas and where irrigation is practiced (Manyowa and Miller, 1991), soil B concentrations of just a few mg kg^{-1}, whether naturally occurring in the soil or added through irrigation water, can cause B toxicity to the plant (Ryan et al., 1977). Some examples of B toxicity observed in field crops in the Near East countries are presented in Table 6.8. Uncontrolled application of sewage sludge and coal fly ash can also induce B toxicity.

Phytotoxicity of B was first described in barley a long time ago. Symptoms of toxicity, i.e., leaf necrosis, are specific only for barley, but even then may be confused with fungal disease (Jenkin, 1993). Such symptoms are related to high tissue concentrations of B which, in turn, are closely related to soluble B in the soil (Francois, 1992). This can occur where water-soluble B exceeds as little as 5 mg kg^{-1}; normal soils contain a few mg, while toxic soils may contain up to 100 mg or more. However, toxicity is modified by factors such as temperature (Mahalakshmi et al., 1995) and is accentuated under dry conditions. Unlike B deficiency, B toxicity is defined by a range of values rather than a narrow critical value.

Boron toxicity can cause substantial reductions in cereal grain and straw yield (Cartwright et al., 1984; Moody et al., 1993) and for squash (*Cucurbita spp*) (Francois, 1992). Boron toxicity has been recognized as a serious and widespread problem in the drylands of South Australia, which also have a Mediterranean-type climate (Cartwright et al., 1986). While the problem was only realized in the 1980s (Jenkin, 1993), of particular concern was the impact of B toxicity on barley (Cartwright et al., 1984).

Subsequently, considerable effort was devoted to ameliorating B toxicity, with a focus on the plant rather than the soil. While B deficiency can be easily rectified by application of soluble B fertilizer to the soil or as a spray on the crop, soil B toxicity is a much more complex problem to manage. Since treating the soil to remove or reduce excess B, e.g., by leaching (Prather, 1977), is not economically feasible, selecting or breeding crop cultivars with high tolerance/resistance to B toxicity is the most promising approach. For wheat and barley, the main field crops in Australia, considerable genetic differences in response to B toxicity have been identified (Nable, 1988; Paull et al., 1988, 1992).

Table 6.8 Some examples of boron toxicity in field crops in Near East countries

Species	Region/Country	Reference
Barley (*Hordeum vulgare*)	Syria	Yau (1997); Mahalakshmi et al. (1995)
Bread wheat (*Triticum aestivum*)	Konya, Turkey	Kalayci et al. (1998)
	Syria	Yau (1997); Yau et al. (1994)
Durum wheat (*Triticum durum Desf.*)	Konya, Turkey	Kalayci et al. (1998)
	Syria	Yau (1997), Yau et al. (1994,1995, 1997a, b)
Lentil (*Lens culinaris*)	Iraq, Syria	Yau and Erskine (2000)

Notwithstanding the information that was available at the global level and indeed indications that high levels of soluble B were present in Iraq and in the Anatolian plateau of Turkey (Sillanpää, 1982), it was not until the late 1980s that B toxicity was even suspected in the Near East region. Therefore, it is important to examine B toxicity, primarily on the basis of work done in Syria and Turkey. While most studies dealt with crops and adaptation to high soil B, there were some exclusive soil studies.

The assessment of micronutrients, including B, with soil depth was done at several of ICARDA's experiment stations and sites (Ryan et al., 1997). Although none of the surface layer samples indicated excess levels of B, as most were in the 0.5–1.0 mg kg^{-1} range; there were substantial increases in subsoil B at 1–2 m depth. The patterns of B accumulation varied with the station, being higher and at shallower depth in the lower rainfall zones. The subsoil pattern of B accumulation would be undetected by normal surface (0–20 cm) sampling. At the Bouiedar site, where B values were highest, a grid-sampling survey indicated that B was highly spatially variable (Ryan et al., 1998), making representative sampling difficult. A survey of farms across the rainfall transect in northern Syria basically showed a pattern of increase in B with decreasing rainfall (Ryan et al., 1996b), but there was little indication of toxicity based on surface samples.

One of the earliest plant studies (Mahalakshmi et al., 1995) showed that barley lines differed in relation to B toxicity symptoms and B concentration in plant tissue; symptoms were influenced by soil temperature in the early growth stage, but not later. Subsequently, Yau and co-workers demonstrated the negative effects of a high level of soluble B (17.4 mg kg^{-1}) on growth and grain yield of durum wheat and increased toxicity symptoms and shoot B concentration (Yau and Saxena, 1997). Varietal differences were evident (Yau et al., 1997a) with lines from Southern Australia, where toxicity is a field-scale problem (Cartwright et al., 1984). Yau and co-workers used soil boxes with varying levels of added soluble B (0, 25, 50 mg kg^{-1}) with seeds of the crops grown in lines under greenhouse conditions. Similar observations on the negative effects of soluble B were made by Yau et al. (1997b) following experimentation under these conditions.

The effect of high soil B on barley lines was also shown by Yau (2000), but straw quality appeared to be improved by increasing levels of B. Similarly, the screening technique was extended to lentil (77 lines), with accessions from Afghanistan being most tolerant to B toxicity, followed by those from India, Syria, Europe, Ethiopia, and Nepal. In Turkey, where the B toxicity problem was first identified, Kalayci et al. (1998) also showed differential varietal responses to high levels of soil B.

The work of Yau at ICARDA has come to a halt following his departure to the American University of Beirut. However, his work clearly showed that screening of tolerant germplasm is a simple, inexpensive technique, and one that set the basis for incorporation of resistant genes into improved varieties in a crop improvement program.

6.4.4 Other Micronutrients

Based on the survey of Sillanpää (1982) and various subsequent reports (e.g., Rashid and Ryan, 2004) from countries of the Near East region, there is little evidence to indicate that micronutrients other than Fe, Zn and B cause growth constraints in crops. Despite the fact that Mn tends to be deficient in calcareous and high-pH soils (Mortvedt et al., 1991), there was little indication of problems of Mn deficiency for crops in the Near East region, and consequently few studies dealing with Mn in soil–plant systems.

The survey of micronutrients in Lebanese soils previously referred to (Bhatti et al., 1982; Khan et al., 1979) indicated no symptoms in the 33 crops that could be attributed to Mn deficiency (Khan and Ryan, 1978). While Mn availability was least in soils where organic matter was low and $CaCO_3$ relatively high, none of the samples in the survey was less than the critical DTPA criterion, but some values were lower than some other conventional indices. One laboratory study elucidated factors related to Mn reactions in soils and therefore indirectly related to plant nutrition (Curtin et al., 1980). Adsorption of Mn was shown to be related to cation exchange capacity, but not to other properties; desorption isotherms showed that Mn was tightly held by the soil-adsorbing phase. A similar approach was used by Curtin et al. (1993) to examine retention of Zn and Cu by soil constituents.

Based on soil analyses from various surveys in Syria (Ryan et al., 1996b) and Morocco (Abdel Monem et al., 1990), there was no indication that any sample had DTPA-Mn values below the critical level (0.2 mg kg^{-1}) of Lindsay and Norvell (1978). If Cu deficiency did exist, it is likely to be mild and localized (Anonymous, 1998); indeed mycorrhizae, which are common in legumes, can contribute to Cu (and Zn) uptake by plants, making crop deficiency even more unlikely. As molybdenum deficiency is only found in non-calcareous soils, its occurrence in the Near East has not been noted. Similarly, information is scant or non-existent on cobalt (Co) and nickel (Ni) in the Near East region.

6.5 Managing Micronutrient Deficiencies

Since the significance of micronutrients as growth-limiting factors for crops has been recognized for the past half century, materials that supply these nutrients in fertiliser form have been developed and are now in common use, as are techniques for applying micronutrients to crops (Mortvedt et al., 1991). Micronutrient fertilisers can be broadly categorized as inorganic materials, mainly water-soluble, synthetic chelates and naturally occurring organic complexes. The most common soluble form of micronutrients is the sulphate form. Of the insoluble inorganic materials, ZnO and $ZnCO_3$ are commonly used. Of the chelates that are formed by coordinate bonding of a chelating agent with a metal cation, the most common are EDTA (ethylene diamine tetra acetic acid), HEDTA (N-hydroxyethyl-ethlylenediamine triacetic acid), and EDDHA (ethylenediamine di-o-hydroxyphenylacetic acid). These materials vary in stability depending on the metal in question; Fe-EDDHA and Zn-EDTA are common for these elements, chelated micronutrients are sold in either liquid or powder formulations. The naturally occurring organic complexes are highly variable in terms of stability and effectiveness and include such compounds as lingo-sulphonates, phenols, and polyflavonoids.

Most micronutrient fertilising materials are applied to the soil, while some are applied to the growing crop as a foliar spray. As soil applications of micronutrients involve small amounts of the material, usually in the order of 10 kg ha^{-1} or less, the problem of uniform application arises. This can be overcome to a large degree by incorporating the micronutrient fertiliser in granular or liquid formulations of N, P, K fertilisers. Foliar sprays are generally in the order of 0.1–1.0% of the metal in aqueous solution. Against these generalisations on micronutrient fertilisers and their methods of application to rectify deficiency in crops, brief mention is made of representative studies in the Near East area that are related to micronutrient deficiency management strategies.

In view of the importance of DTPA as an extractant for available micronutrient metals (Lindsay and Norvell, 1978), the test procedure for available Fe, Zn, Mn, and Cu was compared with common availability indices ($MgCl_2$, $NH_4H_2PO_4$, EDTA-NH_4OAc, HCl) for these metals in a greenhouse study involving wheat and tomato grown in 20 Lebanese soils (Ryan et al., 1985d). In general, the DTPA test proved to be superior to the other tests, but the correlation was only slightly improved by sequential inclusion of soil properties ($CaCO_3$, pH, organic matter, cation exchange capacity) in the regression

As the agronomic efficiency of soluble micronutrient fertilisers is negatively influenced by soil properties, especially $CaCO_3$ and pH, specialized micronutrient materials, such as sulphur-coated granules, were developed to overcome such reactions with $CaCO_3$ and clay particles. Hence, soluble sulphate forms of Fe, Zn, and Mn enveloped in S-coating were tested in incubation studies under varying conditions of moisture, temperature, and $Ca(OH)_2$ concentrations. However, the patterns between uncoated and coated materials were similar, thus indicating that the coatings broke down quickly and were ineffective at controlling the dissolution of the elements (Ryan and Prasad, 1979). Because of such lack of effectiveness, these novel materials never became a commercial success and were not adopted in crop nutrition.

Notwithstanding the advantage of synthetic chelates to improve micronutrient fertilising efficiency, an important consideration was the extent and duration of stability in calcareous soils. Hence, in a laboratory study (168 days), chelates of Fe (EDDHA) and Zn, Mn and Cu (EDTA) were incubated under conditions of varying $CaCO_3$ (0–10%), temperature (5–35 °C) and wetting–drying cycles and destructively sampled and extracted for water-soluble and DTPA-extractable micronutrient cations (Ryan and Hariq, 1983). At the end of the incubation period, 32–56% of the Fe in the EDDHA was still extractable (and the remainder presumably precipitated as insoluble Fe), less than 13% of the Cu and Zn was extractable, while none of the Mn from EDTA was extractable after 2 days. Extractability of the metals was depressed with increasing $CaCO_3$, temperature, and wet–dry cycles.

Examples of practical methods for alleviating micronutrient deficiencies in the field are provided by Cakmak (1998) and Yilmaz et al. (1996) for Zn. In these studies, $ZnSO_4.7H_2O$ was soil-applied at 0, 7, 14, and 21 kg Zn ha^{-1} to wheat under irrigated and rainfed conditions while there were large differences in sensitivity of wheat varieties to Zn deficiency, the lowest application rate (7 kg Zn ha^{-1}) was generally sufficient to alleviate deficiency and overcome yield reductions. In a related study aimed at Zn bioavailability, these authors compared soil, seed, leaf, soil + leaf, and seed + leaf applications of Zn; they concluded that leaf application in addition to the soil one was needed to reduce phytic acid which reduced Zn bioavailability.

Though most studies of micronutrient fertilisers considered direct influences on availability, some were indirect. The study of the behaviour of an N-P fertiliser, urea phosphate, examined effects on micronutrient availability (Ryan et al., 1986). This potentially acidifying material induced a net increase in Mn and Zn solubility. However, the outcome of such effects in terms of crop yields was not established.

6.6 Future Research Needs

Given the vastness of the Near East countries and the wide agro-ecological variations within the region, it is clear that the relatively few studies conducted in a small number of countries can only give a sketchy impression as a whole. Thus, there is need for rapid surveys of countries in the Near East, as was done in Lebanon in the 1970s (Bhatti et al., 1982; Khan et al., 1979) and in Pakistan in 1980s and 1990s (Anonymous, 1998; Rashid, 2006a, b) to establish to what extent, either micronutrient deficiencies or toxicities, might pose a problem for crop production, or indeed by extension to human and animal health. The work of Cakmak (1998) forcefully demonstrated the linkage between micronutrients in the soil and the food chain. Mapping of micronutrients at the country level, based on geostatistical kriging from actual sample values, could help policymakers and scientists respond to perceived issues involving micronutrients, and identify relationships with specific agro-ecological zones. Such geographical surveys are illustrated from Pakistan (Rafique et al., 2002; Rashid et al., 2002d; Rashid, 2006a,b; Rafique et al., 2006), as well as the examples cited from Lebanon (Khan et al., 1979) and Australia (Paull et al., 1988).

Given the likelihood of crop intensification in the future, especially with increased irrigation, improved varieties, and fertiliser use, the issue of micronutrients will be more important, in essence, as factors that restrict potential crop production are eliminated and micronutrients are being removed from the soil in increasing amounts in crop offtake. Without replenishment, deficiency of some micronutrients will inevitably increase; indeed as fertilisers become more concentrated, they will contain lower concentrations of micronutrients as contaminants. This is especially true for micronutrients whose total supply in soils is limited, i.e., Zn and B. Similarly, as the response of crops to micronutrients has genetic variability, new crop cultivars need to be screened for micronutrient efficiency.

Another development that could impinge on micronutrients is the use of wastewater for irrigation (Ryan et al., 2006). Wastewater is the only growing water source in the Near East and the scarcity of water in the region will dictate the need for recycling water from the major cities. It is therefore possible that toxic effects of some elements might become a reality, especially in areas of heavy industry. Though cultural norms of Near Eastern society do not favour use of urban wastes, there is no alternative to reusing wastewater and solid-waste sludge. Developments that have occurred in the West will inevitably occur in the Near East. In the use of urban and industrial wastes, particular attention will need to be given to metals such as lead (Pb) and cadmium (Cd) in view of the potential health hazards involved. However, with few exceptions there is no large-scale awareness of toxic metals in the Near East

Soil micronutrient deficiencies not only result in crop productivity losses but also lead to low-micronutrient concentrations in livestock feed and food stuffs causing animal and human malnutrition (Graham and Welch, 2000). The work of Cakmak (1998) demonstrated that soil micronutrient deficiencies can impact at all levels of the food chain. For example, a recent study in Pakistan (Anonymous, 2004), has revealed the incidence of Zn deficiency in 40–50% of resource-poor women (especially those who are pregnant) and children (<5 years old) dependent entirely on staple cereals (i.e., wheat and rice) produced in the Zn-deficient soils of the country. This alarming situation calls for epidemiological studies on micronutrients in human health, as in the case of Zn. Factors that influence bioavailability of micronutrients will feature in future research. Some countries of the region, like Pakistan, are participating in the CGIAR-led "Future Harvest" research for Zn and Fe biofortification of wheat and rice grain by genetic manipulation of plants. Research in the Near East has adequately established that grain Zn concentrations can be more than doubled with Zn fertilisation (Rafique et al., 2006). However, a slightly higher Zn dose, in excess of the rate required to maximize crop yield, is required to attain adequate Zn-enrichment of grain. As Zn toxicity per se is not a problem and a larger dose of Zn remains effective for a number of subsequent crops, this option appears safe as well as cost-effective. Thus, loading of Zn into grain, by tailoring the soil to fit the plant, is a viable alternative to Zn biofortification for combating Zn deficiency-induced malnutrition in the poorer segments of the society, particularly in less-developed countries of the world.

While methods of micronutrient fertilisation are well established, a new area is the assessment of the effectiveness of micronutrients incorporated in N, P, K compound fertilisers as well as seed-soaking (Cakmak, 1998). The longevity of these effects needs further evaluation.

What emerged from the work of Yau was the importance of manipulating the plant in cases where soil manipulation is neither feasible nor practical. Thus, calls for greater harmony between soil scientists/agronomists and plant breeders, especially for addressing nutrient toxicity, warrant better attention. The studies on Fe and Zn showed the greater potential of breeding for greater efficiency of uptake and physiological use as an alternative to conventional soil or foliar fertilization.

Analysis of soils, plants, and water for micronutrients and other parameters that affect crop production and environmental protection is crucial to the development of agriculture in the Near East (Ryan and Rashid, 2006). Regardless of whether the concern is with micronutrients in water, soil, plant, animal or humans, a critical concern is the reliable and accurate measurement of such elements and the ease of availability of the requisite analytical facilities (Ryan et al., 1999). As this brief review indicated, services for analysis are comparatively rare, and even where they do exist, lack of standardization and quality control is an issue (Ryan et al., 2002).

6.7 Conclusions

In contrast to major nutrients such as N and P, whose effects are generally recognized in the agriculture of the Near East, as well as everywhere else, it is difficult to generalize regarding the importance of micronutrients. Though the Near East has broad agro-ecological features that are likely to induce deficiencies of these nutrients, the extent of deficiency is highly variable from country to country and within any country. Where some deficiencies occur, they can be a severe constraint to crop production as in central Anatolia, in Turkey. It is difficult to draw any conclusions from countries where no information on micronutrients exists. However, in consideration of cropping intensification, it is safe to conclude that micronutrient deficiencies are likely to be more important in the future in Near Eastern agriculture. Not all developments will exacerbate deficiencies of micronutrients; the use of nutrient-rich wastewater and conservation tillage that builds up soil organic matter, are likely to have beneficial effects on micronutrient concentrations as much as using traditional soil approaches. In an age when the broad environment is a societal consideration, the concerns about micronutrients inevitably have to extend beyond the crop in the field to consider what is in the crop and how it affects the humans and animals which consume the crop.

References

Abdel Monem, M., Ryan, J., El-Gharous, M. (1990). Preliminary assessment of the soil fertility status of the mapped area of Chaouia, Morocco. Al-Awamia, 72: 85–107.

Al-Falahi, A. A., Al-Zubaidi, A., Ryan, J. (2001). Boron and its species in saline soils of Iraq. 7th International Meeting of Soils with Mediterranean-type Climate (pp. 88–89). Bari, Italy. Sept. 23–28, 2001. Extended Abstracts.

Al-Khafagi (1997). Boron in soils and waters of Iraq. In J. Ryan (Ed.), Accomplishments and Future Challenges in Dryland Soil Fertility Research in the Mediterranean Area (pp. 204–208). Aleppo, Syria: ICARDA.

Anonymous (1998). Micronutrients in Agriculture: Pakistan Perspective (51 pp.). Islamabad, Pakistan: National Fertilizer Development Center.

Anonymous (2004). National Nutrition Survey, 2001–2002. Islamabad, Pakistan: Planning Commission, Government of Pakistan.

Arshad, M., Ryan, J., Paeth, R. C. (1980). Influence of organic matter, free iron, and calcium carbonate on soil particle size distribution. Agrochimica, 24: 470–477.

Bejiga, G., Singh., K. B., Saxena, M. C. (1996). Evaluation of world collection of Kabuli chickpea for resistance to iron-deficiency chlorosis. Genetic Resources and Crop Evolution, 43: 257–259.

Berger, K.C., Truog, E. (1944). Boron tests and determination for soils and plants. Soil Science, 57: 25–26.

Berger, K. C. (1962). Micronutrient deficiencies in the United States. Journal of Agriculture and Food Chemistry, 10: 178–182.

Bhatti, M. G. S., Ryan, J., Berger. K. C. (1982). Available zinc status of Lebanese soils. Iran Journal of Agricultural Research, 1: 41–47.

Borlaug, N. E. (2003). Feeding a world of 10 billion people: The IFDC/TVA legacy. Travis P. Hignett Memorial Lecture. March 14, 2003. Muscle Shoals, AL.

Cakmak, I. (1998). Selection and characterization of cereal genotypes with high zinc efficiency and evaluation of bioavailability of zinc in wheat for the central Anatolian region. Science for Stability Programme. Adana, Turkey: Cukurova University.

Cakmak, I. (2005). Effect of micronutrients on seed quality. In C. L. Li et al. (Eds.), Plant Nutrition for Food, Security, Human Health, and Environmental Protection (pp. 34–35). Beijing, China: Tsinghua University Press.

Cakmak, I., Atli, M., Kaya, R., Evliya, H., Marschner, H. (1995). Association of high light density in cold-induced leaf chlorosis in grapefruit and mandarin trees. Journal of Plant Physiology, 146: 355–360.

Cakmak, I., Yilmaz, A., Kalayci, M., Ekiz, H., Torum, B., Erenoglu, B., Braun. H. J. (1996a). Zinc deficiency as a critical problem in wheat production in Central Anatolia. Plant and Soil, 180: 165–172.

Cakmak, I., Sari, N., Marschner, H., Kalayci, M., Yilmaz, A., Eker, S., Gulut, K. Y. (1996b). Dry matter production and distribution of zinc in bread and durum wheat genotypes differing in zinc efficiency. Plant and Soil, 180: 173–181.

Cakmak, I., Sari, N., Marschner, H., Kalayci. M., Yilmaz. A., Braun, H. J. (1996c). Phytosiderophore release in bread and durum wheat genotypes differing in zinc efficiency. Plant and Soil, 180: 183–189.

Cakmak, I., Ozturk, L., Marschner, H., Karanlik, S., Ekiz, H. (1996d). Zinc-efficient wild grasses enhance release of phytosiderophores under zinc deficiency. Journal of Plant Nutrition, 19: 551–563.

Cakmak, I., Derici, R., Torum, B., Tolay, I., Braun, H. J., Schlwegel, R. (1997) Role of rye chromosomes in improvement of zinc efficiency in wheat and triticale. Plant and Soil, 196: 249–253.

Cakmak, I., Erenoglu, B., Guilit, K. Y., Derici, R., Romheld, V. (1998). Light-meditated release of phytosiderophores in wheat and barley under iron or zinc deficiency. Plant and Soil, 202: 309–315.

Cakmak, I., Torum, B., Erenoglu, B., Ozturk, L., Marschner, H., Kalayci, M., Ekiz, H., Yilmaz, A. (1998). Morphological and physiological differences in the response of cereals to zinc deficiency. Euphytica, 100: 349–357.

Cakmak, I., Cakmak, O., Eker, S., Ozdemir, A., Watanabe, N., Braun, H. J. (1999). Expression of high zinc efficiency of Aegilops tauschii and Triticum monococcum in synthetic hexaploid wheats. Plant and Soil, 215: 203–209.

Cakmak, I., Tolay, I., Ozdenir, A., Ozkan, H., Ozturk, L., Kling, C. I. (1999). Differences in zinc efficiency among and with diploid, and hexaplaid wheats. Annals of Botany, 84: 163–171.

Cartwright, B., Zarcinas, B. A., Mayfield, A. H. (1984). Toxic concentrations of B in a red-brown earth at Gladstone, South Australia. Australian Journal of Soil Research, 22: 261–272.

Cartwright, B., Zarcinas, B. A., Spoucer, L. R. (1986) Boron toxicity in South Australia barley crops. Australian Journal of Agricultural Research, 37: 351–359.

Chaudhry, F. M., Alam, S. M., Rashid, A., Latif, A. (1977). Mechanism of differential susceptibility of two rice varieties to zinc deficiency. Plant and Soil, 46: 637–642.

Cooper, P. J. M., Gregory, P. J., Tully, D., Harris, H. C. (1987). Improving water use efficiency in the rainfed farming systems of West Asia-North Africa. Experimental Agriculture, 23: 113–158.

Curtin, D., Ryan, J., Chaudhary, R. A. (1980). Manganese adsorption and desorption in calcareous Lebanese soils. Soil Science Society of America Journal, 44: 947–950.

Curtin, D., Ryan, J., Ahmed, M., Piracha, I. (1993). Retention and extractability of copper and zinc in calcareous Lebanese soils. Lebanese Science Bulletin, 6(1): 7–16.

Damania, A. B., Valkoun, J., Wilcox, G., Qualset, C. O. (Ed.) (1998). The origins of agriculture and crop domestication. The Harlan Symposium, Aleppo, Syria: International Center for Agricultural Research in the Dry Areas and Rome, Italy: International Plant Genetics Resource Institute.

Erenoglu, B., Cakmak, I., Romheld, V., Derici, R., Rengel, Z. (1999). Uptake of zinc by rye, bread wheat and durum wheat cultivars differing in zinc efficiency. Plant and Soil, 209: 245–252.

Erskine, W., Saxena, N. P., Saxena, M. C. (1993). Iron deficiency in lentil: Yield loss and geographic distribution in a germplasm collection. Plant and Soil, 151: 249–254.

Eyupoglu, F., Kurucu, N., Talaz, S., Canisag, V. (1998). Plant available iron contact of Turkish soils (pp. 167–173). In International Symposium on Arid Region Soils, Menemen, Izmir, Turkey, Sept. 21–24, Soil Science Society of Turkey, Ankara.

Fageria, N. K., Baligar, V. C., Clark, R. B. (2003). Micronutrients in crop production. Advances in Agronomy, 77: 185–268.

Francois, F. E (1992).Effect of excess boron on summer and winter squash. Plant and Soil, 147: 163–170.

Graham, R. D., Asher, J. S., Hynes, S. C. (1993). Selecting zinc-efficient cereal genotypes for soils of low zinc status. In Randall, P. J. et al. (Eds.), Genetic Aspects of Plant Mineral Nutrition (pp. 349–358), Proceedings, the 4th International Symposium on Genetic Aspects of Plant Mineral Nutrition, Sept. 30–Oct. 4, 1991. Canberra, Australia and Dordrecht, The Netherlands: Kluwer.

Graham, R. D., Welch, R. M. (2000). Plant food micronutrient composition and human nutrition. Communicational in Soil Science and Plant Analysis, 31: 1627–1640.

Gupta, U. C. (1979). Boron nutrition of crops. Advances in Agronomy, 31: 273–307.

Hamze, M., Ryan, J., Shwayri, R., Zaabout, M. (1985). Iron treatment of lime-induced chlorosis: Implications for chlorophyll, Fe^{2+}, Fe^{3+}, and K content in leaves. Journal of Plant Nutrition, 8: 437–448.

Hamze, M., Ryan, J., Zaabout, M. (1986). Screening of citrus rootstocks for lime-induced chlorosis tolerance. Journal of Plant Nutrition, 9: 459–469.

Hamze, M., Ryan, J., Mikdashi, R., Solh, M. (1987). Evaluating of chickpea (*Cicer arietinum L.*) genotypes for lime-induced chlorosis. Journal of Plant Nutrition, 10: 1031–1039.

Jenkin, M. J. (1993). The Genetics of Boron Toxicity in Barley. Ph.D. thesis, University of Adelaide, Adelaide, South Australia.

Jones, J. B., Jr. (1991). Plant Analysis Handbook – A Practical Sampling, Preparation, Analysis and Interpretation Guide. Athens, GA: Micro-Macro.

Kalayci, M., Alkan, A., Cakmak, I., Bayramoglu, O., Yilmaz, A., Aydin, M., Ozbek, V., Ekiz, H., Ozberisoy, F. (1998). Studies on differential response of wheat cultivars to boron toxicity. Euphytica, 100: 123–129.

Kassam, A. H. (1981). Climate, soil and land resources in West Asia-North Africa. Plant and Soil, 58: 1–28.

Khan, M. A., Ryan, J. (1978). Manganese availability of calcareous soils of Lebanon. Agronomy Journal, 70: 79–82.

Khan, Z. D., Ryan, J., Berger, K. C. (1979). Available boron in calcareous soils of Lebanon. Agronomy Journal, 71: 688–690.

Katyal, J. C., Vlek, P. L. G. (1985). Micronutrient problems in tropical Asia. In P. G. L. Vlek (Ed.), Micronutrients in Tropical Food Crop Production (pp. 69–94). Developments in Plant and Soil Sciences, Vol. 14. Dordrecht, The Netherlands: Martinus Nijhoff/Dr. W. Junk.

Loneragan, J. F. (1997). Plant nutrition in the 20th and perspectives for the 21st century. Plant and Soil, 196: 163–174.

Lindsay, W. L., Norvell, W. A. (1978). Development of a soil test for zinc, iron, manganese, and copper. Soil Science Society of America Journal, 42: 421–428.
Matar, A., Torrent, J., Ryan, J. (1992). Soil and fertilizer phosphorus and crop responses in the dryland Mediterranean zone. Advances in Soil Science, 18: 81–146.
Mahalakshmi, V., Yau, S. K., Ryan, J., Peacock, J. M. (1995). Boron toxicity in barley (*Hordeum vulgare* L.) seedlings in relation to soil temperature. Plant and Soil, 177: 151–156.
Malakouti, M. J., Tehrani, M. M., Savagheli, G. R., Lotfallahi, M., Ziaeyan, A., Balali, M. R., Bylordi, A., Majidi, A., Gheibi, M. N., Bybordi, A., Majidi, A., Gheibi, M. N. (2005). The effect of zinc-fortified seed on improving wheat yield and quality. In C. L. Li et al (Eds.), Plant Nutrition for Food, Security, Human Health, and Environmental Protection (pp. 376–377). Beijing, China: Tsinglua University Press.
Manyowa, N. M., Miller, T. E. (1991).The genetics of tolerance to high mineral concentrations in the tribe *Triticae* – a review and update. Euphytica, 57: 175–185.
Materon, L. A., Ryan, J. (1995). Rhizobial inoculation and phosphorus and zinc nutrition for annual medics (*Medicago* spp.) adapted to Mediterranean-type environments. Agronomy Journal, 87: 692–698.
Materon, L., Ryan, J. (1996). Effects of rhizobial inoculation and mineral nutrition of common Mediterranean pasture and forage legumes. Agricultural Mediterranae, 126(1): 64–74.
Moody, D. B., Rathjen, A. J., Carthwright, B. (1993). Yield evaluation of a gene for boron tolerance using backcross-derived lines. In P. J. Randall et al. (Eds.), Genetic Aspects of Plant Mineral Nutrition (pp. 363–366). The Netherlands: Kluwer.
Mortvedt, J. J., Giordano, P., Lindsay, W. (Eds.) (1972). Micronutrients in agriculture. Soil Science Society of America, Madison, Wisconsin, USA. 666 p.
Mortvedt, J. J., Cox, F. R., Shuman, L, M., Welch, R. M. (Eds.) (1991). Micronutrients in Agriculture (2nd edn., 760 p.). Soil Science Society of America, Madison, WI.
Nable, R. O. (1988). Response to boron toxicity amongst several barley and wheat cultivars: a preliminary examination of the resistance mechanism. Plant and Soil, 112: 45–47.
Paull, J. G., Cartwright, B., Rathjen, A. J. (1988). Response of wheat and barely genotypes to toxic concentration of soil boron. Euphytica, 39: 137–144.
Paull, J. G., Nable, R. O., Lake, A. W. H., Materne, M. A., Rathjen, A. J. (1992). Response of annual medics (*Medicago* spp.) to high concentrations of boron: Genetic variation and the mechanism of tolerance. Australian Journal of Agricultural Research, 43: 203–213.
Prather, R. J. (1977). Sulphuric acid as an amendment for reclaiming soils high in boron. Soil science society of Americana Proceedings, 41: 1098–1101.
Rafique, E., Rashid, A., Bhatti, A. U., Rasool, G., Bughio, N. (2002). Boron deficiency in cotton grown in calcareous soils of Pakistan. I. Distribution of B availability and comparison of soil testing methods. In H. E. Goldbach, P. H. Brown, B. Rerkasem, M. Thellier, M. A. Wimmer, R.W. Bell. (Eds.), Boron in Plant and Animal Nutrition (pp. 349–356). New York: Kluwer/Plenum.
Rafique, E., Rashid, A., Ryan, J., Bhatti, A. U. (2006). Zinc deficiency in rainfed wheat in Pakistan: Magnitude, spatial variability, management, and plant diagnostic norms. Communications in Soil Science and Plant Analysis, 37: 1–17.
Rao, S., Ryan, J. (Eds.) (2004). Challenges and Strategies in Dryland Agriculture. Madison, WI, USA: Crop Science Society of America Special Pub. No. 32.
Rashid, A. (1996). Secondary and micronutrients. In A. Rashid, K. S. Memon (Managing Authors), Soil Science (pp. 341–386). National Book Foundation: Islamabad, Pakistan.
Rashid, A. (2006a). Incidence, diagnosis and management of micronutrient deficiencies in crops: success stories and limitations in Pakistan: Invited review. Proceedings, IFA International Workshop on Micronutrients, Feb.27–March 2, 2006, Kunming, China. [online:www.fertilizer.org/ifa/publicat/PDF/2006_ag_kunming_rashid.pdf]
Rashid, A. (2006b). Boron deficiency in soils and crops of Pakistan: Diagnosis and Management (40 pp.). Pakistan Agricultural Research Council, Islamabad.
Rashid, A., Din, J. (1992). Differential susceptibility of chickpea cultivars to iron chlorosis grown on calcareous soils of Pakistan. Journal of Indian Society of Science, 40: 488–492.

Rashid, A., Fox, R. L. (1992). Evaluating internal zinc requirement of grain crops by seed analysis. Agronomy Journal, 84: 469–474.

Rashid, A., Rafique, E. (2002). Boron deficiency in cotton grown in calcareous soils of Pakistan. II. Correction and criteria for foliar diagnosis. In H. E. Goldbach, P. H. Brown, B. Rerkasem, M. Thellier, M. A. Wimmer, R. W. Bell (Eds.), Boron in Plant and Animal Nutrition (pp. 357–362). New York: Kluwer/Plenum.

Rashid, A., Rafique, E. (2005). Internal boron requirement of young sunflower plants: Proposed diagnostic criteria. Communications in Soil Science and Plant Analysis, 36: 2113–2119.

Rashid, A., Ryan, J. (2004). Micronutrient constraints to crop production in soils with Mediterranean-type characteristics: A review. Journal of Plant Nutrition, 27: 959–975.

Rashid, A., Bughio, N., Rafique, E. (1994a). Diagnosing zinc deficiency in rapeseed and mustard by plant analysis and soil testing. Communications in Soil Science and Plant Analysis, 25: 3405–3412.

Rashid, A., Rafique, E., Bughio, N. (1994b). Diagnosing boron deficiency in rapeseed and mustard by plant analysis and soil testing. Communications in Soil Science and Plant Analysis, 25: 2883–2897.

Rashid, A., Rafique, E., Din, J., Malik, S. N., Arain, M. Y. (1997a). Micronutrient deficiencies in rainfed calcareous soils of Pakistan. I. Iron chlorosis in peanut. Communications in Soil Science and Plant Analysis, 28: 135–148.

Rashid, A., Rafique, E., Bughio, N. (1997b). Micronutrient deficiencies in rainfed calcareous soils of Pakistan. II. Boron nutrition of the peanut plant. Communications in Soil Science and Plant Analysis, 28: 149–159.

Rashid, A., Rafique, E., Bughio, N. (1997c). Micronutrient deficiencies in rainfed calcareous soils of Pakistan. III. Boron nutrition of sorghum. Communications in Soil Science and Plant Analysis, 28: 444–454.

Rashid, A., Rafique, E., Bughio, N. Yasin, M. (1997d) Micronutrient deficiencies in rainfed calcareous soils of Pakistan. IV. Zinc nutrition of sorghum. Communications in Soil Science and Plant Analysis, 28: 455–467.

Rashid, A., Kausar, M. A, Hussain, F., Tahir, M. (2000a). Managing zinc deficiency in transplanted flooded rice grown in alkaline soils by nursery enrichment. Tropical Agriculture (Trinidad), 77: 156–162.

Rashid, A., Muhammad, S., Rafique, E. (2000b). Genotypic variation in rice susceptibility to boron deficiency. International Rice Research Notes, 25(3): 29–30.

Rashid, A., Muhammad, S., Rafique, E. (2002a). Genetic variation in boron uptake and utilization by rice and wheat. In H. E. Goldbach, P. H. Brown, B. Rerkasem, M. Thellier, M. A. Wimmer, R. W. Bell (Eds.), Boron in Plant and Animal Nutrition (pp. 305–310). New York: Kluwer/Plenum.

Rashid, A., Rafique, E., Bughio, N. (2002b). Boron deficiency in rainfed alkaline soils of Pakistan: Incidence and boron requirement of wheat. In H. E. Goldbach, P. H. Brown, B. Rerkasem, M. Thellier, M. A. Wimmer, R. W. Bell (Eds.), Boron in Plant and Animal Nutrition (pp. 371–379). New York: Kluwer /Plenum.

Rashid, A., Rafique, E., Muhammad, S., Bughio, N. (2002c). Boron deficiency in rainfed alkaline soils of Pakistan: Incidence and genotypic variation in rapeseed-mustard. In H. E. Goldbach, P. H. Brown, B. Rerkasem, M. Thellier, M. A. Wimmer, R. W. Bell (Eds.), Boron in Plant and Animal Nutrition (pp. 363–370). New York: Kluwer/Plenum.

Rashid, A., Rafique, E., Ryan, J. (2002d). Establishment and management of boron deficiency in field crops in Pakistan: A country report. In H. E. Goldbach, P. H. Brown, B. Rerkasem, M. Thellier, M. A. Wimmer, R. W. Bell (Eds.), Boron in Plant and Animal Nutrition (pp. 339–348). New York: Kluwer/Plenum.

Rashid, A., Yasin, M., Ashraf, M., Mann, R. A. (2004). Boron deficiency in calcareous soils reduces paddy yield and impairs grain quality. International Rice Research Notes, 29: 58–60.

Rerkasem, B., Jamjod, S. (1997). Genotypic variation in plant response to low boron and implications for plant breeding. Plant and Soil, 193: 169–180.

Reuter, D. J., Robinson, J. B. (Eds.) (1986). Plant Analysis – An Interpretation Manual. Sydney, Australia: Inkata.

Reuter, D. J., Robinson, J. B. (Eds.) (1997). Plant Analysis – An Interpretation Manual. Melbourne, Australia: CSIRO.

Robson, A. D. (Ed.) (1993). Zinc in Soils and Plants. Dordrecht, The Netherlands: Kluwer,. 208 p.

Ryan, J. (Ed.) (1997). Accomplishments and future challenges in dryland soil fertility research in the Mediterranean area. Proceedings, International Soil Fertility Workshop, Nov. 19–23, 1995, Aleppo, Syria: ICARDA. 364 p.

Ryan, J. (2002). Available soil nutrients and fertilizer use in relation to crop production in the Mediterranean area. In K. R. Krishna (Ed.), Soil Fertility and Crop Production (pp. 213–246). Enfield, NH: Science.

Ryan, J., Matar, A. (Eds.) (1990). Soil test calibration in West Asia and North Africa. Proceedings, Third Regional Soil Test Calibration Workshop. Amman, Jordan, Sept. 3–9, 1988 (243 p.). Aleppo, Syria: International Center for Agricultural Research in the Dry Areas.

Ryan, J., Matar, A. (Eds.) (1992). Fertilizer use efficiency under rainfed agriculture. Proceedings, Fourth Regional Soil Test Calibration Workshop. Agadir, Morocco. May 5–11, 1991 (286 p.). Aleppo, Syria: International Center for Agricultural Research in the Dry Areas.

Ryan, J., Prasad, J. D. (1979). Factors affecting release and plant availability of sulfur-coated micronutrients. Soil Science Society of America Journal, 43(5): 1039–1043.

Ryan, J., Rashid, A. (2006). Application of soil and plant analysis for applied research in West Asia-North Africa: An international center's perspective. Communications in Soil Science and Plant Analysis, 37: 2185–2198.

Ryan, J., Miyamoto, S., Stroehlein, J. L. (1977). Relation of solute and soluble boron to the boron hazard of irrigation water. Plant and Soil, 47: 253–256.

Ryan, J., El-Fattal, L., Nuruddin, H. (1981a). Micronutrient Studies of Lebanese Soils Bulletin, No. 65, Beirut, Lebanon: American University of Beirut.

Ryan, J., Curtin, D., Safi, I. (1981b). Ammonia volatilization as influenced by calcium carbonate particle size and iron oxides. Soil Science Society of America Journal, 45: 338–341.

Ryan, J., Hariq, S. N. (1983). Transformation of micronutrient chelates in calcareous soil as influenced by selected soil properties, temperature, and wetting-drying during incubation. Soil Science Society of America Journal, 47: 806–810.

Ryan, J., Hamzé, M., Shwayri, R., Hariq, S. N. (1985a). Behaviour of iron-supplying materials in incubated calcareous soils. Journal of Plant Nutrition, 8 (5): 425–436.

Ryan, J., Curtin, D., Cheema, M. A. (1985b). Significance of iron oxides and calcium carbonate particle size in phosphorus sorption and desorption in calcareous soils. Soil Science Society of America Journal, 49: 74–76.

Ryan, J., Hassan, H., Bassiri, M., Tabbara, H. S. (1985c). Availability and transformation of applied phosphorus with time in calcareous Lebanese soils. Soil Science Society of America Journal, 49: 1215–1220.

Ryan, J., Ayubi, A. G., Barsumian, A. (1985d). Comparison of DTPA with individual micronutrient extractants in calcareous Lebanese soils. Lebanese Science Bulletin, 1(2): 163–179.

Ryan, J., Hamzé, M., Daroub, S., Hariq, S. N. (1986). Nutrient availability in incubated urea phosphate-treated calcareous soil. Journal of Fertilizer Issues, 3(4): 146–150.

Ryan, J., Hamzé, M., Harik, S. N., Darub, S. (1988). Influence of urea phosphate on iron-inefficient soybeans. Journal of Fertilizer Issues, 5(2): 56–60.

Ryan, J., Abdel Monem, M., El-Gharous, M. (1990). Soil fertility assessment at agricultural experiment stations in Chaouia, Abda, and Doukkala. Al-Awamia, 72: 1–47.

Ryan, J., Abdel Monem, M., Dafir, M., Mergoum, M., Sali, B. (1995). Response of local and improved corn varieties in Morocco to phosphorus and zinc. Al-Awamia, 90: 69–80.

Ryan, J., Garabet, S., Harmsen, K., Rashid, A. (1996a). A soil and plant analysis manual adapted for the West Asia-North Africa region (126 p.). ICARDA Technical Bulletin, Aleppo, Syria: International Center for Agricultural Research in the Dry Areas (ICARDA).

Ryan, J., Masri, S., Garabet, S. (1996b). Geographical distribution of soil test values in Syria and their relationship with crop response. Communications in Soil Science and Plant Analysis, 27(5–8): 1579–1593.

Ryan, J., Masri, S., Garabet, S., Diekmann, J., Habib, H. (1997). Soils of ICARDA's agricultural experimental stations and sites: Climate, classification, physical-chemical properties and land use (113 p.). ICARDA Technical Bulletin.

Ryan, J., Singh, M., Yau, S. K., Masri, S. (1998). Spatial variability of soluble boron in Syrian soils. Soil and Tillage Research, 45(3, 4): 407–417.

Ryan, J., Garabet, S., Rashid, A., El Gharous, M. (1999). Soil laboratory standardization in the Mediterranean region. Communications in Soil Science and Plant Analysis, 30: 885–894.

Ryan J., Estefan, G. Rashid, A. (2001). Soil and Plant Analysis Laboratory Manual. Aleppo, Syria: International Center for Agricultural Research in the Dry Areas (2001, in English, x + 172 pp.; 2002, in Russian, viii + 122 pp.; 2003, in Arabic, ix + 172 pp.).

Ryan, J., P. Smithson, B. Mandac, Uponi, J. (2002). Role of international research center soil laboratories in agricultural development. Communications in Soil Science and Plant Analysis, 33: 3213–3225.

Ryan, J., Masri, S., Qadir, M. (2006). Nutrient monitoring of sewage water irrigation: Impacts for soil quality and crop nutrition. Communications in Soil Science and Plant Analysis, 37: 2513–2521.

Saxena, M. C., Malhotra, R. S., Singh, K. B. (1990). Iron deficiency in chickpea in the Mediterranean region and its control through resistant genotypes and nutrient application. Plant and Soil, 123: 251–254.

Shorrocks, V. M. (1992). Boron – A global appraisal of the occurrence of boron deficiency. In S. Portch (Ed.), Proceedings International Symposium on Role of Sulphur, Magnesium, and Micronutrients in Balanced Plant Nutrition (pp. 39–53). Washington, DC: The Sulphur Institute.

Shorrocks, V. M. (1997). The occurrence and correction of boron deficiency. Plant and Soil, 193: 121–148.

Sillanpää, M. (1982). Micronutrients and nutrient status of soils: A global study. FAO Soil Bulletin No. 48. Rome, Italy: Food and Agriculture Organization.

Sillanpää, M. (1990). Micronutrient assessment at the country level: An international study. Soils Bulletin No. 63. Rome, Italy: Food and Agriculture Organization.

Sillanpää, M., Velk, P. L. G. (1985). Micronutrients and agroecology of tropical and Mediterranean regions. In P. L. G. Vlek (Ed.) (1985), Micronutrients in Tropical Food Crop Production. Dordrecht, The Netherlands: Martinus Nijhoff/Dr. W. Junk.

Srivastava, S. P., Bhandhari, T. M. S., Yadav, C. R., Joshi, M., Erskine, W. (2000). Boron deficiency in lentil: Yield loss and geographic distribution in a germplasm collection. Plant and Soil, 219: 147–151.

Soltanpour, P. N., Workman, S. (1979). Modification of the NH_4HCO_3-DTPA soil test to omit carbon black. Communications in Soil Science and Plant Analysis, 10: 1411–1420.

Stewart, W. M., Dibb, D. W., Johnston, A. E., Smyth, T. J. (2005). The contribution of commercial fertilizer nutrients to food production. Agronomy Journal, 97: 1–6.

Tandon, H. L. S. (1995). Micronutrients in Soils, Crops, and Fertilizers. New Delhi, India: Fertilizer Development and Consultation Organization.

Uygur, V. (1998). The role of iron oxides in Zn adsorption by calcareous soils (pp. 284–289). Yesilsoy International Symposium on Arid Region. Ankara, Turkey: Turkish Soil Science Society.

Vlek, P. L. G. (Ed.) (1985). Micronutrients in Tropical Food Crop Production. Dordrecht, The Netherlands: Martinus Nijhoff/Dr. W. Junk.

Wei, Y. Z., Bell, R. W., Yang, Y., Xue, J. M., Wang, K., Huang, L. (1998). Prognosis of boron deficiency in oilseed rape (*Brassica napus* L.) by plant analysis. Austrian Journal of Agricultural Research, 49: 867–874.

Welch, R. M., Allaway, W. H., House, W. A., Kubota, J. (1991). Geographic distribution of trace element problems. In J. J. Mortvedt, F. R Cox, L. M. Shuman R. M. Welch (Eds.), Micronutrients in Agriculture (2nd edn., pp. 31–58)., Madison, WI: Soil Science Society of America.

Yaalon, D. H. (1997). Soils in the Mediterranean region: What makes them different? Catena, 28: 157–169.
Yau, S. K. (1997). Differential responses of barley, durum and bread wheat to high levels of soil boron. In J. Ryan (Ed.), Accomplishments and Future Challenges in Dryland Soil Fertility Research in the Mediterranean Area (pp.209–216). Aleppo, Syria: International Center for Agricultural Research in the Dry Areas.
Yau, S. K. (2000). Soil-boron effects straw quality and other agronomic traits in two cultivars of barley. Communications in Soil Science and Plant Analysis, 31: 591–604.
Yau, S. K., Erskine, W. (2000). Diversity of boron-toxicity tolerance in lentil growth and yield. Plant and Soil, 47: 55–61.
Yau, S. K., Saxena, M. C. (1997). Variation in growth, development, and yield of durum wheat in response to high soil boron. I. average effects. Australian Journal of Agricultural Research, 48: 945–949.
Yau, S. K., Hamblin, J., Ryan, J. (1994). Phenotypic variation in boron toxicity tolerance in barley, durum and bread wheat. Rachis, 13, 20–25.
Yau, S. K., Nachit, M. M., Ryan, J., Hamblin, J. (1995). Phenotypic variation of boron toxicity tolerance in durum wheat at seedling stage. Euphytica, 83: 185–191.
Yau, S. K., Nachit, M., Ryan, J. (1997a). Variation in growth, development and yield of durum wheat in response to high soil boron. II. Differences between genotypes. Australian Journal of Agricultural Research, 48: 951–957.
Yau, S. K., Nachit M. M., Ryan, J. (1997b). Variation in boron-toxicity tolerance in a durum wheat core collection. In R. W. Bell, B. B Rerhasem (Eds.), Boron in Soils and Plants (pp. 117–120). Dordrecht, The Netherlands: Kluwer.
Yilmaz, A., Ekiz, H., Torum, B., Gultekin, I., Karanlik, S., Bagci, S. A., Cakmak, I. (1997 OR 1996). Effect of different zinc application methods on grain yield and zinc concentration in wheat cultivars grown on zinc-deficient calcareous soils. Journal of Plant Nutrition, 20: 461–471.
Zahid, M. A., Rashid, A., Din, J. (2000). Balanced nutrient management in chickpea. International Chickpea and Pigeonpea Newsletter (ICRISAT), 7: 24–26.

Chapter 7
Zinc Deficiency in Wheat in Turkey

Ismail Cakmak

Abstract Zinc (Zn) deficiency in soils and crop plants occurs nearly in all countries, particularly in cereal-growing areas. Turkey is among the countries with the most severe Zn-deficient soils. Previously, wheat grown in Central Anatolia was low yielding over many years, but the reason was unclear. In 1991/1992 a field experiment was conducted to determine whether micronutrients were a possible cause of the problem. Only with Zn application there was an impressive increase in growth and yield of wheat (T. aestivum L.). Based on these initial observations, a multi-institutional, long-term project on Zn deficiency in crop production in Central Anatolia was prepared and supported by the NATO "Science for Stability" Program. There were spectacular increases in grain yield of wheat with Zn fertilisation. In certain areas with very low yields (0.25 t ha^{-1}), Zn application enhanced grain yield by a factor of 6–8. The total amount of soil Zn was fairly high, between 40 and 80 mg kg^{-1}, but the level of plant available Zn was extremely low (DTPA-Zn: around 0.1 mg kg^{-1} soil). Field experiments at different locations revealed that in soils containing less than 0.4 mg kg^{-1} DTPA-extractable Zn, wheat, particularly durum wheat, responded significantly to Zn applications. Compared to irrigated conditions, wheat was more sensitive to Zn deficiency under rainfed conditions. Plants emerging from seeds with very high Zn concentrations (up to 55 mg kg^{-1} dry weight) had increased seedling vigour and pathogen resistance, as well as yield. In addition, enhancing Zn concentration in seeds reduced seeding rate, with consequent economic benefits. The impressive effects of Zn fertilisers on crop yield evoked considerable interest by fertiliser companies, which in 1995 started producing Zn-containing compound fertilisers. Today, 12 years after the Zn deficiency problem was diagnosed as a critical problem for wheat production in Turkey, the total amount of Zn-containing compound fertilisers applied in Turkey is at a record level of 300,000 t. Ministry of Agriculture estimates put annual economic benefits from Zn fertilisation at US$100 million. As Zn deficiency is an important micronutrient deficiency in humans in Turkey, increases in grain Zn concentration by Zn fertilisation have obvious implications for human health.

B.J. Alloway (ed.), *Micronutrient Deficiencies in Global Crop Production.*

7.1 Introduction

Zinc (Zn) deficiency is a particular micronutrient deficiency resulting in serious decreases in crop production and nutritional quality of edible parts of crop plants. It is believed that Zn deficiency is the most commonly occurring micronutrient deficiency on the world, especially in cereal growing areas (Graham and Welch, 1996; Cakmak, 2002; Alloway, 2004). Nearly, 50% of the cultivated area for cereal production globally is low in plant-available Zn in the soil. Many reports at country level indicate that Zn deficiency occurs in almost all countries and in different climatic conditions. Generally, soils affected by Zn deficiency are high in calcium carbonate ($CaCO_3$), pH and iron (Fe) and aluminium (Al) oxides and low in organic matter and moisture (Marschner, 1993). These soil factors reduce the solubility and mobility of Zn in soils by stimulating its adsorption to soil constituents and limiting its diffusion to plant roots. Occurrence of Zn deficiency in plants due to its absolute low level in soil is very rare and can be observed only in certain soil conditions (e.g., sandy soils). Countries with a very high distribution of Zn dcficiency problem soils include India (half of the cultivated land affected), China (one-third of soils affected), Western Australia (8 Mha of cultivated land affected) and Bangladesh (2 Mha of land affected) (White and Zasoski, 1999). There are an increasing number of examples showing that application of Zn fertilisers on such Zn-deficient soils greatly improves crop production as shown in field experiments in India (Takkar et al., 1989) and Australia (Graham et al., 1992).

Generally, the regions with the most severely Zn-deficient soils are also the regions where Zn deficiency in human beings is very common, especially in developing countries. According to a recent report, one-third of the world's population is at risk of Zn deficiency due to low concentrations and bioavailability of Zn in the diet (Hotz and Brown, 2004). It is generally believed that Zn deficiency is as widespread as Fe deficiency, which affects half of the world's population (Welch and Graham, 1999; Cakmak et al., 2002). When they occur, Zn and Fe deficiencies induce the development of several health complications in human beings, such as impairments in physical growth, immune system, mental and cognitive development, reproductive performance and increases in anaemia and maternal mortality (Black, 2003; Hotz and Brown, 2004; Welch and Graham, 2004). Extensive and monotonous consumption of cereal-based foods seems to be the major reason for the widespread occurrence of Zn deficiency in human beings. In a large number of countries, wheat is the predominant source for the daily calorie intake (Fig. 7.1) and wheat, like other cereal species, is very low in concentration and availability of Zn in grain. Consequently, any effort aiming at improving Zn concentration in grain can greatly contribute to human health and nutrition.

In Turkey, Zn deficiency is a common problem both in crop plants (Cakmak et al., 1996, 1999) and human beings (Cavdar et al., 1983). According to the survey reports at country level by Sillanpää (1982, 1990), Zn concentrations in Turkish soils are some of the lowest ever recorded. Based on the results obtained from 1,511 soil samples by using the soil-DTPA test (Lindsay and Norvell, 1978), 49.8% of the cultivated soils in Turkey were classified as Zn deficient (Eyupoglu et al., 1994).

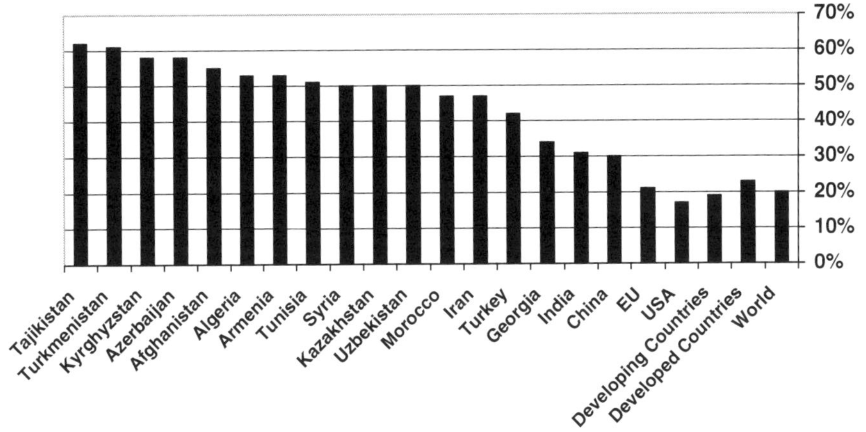

Fig. 7.1 Daily calorie intake from wheat in different countries and regions (FAO Database 2003; compiled by H.J. Braun, CIMMYT-Turkey)

The prevalence of Zn deficiency in Turkish soils is increased by increasing pH value and decreasing level of organic matter of soils collected from different regions. As reviewed by Kacar (1998), in several pot experiments under greenhouse conditions using soils from different parts of Turkey, it was shown that the soils are generally deficient in Zn and this is reflected in poor growth of plants under greenhouse conditions. Despite so many reports, the role of Zn application on the yield of crop plants under field conditions has very rarely been studied, especially in semiarid regions, e.g., the Central Anatolia region (Fig. 7.2). Central Anatolia is the major wheat growing area (4.5 Mha) where a wheat/fallow production system dominates and the soils are extremely poor in their level of soil moisture and organic matter and very high in pH and $CaCO_3$. The long-term average precipitation in the region varies between 300 and 400 mm.

Historically, wheat grown in several parts of Central Anatolia produced low yields and poor growth. Often plants showed necrotic and chlorotic patches on leaves and reduced shoot elongation, but the reason for such growth disorders was not clear. At that time, several biotic and abiotic stress factors (such as drought, heat, pathogens, and boron (B) toxicity, or micronutrient deficiencies) were considered and discussed as possible causes for the development of the leaf symptoms and poor growth. Before the 1990–1991 cropping season there was no scientific evidence that the poor growth and leaf symptoms observed in wheat under field conditions were caused by low concentrations of plant-available Zn in soil. To identify the underlying cause of the unknown disorders in wheat and to clarify its relationship to micronutrient nutrition of plants, a field experiment was conducted at the Transitional Zone Agricultural Research Institute in Eskisehir Province by agronomist M. Kalayci and his colleagues together with representatives of CIMMYT-Turkey. In parallel to this experiment, wheat germplasm from the University of Adelaide having genotypes with wide tolerance to different micronutrient deficiencies (developed by

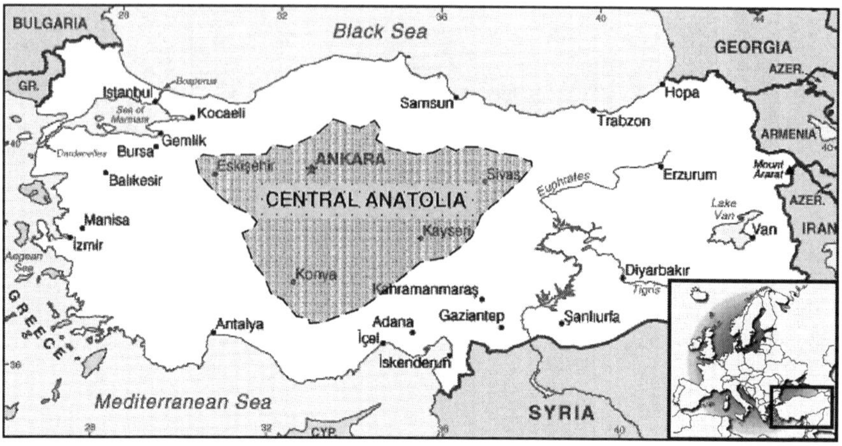

Fig. 7.2 Map showing the location of Central Anatolia in Turkey

R.D. Graham) was grown at the research farms of the Transitional Zone Agricultural Research Institute and the Bahri Dagdas International Agricultural Research Centre in Konya. These experiments showed that, of the micronutrients applied to wheat, only Zn was effective in increasing yield and, within the germplasm examined, the genotypes showing high sensitivity to Zn deficiency were particularly damaged.

Based on these initial observations a long-term and multi-institutional project was developed and submitted to the NATO-Science for Stability (NATO-SFS) Programme in 1992. The institutions involved were Cukurova University in Adana, the Transitional Zone Agricultural Research Institute in Eskisehir, the Bahri Dagdas International Agricultural Research Center in Konya and CIMMYT-Ankara. In 1992, the project was approved by the NATO-SFS program for the of study different aspects of Zn deficiency in wheat production in Turkey, especially in Central Anatolia. The results obtained from this NATO-SFS project and from the subsequent studies following the NATO project are described and discussed below. This chapter is based on a paper that was presented to the International Fertiliser Society at its Conference in Cambridge in December 2004.

7.2 Results and Discussion

7.2.1 Soil Properties and Zinc Status of Soils

In order to obtain a general picture of the chemical and physical properties of the soils, 76 topsoil samples (0–30 cm depth) were collected from different parts of Central Anatolia. All soils sampled had high pH values ranging between 7.5 and 8.1 with a mean value of 7.9. Almost all soil samples were found to be rich in

$CaCO_3$. More than 65% of the soil samples contained greater than 20% $CaCO_3$. The content of organic matter was very low and averaged 1.5%. There was a wide variation in soil texture; on average, soils were clay-loam textured.

In soils with high pH and $CaCO_3$ and very low contents of organic matter as described above, availability of Zn to plant roots is extremely low (Marschner, 1993). Among these factors, pH plays a decisive role in decreasing Zn availability to plant roots by stimulating its adsorption to soil constituents (e.g., clay minerals and Fe/Al oxides) (Barrow, 1993; Loosemore et al., 2004). A continuous supply of Zn to plant roots is critical for the maintenance of plant growth without Zn deficiency stress. Release (desorption) of Zn sorbed/fixed to soil constituents is, therefore, important for maintaining the Zn supply to plant roots. In order to collect information on the Zn desorption capacity of soils, several Zn desorption experiments were carried out under laboratory conditions. The studies showed that Zn added or present in soils is very strongly adsorbed and held on soil constituents. For example, with a Zn addition of 521 µg g^{-1} soil, Zn desorption with $CaCl_2$ was around 5 µg in severely Zn-deficient soils, while in soils without Zn deficiency stress, up to 100 µg Zn could be desorbed by $CaCl_2$ (Erenoglu, 1995). These results indicate that Zn-deficient soils in Central Anatolia have a very high Zn-adsorption capacity, and that the Zn adsorbed in soils is held with high bonding energy, and consequently, Zn is only poorly available to plants.

In close agreement with the results of the Zn-adsorption/desorption experiments, the concentration of plant-available (DTPA-extractable) Zn in soils was found to be very low, and ranged between 0.08 and 1.20 mg kg^{-1} with a mean value of 0.29 mg kg^{-1} (Table 7.1). With the exception of a few samples, almost all the samples contained less than 0.5 mg kg^{-1} soil DTPA-extractable Zn, which is widely considered to be the critical deficiency concentration of Zn for plants grown in alkaline soils (Lindsay and Norvell, 1978; Armour and Brennan, 2001). Half of the collected soil samples seemed to be extremely Zn deficient (DTPA-extractable Zn: <0.25 mg kg^{-1}). In contrast to extractable Zn, Zn-deficient soils were found to be fairly high in total Zn. The total amount of Zn in severely Zn-deficient soils varied from 39.6 to 62.4 mg kg^{-1}. Of the Zn fractions, the lowest amounts of Zn were found in exchangeable and organically bound Zn (Karanlik, 1995).

Table 7.1 Concentrations of Zn, Fe, Mn and Cu in 76 topsoils (0–30 cm) collected from different locations in Central Anatolia. Soil concentrations of micronutrients were measured by DTPA extraction

	Soil		Critical level	Samples
Element	Mean	Range	(mg kg^{-1} soil)	below critical level (%)
Zn	0.29	0.1–1.2	0.5	92
Fe	3.04	1.2–9.7	2.5	25
Mn	8.73	1.1–18.4	0.5	0
Cu	1.14	0.5–2.0	0.4	0

Contrary to Zn, concentrations of DTPA-extractable manganese (Mn) and copper (Cu) in Central Anatolia were in the adequate range, while in the case of Fe about 25% of the soils were potentially deficient (e.g., <2.5 mg kg^{-1} DTPA-extractable Fe).

7.2.2 *Response of Wheat to Zinc Fertiliser Application*

Application of Zn to wheat and also to most of other cereal species grown in Central Anatolia resulted in significant increases in growth and grain yield. In the first field experiment conducted in Eskisehir province in the 1991–1992 cropping season, the effect of soil application of different micronutrients was studied by using two bread wheat (*Triticum aestivum* L.), one durum wheat (*Triticum turgidum* L.) and one barley (*Hordeum vulgare* L.) cultivar. As presented in Table 7.2, only in the absence of Zn application was the grain yield depressed. On average, Zn application enhanced the grain yield of the wheat and barley cultivars by 58%. Also foliar application of Zn to plants rapidly improved growth and eliminated development of necrotic and chlorotic symptoms on leaves.

In further field experiments conducted at six different locations, application of Zn fertilisers eliminated leaf symptoms and resulted in very high increases in shoot growth and grain yield of wheat. There was a close relationship between the increases in grain yield by Zn application and the concentration of DTPA-extractable Zn in soils (Table 7.3). The highest increases in grain yield by Zn application (more than 100%) were found in locations where DTPA-extractable Zn was around 0.1 mg Zn kg^{-1} soil. The results in Table 7.3 indicate that wheat grown in Anatolia can positively respond to Zn application when soils contain less than 0.4 mg kg^{-1} DTPA-extractable Zn (Fig. 7.3 in Colour Section). Based on 20 field experiments on farmers fields in India, Dwivedi and Tiwari (1992) showed that Zn fertiliser application to wheat at the rate of 5 kg Zn ha^{-1} improved grain yield when the soil DTPA-extractable Zn concentration was below 0.6 mg kg^{-1} soil. In Canada, in

Table 7.2 Effect of soil applications of different micronutrients on grain of the bread wheat cultivars Bezostaja-1 and Gerek-79, durum wheat cultivar Kunduru-1149 and barley cultivar Tokak grown in a Zn-deficient soil under field conditions in Central Anatolia (Eskisehir Province). Micronutrients were applied at the rate of 5 kg ha^{-1} (Cakmak et al., 1999)

Application	Bezostaja-1	Gerek-79	Kunduru-1149	Tokak	Mean
Control (all micro nutrients)	4.64	4.76	2.65	4.73	4.19
–Fe	4.36	4.37	2.38	4.53	3.91
–Mn	4.54	4.71	1.93	5.38	4.14
–Cu	3.93	4.30	2.17	4.91	3.62
–B	4.20	4.37	2.63	4.81	4.00
–Zn	2.96	3.00	1.58	3.07	2.65

Table 7.3 Effect of soil Zn application (+Zn: 23 kg Zn ha^{-1}) on the concentration of Zn in flag leaves and grain yield of wheat (cv. Gerek-79) grown under field conditions in various locations having different levels of DTPA-extractable Zn in soils (Cakmak et al., 1996)

	DTPA-extractable Zn	Leaf Zn concentration		Grain yield		Yield increase by +Zn
		–Zn	+Zn	–Zn	+Zn	
Location	(mg kg^{-1} soil)	(mg kg^{-1} DM)		(t ha^{-1})		(%)
Konya (Centrum)	0.13	8 ± 1	13 ± 1	2.8 ± 0.7	5.9 ± 0.5	109
Konya (Comakli)	0.11	6 ± 1	16 ± 1	0.2 ± 0.1	1.4 ± 0.2	554
Eskisehir	0.15	7 ± 1	16 ± 2	2.5 ± 0.9	3.3 ± 0.2	31
Cesmelisebil	0.25	7 ± 1	15 ± 2	2.0 ± 0.5	2.3 ± 0.5	16
Gözlü	0.38	10 ± 2	18 ± 1	1.1 ± 0.1	1.5 ± 0.3	27
Cumra	0.64	10 ± 1	16 ± 1	5.4 ± 0.4	5.6 ± 1.1	5

Fig. 7.3 Growth of wheat plants on a Zn-deficient calcareous soil (DTPA-Zn 0.1 mg kg^{-1}) in Eskisehir Province, Central Anatolia showing details of plants with (+Zn) and without (–Zn) fertilisation with 50 kg $ZnSO_4$ ha^{-1} (H. Braun, CIMMYT) (*See Colour Plates*)

bean-growing areas, there were significant increases in seed yield in response to Zn application when the concentration of DTPA-extractable Zn in soil was equal or less than 0.5 mg Zn kg^{-1} (Goh and Karamanos, 2004).

In Central Anatolia, drought stress is the most important constraint to crop production. By affecting diffusion/mobility of Zn in soils, low soil moisture can also affect development of Zn deficiency stress in plants. To understand the relationship between Zn nutrition and drought stress, several field experiments have been carried out. In a field experiment with 25 wheat cultivars, it was shown that enhancements in grain yield brought about by irrigation were more significant when Zn was applied. Relative increases in average grain yield of 25 wheat cultivars by Zn fertiliser application were much higher under rainfed than irrigated conditions indicating greater susceptibility of genotypes to Zn deficiency under rainfed conditions. In accordance with these results, we often observed that the wheat cultivars developed/bred for irrigated conditions had greater susceptibility to Zn deficiency than the cultivars improved for rainfed conditions (Kalayci et al., 1999). Similar observations were also made in experiments examining the effect of Zn application on the grain yield of different cereal species with and without irrigation. With the exception of rye, Zn application enhanced grain yield of all cereal cultivars both under rainfed and irrigated conditions (Table 7.4). Interestingly, rye was not affected by Zn deficiency in soil; there was even a tendency for a decrease in yield with Zn application. Among the cereal species, yields of durum wheat cultivars were particularly depressed under Zn deficiency, especially under rainfed conditions. On average, increases in grain yield of cereal species by Zn application were

Table 7.4 Effects of Zn applications on grain yield of cereal species and cultivars grown in a Zn-deficient calcareous soil in Central Anatolia with and without irrigation. Since the differences between 7, 14, 21 kg ha^{-1} Zn applications were not different, average values are presented (Ekiz et al., 1998)

	Rainfed			Irrigated		
Species and cultivares	0 kg Zn ha^{-1} (kg ha^{-1})	Mean of 7,14, 21 kg Zn ha^{-1}	Increase in yield (%)	0 kg Zn ha^{-1} (kg ha^{-1})	Mean of 7,14, 21 kg Zn ha^{-1}	Increase in yield (%)
Bread wheat						
Gerek 79	3,100	4,460	44	4,070	5,250	29
Bezostaya-1	1,900	4,150	118	3,190	5,080	59
Durum wheat						
Kunduru-1149	330	2,470	648	1,550	3,580	130
Cakmak-79	170	760	347	630	2,870	355
Triticale						
Presto	3,380	4,280	26	3,870	5,290	37
BDMT-19	4,000	4,240	6	4,350	5,320	22
Barley						
Tokak-157/37	2,240	4,470	113	2,580	5,100	98
Erginel-91	1,920	4,360	127	2,650	5,820	120
Rye						
Aslim	3,560	3,340	−7	4,180	4,190	0
Oat						
Chekota	810	1,880	132	1,190	3,270	174

greater under rainfed (156%) than the irrigated conditions (102%). Based on the severity of leaf symptoms observed and the decreases in grain yield by Zn deficiency, tolerance to Zn deficiency of cereal species declined in the following order: durum wheat < oats < barley ≤ bread wheat < triticale < rye (Ekiz et al., 1998; Cakmak et al., 1997). This same order was also found in greenhouse experiments with pot experiments (Cakmak et al., 1998). Enhanced tolerance of triticale to Zn deficiency indicates that the genes affecting high Zn deficiency tolerance in rye are transferable to wheat. Triticale is a hybrid cross made between wheat and rye.

Despite dramatic differences in Zn deficiency tolerance between and among the cereal species, leaf or shoot Zn concentrations of plants were not related to their differential tolerance to Zn deficiency. For example, in 20 wheat cultivars grown in a Zn-deficient field, differences in Zn deficiency tolerance did not correlate with shoot or grain Zn concentration (Fig. 7.4). There was even a negative trend between these parameters. This indicates that differences in internal Zn use efficiency may play an important role in differential expression of Zn deficiency tolerance between cereal species or genotypes of a given species. It is also possible that the mechanisms affecting Zn deficiency tolerance and Zn concentrations in shoot (or grain) are not same and are affected by different genetic systems.

In Central Anatolia, the occurrence of spike (ear) sterility is a common problem in cereals, especially under rainfed conditions in combination with high day temperatures. Spike sterility becomes more pronounced when plants suffer from Zn deficiency. Therefore, under drought and heat stress conditions, improving the Zn nutritional status of plants by soil or foliar applications can be very helpful in the protection of plants from spike sterility. The reason why a better Zn nutritional status is important for protection of plants from drought and heat stress is not well understood. Possibly, Zn protects plants from damage by reactive oxygen species (ROS) under drought and heat stress. Drought and heat represent oxidative stress and cause oxidative damage to cell constituents by inducing production of ROS as found in wheat (Selote et al., 2004; Bartolli et al., 2004). In good agreement

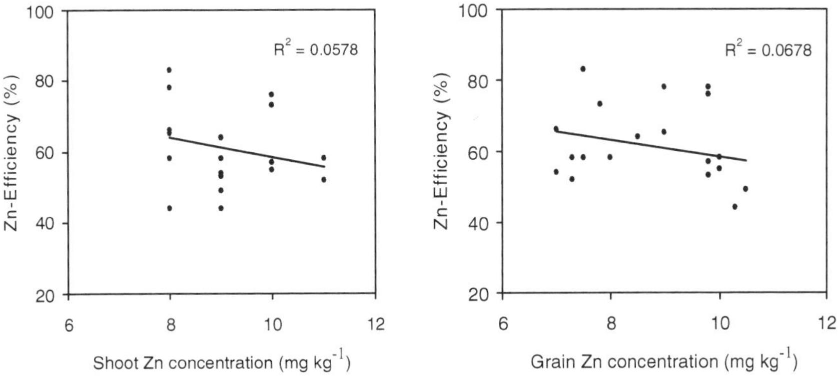

Fig. 7.4 Relationship between Zn efficiency and shoot and grain Zn concentrations of Zn-deficient 20 wheat cultivars grown in Central Anatolia. Zinc efficiency was calculated as grain yield at –Zn to grain yield at +Zn (for further details see Torun et al., 2001; Erdal et al., 2002)

with this, water stress-induced spikelet sterility in upland rice is closely associated with poor antioxidative defence systems against ROS (Selote and Khanna-Chopra, 2004). Since production of ROS is also very high in Zn-deficient plants (Cakmak and Marschner, 1988a; Pinton et al., 1994), ROS-induced cell damage (necrosis, chlorosis) under drought stress can be additionally increased when plants suffer from Zn deficiency. Adequate Zn supply interferes with the generation, and catalyses the detoxification, of the superoxide radical by Zn-containing superoxide dismutase (Cakmak, 2000). As Zn deficiency limits photosynthetic carbon metabolism at different levels, the well-documented photogeneration of ROS during photosynthesis is intensified when plants are exposed simultaneously to Zn deficiency and excessive light (Marschner and Cakmak, 1989; Cakmak et al., 1995). An adequate Zn supply can, therefore, protect plants from photooxidative damage under high light and heat/drought stress which are very common under field conditions in Central Anatolia.

7.2.3 Zinc Application Methods for Correcting Zinc Deficiency

Several application methods are used for the correction of Zn deficiency. The most frequent method is the broadcast application of Zn onto soil. Depending on DTPA-extractable Zn in soil, Zn deficiency can be corrected by broadcast applications of Zn between 4.5 and 34 kg Zn ha^{-1} in the form of $ZnSO_4$ (Martens and Westermann, 1991). Zinc can also be applied to plants as a foliar spray with application rates of 0.5–1.0 kg Zn ha^{-1} as $ZnSO_4$.

In Central Anatolia, the effect of different Zn application methods has been tested on shoot biomass production and grain yield of durum and bread wheat cultivars (Yilmaz et al., 1997). The Zn application methods tested were:

1. Control (no Zn application)
2. 23 kg Zn ha^{-1} as broadcast to soil
3. Seed coating (1 L 30% $ZnSO_4$ sprayed on to 10 kg seeds and then the seeds dried and sown)
4. Foliar application (2 × 220 g Zn ha^{-1} as $ZnSO_4$ in 450 L water at tillering and stem elongation)
5. Combination of the methods (2) and (4)
6. Combination of the methods (3) and (4)

Irrespective of the methods used, application of Zn very significantly enhanced grain yield and shoot biomass. The highest increases in grain yield were found with soil, soil + leaf, or seed + leaf application methods (Fig. 7.5). The increases in biomass production by Zn were much lower than the increases in grain yield (Fig. 7.3) indicating that Zn is of critical importance for generative growth (seed setting) rather than the vegetative growth. Zinc could be important for anther and pollen grain development, possibly by improving the phytohormone status of plants and protein synthesis (Brown et al., 1993). As Zn applied to soil has a significant

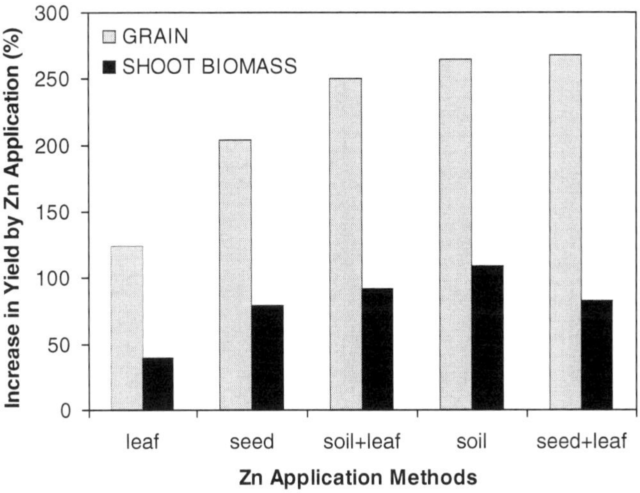

Fig. 7.5 Increases in grain and whole shoot biomass production by different Zn application methods. The results represent the average values of one durum and three bread wheat cultivars grown in Central Anatolia on a severely Zn-deficient calcareous soil. The yield values for control (without Zn) plants: 558 kg ha^{-1} for grain and 3,114 kg ha^{-1} for shoot biomass (Yilmaz et al., 1997)

residual effect for a couple of years, the soil application method alone has been considered as the most effective and long-term economical method for wheat.

With the exception of the seed application method, all Zn application methods clearly enhanced grain Zn concentration (Table 7.6). The most effective method for increasing Zn in shoot and seed was the soil + leaf application method that resulted in 3.5-fold increase in the grain Zn concentration and 7-fold increase in the shoot Zn concentration. Elevated levels of Zn in grain can be very beneficial for human nutrition and health due to the very widespread occurrence of Zn deficiency in humans (see below). When high grain Zn concentration is required besides improving grain yield, combined soil and foliar Zn application should be recommended.

7.2.4 The Role of High Seed Zinc Contents on Wheat Production

As shown in Fig. 7.5, Zn-coating of seed significantly enhanced shoot biomass production and grain yield under Zn-deficient conditions. The increases in grain yield by seed Zn-coating were comparable with the increases in yield caused by soil Zn application. Despite such marked effects of seed-Zn coating on shoot growth and grain yield, Zn concentrations in plant leaves at the shooting stage and in grain

Table 7.5 Effect of different Zn-application methods on Zn concentration in whole shoots and grain, and the increases in shoot biomass and grain yield by Zn applications. Shoots were sampled at the beginning of stem elongation. The values represent the average values of one durum and three bread cultivars grown on a Zn-deficient soil in Central Anatolia (Yilmaz et al., 1997). For Zn-application methods see text

	Zn concentration		Increases in yield by Zn application	
Zn application methods	Whole shoot	Grain (mg kg^{-1})	Whole biomass	Grain (%)
Control	10	10	–	–
Soil	19	18	109	265
Seed	12	10	79	204
Leaf	60	27	40	124
Soil + leaf	69	35	92	250
Seed + leaf	73	29	83	268

at maturity were not affected (Table 7.5). Average concentrations of Zn in shoots and grain of plants derived from seed-coated Zn were same for grain (e.g., 10 mg kg^{-1}) or only slightly higher for shoot (10 versus 12 mg kg^{-1}) of the control plants without Zn treatment (Table 7.5). In the case of soil application, Zn concentrations in plants were increased by a factor of 2. It can be argued that in plants derived from the Zn-coated seeds, Zn could be diluted as a consequence of enhanced shoot and grain production. Increases in biomass production and grain yield by soil Zn application were much greater than the increases caused by seed Zn-coating. But, in the case of soil Zn application, apparently, there was no substantial dilution of Zn as supposed in the case seed Zn-coating. It seems very likely that a high seed Zn concentration affects very positively seed germination and early seedling development by improving the tolerance of seedlings to different biotic and abiotic stress factors (see below). In order to understand more clearly the role of high seed-Zn in wheat production, field experiments were conducted by using seeds differing in Zn content. Seeds with varying Zn contents were obtained through different numbers of foliar applications of $ZnSO_4.7H_2O$ in the previous crop year. Three groups of seed with different Zn contents were obtained and used in field experiments, i.e., seeds containing 355, 800 and 1,465 ng Zn $seed^{-1}$ which were more or less equivalent to Zn concentrations of 8, 19 and 32 mg kg^{-1} seed, respectively.

As expected, increasing seed Zn content was highly effective in improving shoot growth and grain yield under Zn deficient conditions in Central Anatolia. Seedlings derived from low Zn-seeds were very poorly developed, while in the case of high seed-Zn, there was a rapid and uniform emergence and a higher number of tillers. These positive effects of high seed-Zn on seedling vigour and field establishment were reflected in grain yield and shoot dry matter production under Zn-deficient conditions (Table 7.6). Even, with soil Zn application, high seed-Zn tended to increase grain yield. The relative increases in growth and yield caused by high seed-Zn were more pronounced under rainfed than the irrigated conditions (Table 7.6). As found with Zn-coated seeds (Fig. 7.5), despite dramatic effects of high seed-Zn on early seedling growth, shoot dry matter production and grain yield, and

Table 7.6 Effects of seed-Zn content on grain yield of the bread wheat cv. Atay grown under rainfed and irrigated conditions in a Zn-deficient calcareous soil with (+Zn = 23 kg Zn ha^{-1}) and without Zn (–Zn) fertilizer application in Central Anatolia (Yilmaz et al., 1998)

Seed-Zn content	Rainfed		Irrigated	
(ng Zn seed–1)	–Zn	+Zn	–Zn	+Zn
(kg ha^{-1})				
355	480	2,720	5,700	7,170
800	920	3,170	5,930	7,800
1,465	1,040	2,840	6,190	7,450
Mean	810	2,810	5,940	7,470

Table 7.7 Effect of increasing seed-Zn content on Zn concentrations of shoots at shooting stage and grains at harvest of the bread wheat cv. Atay grown under rainfed conditions in a Zn-deficient calcareous soil (Yilmaz et al., 1998)

	Soil Zn applications			
Seed-Zn	Shooting stage		Mature grains	
content	0 kg Zn ha^{-1}	23 kg Zn ha^{-1}	0 kg Zn ha^{-1}	23 kg Zn ha^{-1}
(ng Zn $seed^{-1}$)	(mg Zn kg^{-1} DW)			
355	7.6	16.2	9.8	13.2
800	7.8	16.1	10.1	13.0
1,465	7.2	16.2	9.5	13.3
MEAN	7.5	16.2	9.8	13.2

Zn concentrations in shoot and grain were not changed by increasing seed Zn content (Table 7.7). High seed Zn possibly provides several advantages for better growth and greater tolerance to adverse soil and environmental conditions (Fig. 7.6). Similar results were obtained by Rengel and Graham (1995) in pot experiments. They found marked increases in root and shoot growth in plants derived from seeds with high seed Zn content. According to Rengel and Graham (1995), high seed Zn acts as a "starter Zn fertiliser" and should be considered when genotypes are tested for their ability to tolerate Zn deficiency in soils.

The greater effects of high seed Zn on growth under rainfed than irrigated conditions may indicate that it protects seedlings from drought stress. Susceptibility of plants to Zn deficiency is much higher when grown under drought conditions, or positive effects of irrigation on grain yield is maximised when Zn is adequately supplied (Tables 7.4 and 7.6). Zinc is possibly an important factor contributing to the tolerance of plants to different stress conditions during germination and early seedling growth stages. Zinc has been shown to affect both synthesis and activity of antioxidative defence systems and reduces generation of ROS by the plasma-membrane bound O_2-generating NADPH oxidase (see for more detail see Cakmak, 2000). There is also increasing evidence that Zn has protective effects against root-rotting pathogens, *Fusarium* crown-rot (*F. graminearum*), nematode penetration and take-all (*Gaeumannomyces graminis*) infection (Brennan, 1992;

Fig. 7.6 Benefits of high seed-Zn on plant crop production and human health

Rengel et al., 1996; Streeter et al., 2001; Siddiqui et al., 2002; Kalim et al., 2003). One major reason for the protective role of Zn against diseases and pests could be related to its role in maintaining structural integrity and controlling the permeability of cell membranes. Under Zn deficiency, root exudation of several organic compounds such as amino acids and carbohydrates significantly increases (Table 7.8). When the concentration of carbon-containing compounds is elevated in the rhizosphere of Zn-deficient plants, susceptibility of plants to different root diseases such as *Fusarium graminearum* (Sparrow and Graham, 1988), pests (Siddiqui et al., 2002); *Gaumannomyces graminis* (Brennan, 1992) and *Rhizoctonia solani* (Thongbai et al., 1993) can be enhanced. Protective effects of high Zn against root-rot and nematode invasion became more distinct when Zn was applied in combination with disease-suppressive strains of *Pseudomonas* (Duffy and Defago, 1997, 1999; Shaukat and Siddiqui, 2003). It is proposed that Zn possibly stimulates production of antimicrobial compounds (such as 2, 4- diacethylphloroglucinol and pyoluteorin) by disease-suppressive strains of *Pseudomonas fluorescens*.

Sowing seeds with elevated contents of Zn can be a practical solution to the Zn deficiency problem, and can contribute to reducing very high seeding rates in Central Anatolia (Fig. 7.6). The seeding rate is much higher in Central Anatolia (up to 250–300 kg ha^{-1} under farmer conditions) than the rate applied in other countries with similar climatic conditions. Due to the prevalence of Zn deficiency in soils in Central Anatolia, seed Zn concentrations are very low (e.g., generally below 15–20 mg Zn kg^{-1} seed), and plants derived from such seeds can be susceptible to adverse soil conditions and soil-born pathogens at an early growth stage leading to poor seedling performance and reduced final yield. This might be one major reason why in Central Anatolia a high seeding rate is considered essential in wheat production. As shown under boron (B) deficiency (Bell et al., 1989) and Mn deficiency (Longnecker et al., 1996) seeds obtained from micronutrient-deficient plants have very low vigour and viability causing poor seedling establishment and high susceptibility to diseases. These results indicate that adequate and continuous transport of Zn into seeds during the reproductive growth stage should be ensured to achieve an optimum seed viability and seedling establishment. Under potentially Zn-deficient conditions, late foliar application of Zn is, therefore, highly recommended for increasing seed Zn and thereby for improving stress tolerance of plants at an early growth stage. Efforts aimed at increasing seed Zn concentrations can contribute to reducing the seeding rate with considerable economic savings in Central Anatolia. According to Braun (1999), by using Zn-enriched seeds, seeding rates in Central Anatolia could be reduced up to 40–50%, resulting in economic benefits of around US$100 million year^{-1}.

7.2.5 Development of Zinc-Containing Compound Fertilisers

The spectacular effects of Zn fertiliser application on wheat production at different locations in Central Anatolia evoked a growing interest in the project results from the farmers and the fertiliser industries. The TOROS Fertilizer and Chemical Industry was the first fertiliser producer to respond to the project results and produce NP and NPK fertilisers containing 1% Zn. To test the reaction of farmers, the TOROS Company produced nearly 4,000 t of Zn-containing compound fertilisers in 1994 and distributed it to selected farmers in Central Anatolia without charge. The reaction of the farmers was so positive that, in the following years, increasing amounts of Zn-containing fertilisers were produced and applied in Central Anatolia and also in other regions of Turkey. Besides the TOROS company, other fertiliser companies in Turkey (e.g., Gubretas, Bagfas, Ege Gubre and Türksas) also started to produce different Zn-containing fertilisers. Additionally, several fertiliser distributors imported new Zn-containing foliar fertilisers from various countries to supply plants with Zn by foliar application of Zn. In 1997, the Turkish Government decided to consider making Zn-containing NP and NPK fertilisers eligible for the state subsidy. This was the first state subsidy given for a micronutrient-containing fertiliser so far in Turkey. Today, more than 10 years after the Zn deficiency problem was first diagnosed as critical for wheat production in Central Anatolia, the total amount of Zn-containing compound fertilisers applied in Turkey is at a record level of approximately 300,000 t (Fig. 7.7), representing an annual turnover of about US$70,000. The fertilisers amended with Zn and their production amounts in 2003 were as follows:

20-20-0-Zn (210,000 t)
15-15-15-Zn (42,000 t)
10-25-20-Zn (29,000 t)
10-15-25-Zn (18,000 t)

Besides these compound fertilisers, increased amounts of $ZnSO_4$, ZnO and Zn-containing foliar fertilisers are also being applied, but today there are no available statistics about the amount of these Zn fertilisers applied in Turkey. The economic benefits associated with application of Zn fertilisers are very impressive. The Ministry of Agriculture estimates put annual economic benefits from the Zn fertiliser application at US$100 M.

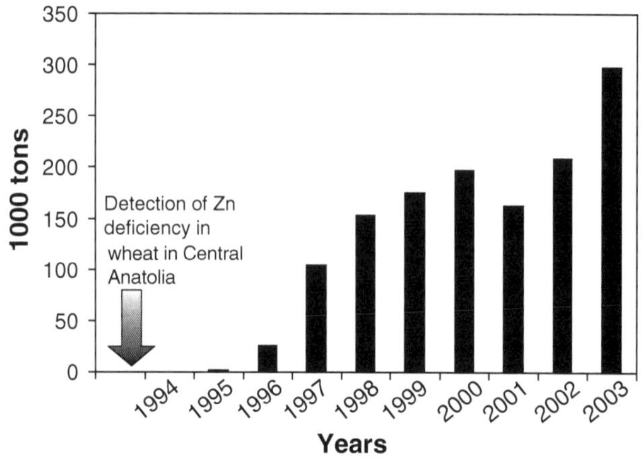

Fig. 7.7 Consumption of Zn-containing fertilisers in Turkey (Turkish Fertilizer Producer Association, 2001; Turkish Ministry of Agriculture, 2004)

7.2.6 Human Nutritional Aspects

Besides the impressive improvements in cereal productivity and increases in farmers' profits, use of Zn fertilisers should also improve the health and productivity of large numbers of Turkish people that consume little Zn from their diets because supplying more Zn to cereal crops results in Zn-enriched grain (Fig. 7.4). As demonstrated in this project, applying Zn to wheat can increase grain Zn concentration by nearly twofold. In addition, an adequate Zn supply can also reduce the concentration of P and phytic acid in grains. By binding Zn, phytic acid hampers the biological utilization of Zn and is an important inhibitor in Zn absorption in the body (Lonnerdal, 2000). It is a well-documented phenomenon that Zn deficiency is associated with enhanced uptake and accumulation of P in plants (Loneragan et al., 1982; Cakmak and Marschner, 1986). In wheat grown under Zn-deficient conditions in Central Anatolia, improving the Zn nutritional status of wheat plants was shown to reduce shoot and grain P concentrations in plants and this reduction was accompanied with corresponding decreases in phytic acid concentration in grain (Erdal et al., 2002; Torun et al., 2001). It appears that improving the Zn nutritional status of plants under Zn-deficient conditions can alleviate the risk of the occurrence of phytic acid-induced Zn deficiency in the body.

In Turkey, wheat is still the major source of daily calorie intake: on average, wheat alone provides nearly 45% of the daily calorie intake at country level (Fig. 7.1). It is estimated that this proportion could be more than 75% in rural regions. Providing Zn-dense grain to the poor in Turkey should lead to improvements in their health and productivity. This NATO-SFS project was one of the first in the world to use an agricultural intervention to address problems in relation to public

Table 7.8 Effect of zinc nutritional status of wheat plants on root exudation of amino acids, sugars, and phenolics. Wheat plants were grown in nutrient solution for 19 days (Cakmak and Marschner, 1988b)

Zn supply	Amino acids	Sugars ($\mu g\ g^{-1}$ root dry weight $6h^{-1}$)	Phenolics
–Zn (deficient)	48 ± 3	615 ± 61	80 ± 6
+Zn (sufficient)	21 ± 2	315 ± 72	34 ± 6

health and crop production, and its success should be a model for other nations to follow that face Zn deficiencies within their populations and in crop production as well. As shown by the success of this project, improving human health through agricultural interventions should become a primary tool in the fight against micronutrient malnutrition worldwide because it is sustainable and reaches all the people at risk of developing micronutrient deficiencies in target populations.

Acknowledgements The author is grateful to Ernest Kirkby, University of Leeds, for valuable suggestions and correction of the original English text. This work was supported by the NATO's Scientific Affairs Division in the framework of the Science for Stability Programme, TUBITAK (the Scientific and Technical Research Council), Cukurova University and Ministry of Agriculture and Rural Affairs of Turkey. The key scientists in this project involved in establishment and organization of the field experiments in Central Anatolia were Müfit Kalayci, Mehmet Aydin and Fahri Altay from the Transitional Zone Agricultural Research Institute in Eskisehir; Hasan Ekiz, Ahmet Yilmaz and Engin Kinaci from the Bahri Dagdas International Agricultural Research Center in Konya and Hans J. Braun from CIMMYT-Ankara. The author also acknowledges permission from the International Fertiliser Society to use the text of his paper for the 2004 conference in this chapter.

References

Alloway, B.J. (2004) Zinc in soils and crop nutrition. International Zinc Association Communications. IZA, Brussels, Belgium.

Armour, J.D., Brennan, R.F. (2001) Zinc. In Peverill, K.I., Sparrow, L.A., Reuter, D.J. (Eds.) Soil Analysis: An Interpretation Manual. CSIRO, Melbourne, pp. 281–285.

Barrow, N.J. (1993) Mechanisms of reaction of zinc with soil and soil components. In Robson, A.D. (Ed.) Zinc in Soils and Plants. Kluwer, Dordrecht, The Netherlands, pp. 15–31.

Bartolli, C.G., Gomez, F., Martinez, D.E., Guiamet, J.J. (2004) Mitochondria are the main target for oxidative damage in leaves of wheat (*Triticum aestivum* L.). J. Exp. Bot., 55: 1663–1669.

Bell, R.W., McLay, L., Plaskett, D., Dell, B., Loneragan, J.F. (1989) Germination and vigor of black gram (*Vigna mungo* (L.) Hepper) seed from plants grown with and without boron. Aust. J. Agric. Res., 40: 273–279.

Black, M.M. (2003) Micronutrient deficiencies and cognitive functioning. J. Nutr., 133: 3927S–3931S.

Braun, H.J. (1999) Prospects of Turkey's Wheat Industry, Breeding and Biotechnology. In Ekiz, H (ed.) Hububat Sempozym, International Winter Wheat Research Centre-Konya, pp1 –744.

Brennan, R.F. (1992) The effect of zinc fertilizer on take-all and the grain yield of wheat grown on zinc-deficient soils of the Esperance region, Western Australia. Fert. Res., 31: 215–219.

Brown, P.H., Cakmak, I., Zhang, Q. (1993) Form and function of zinc in plants. In Robson, A.D. (Ed.) Zinc in Soils and Plants. Kluwer, Dordrecht, The Netherlands, pp. 93–106.
Cakmak, I. (2000) Role of zinc in protecting plant cells from reactive oxygen species. New Phytol., 146: 185–205.
Cakmak, I. (2002) Plant nutrition research: Priorities to meet human needs for food in sustainable ways. Plant Soil, 247: 3–24.
Cakmak, I., Marschner, H. (1986) Mechanism of phosphorus induced zinc deficiency in cotton. I. Zinc deficiency-enhanced uptake rate of phosphorus. Physiol. Plant, 68: 483–490.
Cakmak, I., Marschner, H. (1988a) Enhanced superoxide radical production in roots of zinc deficient plants. J. Exp. Bot., 39: 1449–1460.
Cakmak, I., Marschner, H. (1998b) Increase in membrane permeability and exudation in roots of zinc deficient plants. J. Plant Physiol., 1988b, 132: 356–361.
Cakmak, I., Atli, M., Kaya, R., Evliya, H., Marschner, H. (1995) Association of high light and zinc deficiency in cold induced leaf chlorosis in grapefruit and mandarin trees. J. Plant Physiol., 146: 355–360.
Cakmak, I., Yilmaz, A., Ekiz, H., Torun, B., Erenoglu, B., Braun, H.J. (1996) Zinc Deficiency as a critical nutritional problem in wheat production in Central Anatolia. Plant Soil, 180: 165–172.
Cakmak, I., Ekiz, H., Yilmaz, A., Torun, B., Köleli, N., Gültekin, I., Alkan, A., Eker, S. (1997) Differential response of rye, triticale, bread wheat and durum wheats to zinc deficiency in calcareous soils. Plant Soil, 188: 1–10.
Cakmak, I., Torun, B., Erenoglu, B., Oztürk, L., Marschner, H., Kalayci, M., Ekiz, H., Yilmaz, A. (1998) Morphological and physiological differences in cereals in response to zinc deficiency. Euphytica, 100: 349–357.
Cakmak, I., Kalayci, M., Ekiz, H., Braun, H.J., Yilmaz, A. (1999) Zinc deficiency as an actual problem in plant and human nutrition in Turkey: A NATO-Science for Stability Project. Field Crops Res., 60: 175–188.
Cakmak, I., Graham, R., Welch, R.M. (2002) Agricultural and molecular genetic approaches to improving nutrition and preventing micronutrient malnutrition globally. In Cakmak, I. Welch, R.M. (Eds.) Encyclopaedia of Life Support Systems. UNESCO-EOLSS. UK, ISBN: 0 9542989–0-X, 3490 pp.
Cavdar, A.O., Arcasoy, A., Cin, S., Babacan, S., Gözdasoglu, S. (1983) Geophagia in Turkey: Iron and zinc deficiency, iron and zinc absorption studies and response to treatments with in geophagia cases. In Prasad, A.S., Cavdar, A.O., Brewer, G. J. (Eds.) Zinc Deficiency in Human Subjects. Alan R. Liss, New York, pp. 71–79.
Duffy, B.K., Defago, G. (1997) Zinc improves biocontrol of Fusarium crown and root rot of tomato by Pseudomonas fluorescens and represses the production of pathogen metabolites inhibitory to bacterial antibiotic biosynthesis. Phytopath., 87: 1250–1257.
Duffy, B.K., Defago, G. (1999) Macro- and microelement fertilizers influence the severity of Fusarium crown and root rot of tomato in a soilless production system. HortScience, 34: 287–291.
Dwivedi, B.S., Tiwari, K.N. (1992) Effect of native and fertilizer zinc on dry matter yield and zinc uptake by wheat (Triticum aestivum) in Udic Ustochrepts. Trop. Agric., 69: 357–361.
Ekiz, H., Bagci, S.A., Kiral, A.S., Eker, S., Gultekin, I., Alkan, A., Cakmak, I. (1998) Effects of zinc fertilization and irrigation on grain yield and zinc concentration of various cereals grown in zinc-deficient calcareous soil. J. Plant Nutr., 21: 2245–2256.
Erdal, I., Yilmaz, A., Taban, S., Eker, S., Cakmak, I. (2002) Phytic acid and phosphorus concentrations in seeds of wheat cultivars grown with and without zinc fertilization. J. Plant Nutr., 25: 113–127.
Erenoglu, B. (1995) Zinc Adsorption and Desorption Characteristics in Selected Soils from Central Anatolia, GAP and Cukurova Regions (in Turkish). M.Sc. thesis, Cukurova University, Adana, Turkey.
Eyupoglu, F., Kurucu, N., Sanisa, U. (1994) Status of plant available micronutrients in Turkish soils (in Turkish). Annual Report, Report No: R-118. Soil and Fertilizer Research Institute, Ankara, 25–32.

Goh, T.B. and Karamanos, R.E. (2004) Zinc responses of dry beans in Manitoba. Can. J. Soil Sci., 84: 213–216.

Graham, R.D., Ascher, J.S., Hynes, S.C. (1992) Selection of zinc-efficient cereal genotypes for soils of low zinc status. Plant Soil, 146: 241–250.

Graham, R.D., Welch, R.M. (1996) Breeding for staple-food crops with high micronutrient density: Working Papers on Agricultural Strategies for Micronutrients, No.3. International Food Policy Institute, Washington DC.

Hotz, C., Brown, K.H. (2004) Assessment of the risk of zinc deficiency in populations and options for its control. Food Nutr. Bull., 25: 94–204.

Kacar, B. (1998) Zinc dynamics in Soils (in Turkish). In First National Zinc Congress. 12–16 May, 1997, Eskisehir, Kemal Matbaasi, 47–60.

Kalayci, M., Torun, B., Eker, S., Aydin, M., Oztürk, L., Cakmak, I. (1999) Grain yield, zinc efficiency and zinc concentration of wheat cultivars grown in a zinc-deficient calcareous soil in field and greenhouse. Field Crops Res., 63: 87–98.

Kalim, S., Luthra, Y.P., Gandhi, S.K. (2003) Role in zinc and manganese in resistance of cowpea root rot. J. Plant Dis. Protect., 110: 235–243.

Karanlik, S. (1995) Determination of Total and Plant Available Zinc Concentrations in Selected Soils Collected from Central Anatolia, Cukurova and GAP Regions. M.Sc. thesis, Cukurova University.

Lindsay, W.L., Norvell, W.A. (1978) Development of a DTPA soil test for zinc, iron, manganese and copper. Soil Sci. Soc. Amer. J., 42: 421–428.

Loneragan, J.F., Grunes, D.L., Welch, R.M., Aduayi, E.A., Tengah, A., Lazar, V.A., Cary, E.E. (1982) Phosphorus accumulation and toxicity in leaves in relation to zinc supply. Soil Sci. Soc. Amer. J., 46: 345–351.

Longnecker, N., Crosbie, J., Davies, F., Robson, A. (1996) Low seed manganese concentration and decreased emergence of *Lupinus angustifolius*. Crop Sci., 36: 355–361.

Lonnerdal, B. (2000) Dietary factors influencing zinc absorption, J. Nutr. 130:1378–1383.

Loosemore, N., Straczek, A., Hinsinger, P., Jaillard, B. (2004) Zinc mobilization from a contaminated soil by three genotypes of tobacco as affected by soil and rhizosphere pH. Plant Soil, 260: 19–32.

Marschner, H. (1993) Zinc uptake from soils. In Robson, A.D. (Ed.) Zinc in Soils and Plants, Kluwer, Dordrecht, The Netherlands, pp. 59–77.

Marschner, H., Cakmak, I. (1989) High light intensity enhances chlorosis and necrosis in leaves of zinc potassium, and magnesium deficient bean (*Phaseolus vulgaris* L.) plant. J. Plant Physiol., 134: 308–315.

Martens, D.C., Westermann, D.T. (1991) Fertilizer applications for correcting micronutrients in Agriculture. In Mordvedt, J.J., Cox, F.R., Shuman, L.M., Welch, R.M. (Eds.) Micronutrients in Agriculture, 2nd edn. Soil Science Society of America, Madison, WI, pp. 89–112.

Pinton, R., Cakmak, I., Marschner, H. (1994) Zinc deficiency enhanced NAD(P)H-dependent superoxide radical production in plasma membrane vesicles isolated from roots of bean plants. J. Exp. Bot., 45: 45–50.

Rengel, Z., Graham, R.D. (1995) Importance of seed zinc content for wheat growth on zinc-deficient soil. I. Vegetative Growth. Plant Soil, 173: 259–266.

Rengel, Z., Gutteridge, R., Hirsch, P., Hornby, D. (1996) Plant genotype, micronutrient fertilization and take-all infection influence bacterial populations in the rhizosphere of wheat. Plant Soil, 183: 269–277.

Selote, D.S., Khanna-Chopra, R. (2004) Drought-induced spikelet sterility is associated with an inefficient antioxidant defence in rice panicles. Physiol. Plant., 121: 462–471.

Selote, D.S., Bharti, S., Khanna-Chopra, R. (2004) Drought acclimation reduces O_{2-1} accumulation and lipid peroxidation in wheat seedlings. Biochem. Biophys. Res. Commun., 314: 724–729.

Shaukat, S.S., Siddiqui, I.A. (2003) The influence of mineral and carbon sources on biological control of charcoal rot fungus, *Macrophomina phaseolina* by fluorescent pseudomonads in tomato. Letters App. Microbiol., 36: 392–398.

Siddiqui, I.A., Shaukat, S.S., Hamid, M. (2002) Role of zinc in rhizobacteria-mediated suppression of root-infecting fungi and root-knot nematode. J. Phytopathol., 150: 569–575.

Sillanpää, M. (1982) Micronutrients and the nutrient status of soils: A global study. FAO Soils Bulletin 48. Food and Agriculture Organization of the United Nations, Rome, 75–82.

Sillanpää, M. (1990) Micronutrient assessment at the country level: An international study. FAO Soils Bulletin 63. Food and Agriculture Organization, Rome.

Sparrow, D.H., Graham, R.D. (1988) Susceptibility of zinc-deficient wheat plants to colonization by *Fusarium graminearum* Schw. Group 1. Plant Soil, 112: 261–266.

Streeter, T.C., Rengel, Z., Neate, S.M., Graham, R.D. (2001) Zinc fertilisation increases tolerance to *Rhizoctonia solani* (AG 8) in *Medicago truncatula*. Plant Soil, 228: 233–242.

Takkar, P.N., Chibba, I.M., Mehta, S.K. (1989) Twenty Years of Coordinated Research of Micronutrients in Soil and Plants (1967–1987). Indian Institute of Soil Science, Bhopal, IISS, Bull.

Thongbai, P., Hannam, R.J., Graham, R.D., Webb, M.J. (1993) Interaction between zinc nutritional status of cereals and *Rhizoctonia* root rot severity. Plant Soil, 153: 207–214.

Torun, A., Gültekin, I., Kalayci, M., Yilmaz, A., Eker, S., Cakmak, I. (2001) Effects of zinc fertilization on grain yield and shoot concentrations of zinc, boron and phosphorus of 25 wheat cultivars on a zinc-deficient and boron-toxic soil. J. Plant Nutr., 24: 1817–1829.

Yilmaz, A., Ekiz, H., Torun, B., Gültekin, I., Karanlik, S., Bagci, S.A., Cakmak, I. (1997) Effect of different zinc application methods on grain yield and zinc concentration in wheat grown on zinc-deficient calcareous soils in Central Anatolia. J. Plant Nutr., 20: 461–471.

Yilmaz, A., Ekiz, H., Gültekin, I., Torun, B., Barut, H., Karanlik, S., Cakmak, I. (1998) Effect of seed zinc content on grain yield and zinc concentration of wheat grown in zinc-deficient calcareous soils. J. Plant Nutr., 21: 2257–2264.

Welch, R.M., Graham, R.D. (1999) A new paradigm for world agriculture: Meeting human needs. Productive, sustainable, nutritious. Field Crops Res., 60: 1–10.

Welch, R.M., Graham, R.D. (2004) Breeding for micronutrients in staple food crops from a human nutrition perspective. J. Exp. Bot., 55: 353–364.

White, J.G., Zasoski, R.H. (1999) Mapping soil micronutrients. Field Crops Res., 60: 11–26.

Chapter 8
Micronutrient Deficiencies in Crops in Africa with Emphasis on Southern Africa

Johan H. van der Waals and Michiel C. Laker

Abstract Even though deficiencies of micronutrients in Africa are considered to be widespread, limited research has to date been published on the extent or impact of these deficiencies on crop production and human health and well-being. Published and/or commercially available data pertains mainly to a number of cash crops or to commercial farming sectors. Information on nutrient contents of staples in areas and countries dominated by low-input agriculture is generally lacking. The dominant soils of Africa are inherently infertile due to them being derived from geologically old material or hard rock types. Furthermore, climatic conditions contribute to the low productivity of African soils.

Micronutrient deficiencies have been reported for a range of soils under differing crop production conditions. These deficiencies are readily corrected in the commercial farming sector, but often go unreported or unnoticed in the subsistence farming sector. Variation in micronutrient content of produce, as sold at fresh produce markets, is high and poses challenges to adequate human nutrition. Several strategies exist for the alleviation of micronutrient deficiencies, but many of these will require significant involvement from governments. Governments, such as that of South Africa, have in some cases resorted to the fortification of commercially available staple foods to alleviate malnutrition. This is not a solution for subsistence farming sectors and a significant effort will be required in the future to identify, assess, and correct such deficiencies.

8.1 Introduction

Deficiencies of micronutrients in Africa are considered to be widespread, however limited research has to date been published on the extent or impact of these deficiencies on crop production and human health and well-being. The population of Africa faces several threats that include diseases such as malaria and HIV/AIDS as well as challenges regarding sustainable food production. The severity of the health threats faced is exacerbated by widespread malnourishment that poses a risk to community health, especially in terms of long-term quality of life.

B.J. Alloway (ed.), *Micronutrient Deficiencies in Global Crop Production*.

Most of the agricultural research in sub-Saharan Africa focuses on the management of the macronutrients nitrogen (N), phosphorus (P), and potassium (K), such as the studies by Kang and Balasubramanian (1990), Smaling (1993) and Smaling and Braun (1996). This phenomenon is not restricted to Africa but is considered a major problem worldwide (Brown, 2004). Published data on micronutrient deficiencies in food crops in Africa are scarce and available information pertains mainly to a number of cash crops (Kang and Osiname, 1985). This is also the case for South Africa, where extensive knowledge on micronutrient deficiencies and fertiliser amendments exists for the commercial crop production sector. Information on nutrient contents of staples in areas and countries dominated by low-input agriculture is generally lacking, as was the case reported by Buerkert et al. (2001) for West Africa. The lack of information on low-input agriculture also applies to South Africa. Preliminary investigations indicate that the nutrient contents of crops procured at fresh produce markets are highly variable (Steyn and Herselman, 2005). For this reason Van Heerden and Schönfeldt (2004) called for the generation and compilation of information on food nutrient composition in southern Africa and for this information to be distributed widely. The study by Sika et al. (1995) represents such an exercise specifically for Morocco.

Adequate nutrient supply to crops has always been a constraint in the resource-poor agricultural sector of Africa and has become a serious problem due to high net nutrient removal rates or nutrient mining (Smaling, 1991). Data from 38 African countries indicated that nutrient mining had become a serious problem leading to the non-sustainability of most agricultural systems in these countries (Smaling, 1991). Due to increasing population numbers, many countries in Africa face the challenge of increasing food production through agricultural intensification – a situation that poses many threats considering the low contents of most nutrients in African soils.

In this chapter, the established information and knowledge on micronutrient deficiencies in Africa will be briefly discussed with emphasis on the two widely differing farming sectors. In addition, special emphasis will be placed on the restrictions faced by subsistence farming communities and this will be illustrated with a case study. Although plant micronutrients are addressed in this study, reference will also be made to other nutrient elements that are of importance to human health.

8.2 Micronutrient Deficiencies in Africa

8.2.1 Geology and Soils

Micronutrient deficiencies are often related to low total levels in certain soil parent materials and are more prevalent in highly weathered and pedologically old soils (Kang and Osiname, 1985; Fageria et al., 2002). Laker (2003) distinguishes three major soil regions of the world with two dominating large parts of Africa. These are:

- The high latitude areas of the northern hemisphere
- The humid tropics
- The hot semi-arid zones bordering the humid tropics

In the humid tropics, soils represent the most important and most limiting factor for crop production (Agboola and Akinnifisi, 1993). Agboola and Akinnifisi (1993) point out that these soil limitations are chemical rather than physical. The soils are predominantly extremely infertile, with low nutrient contents and high acidity. Nutrient management is difficult due to the low nutrient storage capacities of the soils. This is mainly due to the clay fractions being dominated by kaolinite and/or sesquioxides, leading to excessive leaching of nutrients under the high rainfall, and due to high fixation capacities for nutrients like P. Due to the low buffering capacities of the soils, nutrient imbalances can easily be caused by injudicious plant nutrient management (Deckers et al., 2000).

The hot semi-arid zones bordering the humid tropics have a low cropping potential due to both poor quality soils and unfavourable climate. In these zones rainfall is not only low, but also erratic, unreliable and very inefficient. The efficiency of rainfall is restricted inter alia by (1) the fact that the rain over most of these zones falls as very aggressive thunderstorms, leading to high runoff, (2) the very high evapotranspiration losses of water due to the prevailing high temperatures and low humidity and (3) the low water storage capacities of the soils. The soils of these zones are predominantly very shallow, due to slow weathering caused by the lack of water and the fact that the soils develop mainly from the weathering of hard rock types. Significant areas of aeolian sands occur, which under these semi-arid conditions are the most favourable soils for rainfed cropping (Laker and Remmelzwaal, 1994). The pH of the soils of the drier parts of these areas is predominantly neutral to alkaline and the soils often contain free lime or are underlain by lime pans. Phosphorus deficiencies are widespread, as are deficiencies of trace elements like zinc (Zn), iron (Fe) and copper (Cu).

Information on the distribution of micronutrient deficiencies as associated with specific soil types in Africa is restricted. Even though generalized soil data for Africa is available in the form of a soil map on a scale of 1:5 million (FAO-UNESCO, 1977), Vlek (1995) states that its compilation involved a minimum of ground truthing. This restricts the accuracy of attempts to predict deficiencies for the continent through the use of available soil data. The existing soil data was used by Eswaran et al. (1997) in the compilation of a map based on the USDA's Soil Taxonomy for the purposes of assessing the soil resources of Africa. A summary of the areal distribution of soil Orders is provided in Table 8.1 and the distribution of soil pH values in Table 8.2 (from Eswaran et al., 1997). As discussed earlier by Laker (2003), arid and young soils dominate in Africa followed by highly weathered pedologically old soils. Around 15% of the continent's soils can be considered acid and it is in these soils that most of the agricultural production is concentrated.

Table 8.1 Distribution of soil orders (USDA Soil Taxonomy) and other forms of land cover for Africa (From Eswaran et al., 1997)

Soil order (Soil taxonomy)	Area ($\times$1,000 km^2)	Percentage of total
Andisols	49	0.16
Histosols	15.3	0.05
Spodosols	30.7	0.10
Oxisols	4,389	14.32
Vertisols	990	3.23
Aridisols	8,076	26.35
Ultisols	1,906	6.22
Mollisols	70	0.23
Alfisols	3,200	10.44
Inceptisols	2,378	7.76
Entisols	7,506	24.49
Dune sands	1,441	4.69
Rock	324	1.03
Salt flats	35	0.12
Inland water	263	0.89

Table 8.2 Soil pH ranges and geographical extent for the soils of Africa (From Eswaran et al., 1997)

Soil pH range	Area ($\times$1,000 km^2)	Percentage of total
<3.5	31	0.1
3.5–4.2	1,193	3.9
4.2–5.5	3,278	10.7
5.5–6.5	4,306	14.0
6.5–8.5	6,997	22.8
>8.5	14,845	48.4

For South Africa specifically, there are no recent comprehensive and up-to-date accounts of its soils and Laker (2000) provides a brief and informative description and general distribution of South African soils. The Agricultural Research Council's Institute for Soil, Climate and Water (ISCW) has published a series of "land-type" maps, with accompanying memoirs (reports) on a scale of 1:250,000 for large parts of the country. Information on profile pits and soil samples collected complement the land data and forms the basis of investigations that will be discussed in more detail further in the chapter. Soil acidity data for the country have not been published but an unpublished report by the ISCW (ARC-ISCW, 2004) indicates a similar trend, albeit less alkaline, as for the rest of the continent (Table 8.3). Although the distribution of acid soils is limited, these soils are often used for crop production and run the risk (as do the more neutral soils) of acidification through extensive use of N-fertilisers and inadequate lime use.

Table 8.3 Soil pH ranges and frequency in South African soils (ARC-ISCW, 2004)

Soil $pH_{(water)}$ range	Percentage of total
<5.5	5.7
5.6–6.4	24.4
6.5–7.4	34.2
7.5–8.4	29.2
>8.5	6.6

8.2.2 Reported Deficiencies – South Africa

Due to the important role of commercial agriculture in South Africa (and Zimbabwe to a lesser degree in the past), micronutrient deficiencies have been addressed to a very large extent. For most crops, threshold values have been determined that usually lead to efficient diagnosis. In most cases, deficiencies have been reported for sandy soils associated with sandstones or aeolian deposits, or with intensive agricultural practices such as fruit production. The bulk of the research was conducted in the 1950s, 1960s and 1970s and was published in technical reports of the different research institutions. Although very few of these results were published in scientific journals, regular communication took place at conferences or through the technical reports as mentioned earlier. In this sense, in his summary on the highlights of soil chemistry research in southern Africa from 1953 to 1978, De Villiers (1980) states that

> More or less routine soil analysis conducted on a continuing basis by many laboratories for fertiliser advisory and soil classification purposes has resulted in a build-up of local experience of chemical characteristics that is normally unpublished and cannot easily be generalised.

The following, except where cited with a different reference, is a summary of the identified deficiencies in a range of different crops as contained in the Fertiliser Society of South Africa's (FSSA) Fertiliser Handbook (Bornman et al., 1989). Province names and regions discussed in the following section are shown in Fig. 8.1.

8.2.2.1 Boron

Boron (B) deficiencies in maize were reported on sandy soils (mainly Acrisols and Lixisols, FAO-UNESCO Classification) of the Eastern Free State and Eastern Highveld. Deficiencies of B were reported for sunflower (Grant, 1980) on well-drained and leached soils and are corrected through soil applications when leaf analysis indicates a deficiency. Deficiencies of B were also indicated for groundnuts and tobacco on sandy soils. In terms of fruit production, deficiencies were reported for citrus, avocados, pecan nuts, and banana and these are usually corrected through foliar sprays (except for banana where soil applications are preferred). Although no B deficiencies have been reported for sugarcane in South Africa, it has been identified in sugar plantations in Malawi (Meyer, 2005).

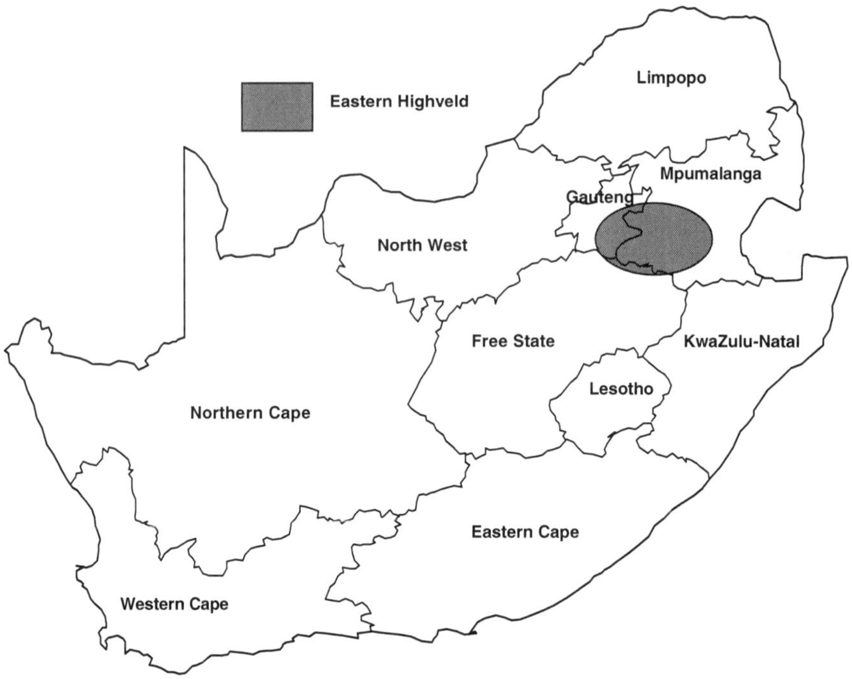

Fig. 8.1 Provinces and regions in South Africa

8.2.2.2 Copper

Copper deficiencies are not widespread and have been reported for wheat on sandy soils of the Western Cape Province. Deficiencies are considered to be associated with sandy soils low in organic matter in Southern Africa (Cooper, 2005). In the Cape coastal strip, correction of deficiencies has resulted in large yield increases in wheat (Beyers and Roach, 1965 as cited by Grant, 1980). Deficiencies have also been identified in citrus, avocados, pecan nuts and banana and are treated in the same way as B. Copper deficiencies were first observed in sugarcane in the fifties and are efficiently corrected through foliar application (Du Toit, 1956).

8.2.2.3 Iron

Iron deficiencies were reported for sugarcane in young cane in soils with pH values above 8, on termite affected soils, and as induced by high or toxic manganese (Mn) concentrations in acid soils under moist environmental conditions (Meyer, 2005). Iron deficiencies for other crops have not been widely reported in South Africa but where they are identified, they are corrected either through foliar sprays or soil application of Fe chelates.

8.2.2.4 Manganese

Deficiencies of Mn have been reported for wheat on sandy soils of the Western Cape Province as well as certain cultivars of peas. Citrus, avocados, and banana have exhibited sporadic deficiencies and treatment is the same as for B. Deficiencies are also often observed during episodes of over-liming in a range of soils. Manganese toxicity is often suspected to be a major contributor to acid soil infertility (De Villiers, 1980).

8.2.2.5 Molybdenum

In the 1960s, molybdenum (Mo) deficiencies were reported on a range of acid soils (De Villiers, 1980), especially in maize (Grant, 1980), and the problem was solved with the coating of seed with Mo and liming of soils. Liming of acid soils is the most common method of alleviating Mo deficiencies, although Grant (1980) indicated that the use of Mo salts was often a much cheaper option. Deficiencies have also been reported for several other crops (sunflower, groundnuts, dry beans, peas) and in all cases seeds are treated.

8.2.2.6 Zinc

In South Africa, widespread Zn deficiencies are known on the sandier soils of the maize belt (Altona, 1975 as cited by Grant, 1980). Deficiencies often associated with liming, were reported in the 1960s (Laker, 1964) on maize and the problem was solved through the application of Zn-containing fertilisers (De Villiers, 1980). Also in the 1960s, deficiencies were reported for sugarcane on sandy soils (Alexander, 1967; Meyer et al., 1971). Deficiencies associated with liming and high P applications were reported for dry beans and were confirmed by Laker (1967) in rye in pot experiments. Zinc deficiencies were also reported for wheat, peas, citrus, avocados, pecan nuts, and banana. In most cases the problems were solved through the use of Zn-containing fertilisers, application of foliar sprays (fruit trees) or soil application (banana).

Grobler and Hugo (1963) and Stanton (1964) found low Zn levels in the sandstone derived soils of the South African Highveld. In the study by Grobler and Hugo (1963), Zn was found to be mainly associated with the weatherable silt fraction, a fraction that is very low in the soils of the Highveld. Adjusting the pH in these soils to optimal levels for Zn plant uptake is not a successful strategy due to the inherent low Zn levels. In the north-western Free State (predominantly a maize producing area), in cases where abundant lime is present in the soils close to the surface, Zn deficiencies are induced in maize through ploughing of the subsurface lime into the plough layer. Other management practices also play a role and Du Preez et al. (2001) indicated that straw burning and conservation tillage, in the long term, increased surface soil Zn levels (along with pH, K and P) on a semi-arid Plinthosol compared to no burning and conventional tillage.

8.3 Reported Deficiencies – Tropical Africa

Kang and Osiname (1985) reviewed the status of micronutrient problems in tropical Africa and provided a map with the distribution of reported field cases (based on published data) to illustrate these problems. Figure 8.2 shows the distribution of the deficiencies as reported by Kang and Osiname (1985) on the African continent. Micronutrient deficiencies were often related to parent material with low total and extractable levels of B, Cu, Mo, and Zn. Iron and Mn deficiencies were locally important and toxicities were experienced more often than deficiencies. More micronutrient problems were encountered in subhumid and arid zones than in humid zones but White and Zasoski (1999) attribute this to a probable concentration of research effort in the former. A summary of low extractable soil micronutrient levels as reported for a number of countries is provided in Table 8.4 (the listed references are from Kang and Osiname, 1985). Deficiencies in crops, of which the

Fig. 8.2 Distribution of micronutrient deficiencies as indicated by Kang and Osiname (1985) for the African continent

Table 8.4 Low extractable micronutrient levels as reported for different parts of Africa (Data and references from Kang and Osiname, 1985)

Area/Country	Reference	Element	Soils
Ivory Coast	Dabin and Leneuf (1960)	Cu	Young peat soils
		Fe	Soils derived from basic rocks
		Mn	Soils of peat and lacustrine alluvial origins
		Mo	All soils except organic and hydromorphic soils
		Zn	Soils derived from tertiary sands, granites, schists, and alluvial materials
Humid zone of southern Nigeria and Togo	Cottenie et al. (1981)	Mn	Soils derived from sandstones and sedimentary material
		B, Mo, Zn	All soils studied
Sierra Leone	Haque et al. (1981a, b)	Mn	Lowland soils
		Zn	All soil samples
Zaire	Vancompernolle et al. (1965)	Zn	Most samples
		Fe, Mn	Some soil samples
Rift Valley region of Kenya	Pinkerton (1967), Nyandat and Ochieng (1976)	Cu	Soils from the Rift Valley region
Sudan (Gezire scheme)	Ibrahim (1982)	Mo	Three major soils from the Gezira scheme

most is known about cash crops, such as banana, groundnuts, cotton, cocoa and oil palm, were also reviewed by Kang and Osiname (1985) and a summary is provided in Table 8.5

The work by Sillanpää (1982) was reviewed by Kang and Osiname (1985) and White and Zasoski (1999), but due to the extensive nature of the work only some of the results will be addressed in further detail. The project was started in 1974 and had the aim of providing a picture of the status of soil micronutrients in agriculture in thirty countries worldwide. The countries in Africa included Egypt, Ethiopia, Ghana, Malawi, Nigeria, Sierra Leone, Tanzania, and Zambia. Representative soil samples were collected from the participating countries and spring wheat (cv "Apu") was grown on samples of the soils under controlled conditions in a greenhouse in Finland. Soil and plant samples were analysed for, amongst others, B, Cu, Fe, Mn, Mo, and Zn.

The products of soil concentration × plant concentration ("concentration products") were calculated for each nutrient. The lowest 5% of these concentration product values were allocated to Zone I and the next lowest 5% to Zone II. The middle 80% of the concentration products were allocated to Zone III, the top 90–95% of concentration products to Zone IV and the uppermost 95–100% of the concentration

Table 8.5 Summary of reported deficiencies in crops in tropical Africa (Data and references from Kang and Osiname, 1985)

Micro-nutrient	Crop	Comment	Country/region	Reference [Country]
B	Oil palm	Depleted sandy soils	Congo, Ivory Coast	Ferrand et al. (1951), Fewerda (1954), Ollagnier and Valverde (1968), Martin (1969)
	Cocoa		Ghana, Nigeria	Asomaning and Kwakwa (1967), Egbe and Omotso (1972)
	Banana	Induced by increasing soil pH	Ivory Coast	Carpentier and Martin-Prevel (1967)
	Cotton	Light textured soils	Tanzania, Zambia, Chad, Cameroon, Benin, Nigeria	Le Mare (1970), Rothwell et al. (1967), Braud et al. (1969), Smithson (1972), Smithson and Heathcote (1974), Heathcote and Smithson (1974)
	Maize, wheat, groundnuts	Ferrallitic soils	Zimbabwe, Madagascar, Zambia	Fenner and Goldring (1968), Velley et al. (1974), Smart (1961)
	Sisal, eucalyptus		Zimbabwe, Zambia, Upper Volta	Dickmahns (1957), Savory (1962), Birot (1967)
Cu	Oil palm	Depleted sandy soils	Congo	Ferrand et al. (1951)
	Banana	Younger peat soil	Ivory Coast	Dabin and Leneuf (1960), Carpentier and Martin-Prevel (1967)
	Wheat	Recent Rift Valley soils of Kenya derived from ash and pumice/ sandy soils	Kenya, Zimbabwe	Pinkerton et al. (1965), Pinkerton (1967), Wapakala (1973), Nyandat and Ochieng (1976) Tanner et al. (1981)
	Maize	Ferrallitic soil	Madagascar	Velley et al. (1974)
Fe	Cacao	Of a local nature after burning	Ghana, Nigeria	Greenwood and Hayfron (1951), Egbe and Omotoso (1972)
	Rice	After extensive burning of plant residues	Nigeria	Kang et al. (1976)
	Sorghum	After irrigation with Ca and Mg containing water	Senegal	Blondel (1970)

(continued)

Table 8.5 (continued)

Micro-nutrient	Crop	Comment	Country/region	Reference [Country]
Mn	Oil palm	Associated with Fusariosis	Congo, Ivory Coast, West Africa	Ferrand et al. (1951), Bachy and Fehling (1957), Prevot (1959)
	Maize	Soil derived from sandy sedimentary material	Nigeria	Okoye (1972), Tinker and Ziboh (1959)
	Pineapple, maize, cassava, cowpeas	Induced by liming on acid ferrallitic soil	Ivory Coast, Swaziland, Nigeria	Marchal et al. (1980), Jones (1979), Edwards and Kang (1978), Kang (1978)
Mo	Groundnuts	Oxisols and other, sandy Aeolian soils	Sierra Leone, Senegal, Ghana, Nigeria, Gambia, West Africa	Haque and Kamara (1978), Martin and Fourier (1965), Stephens (1959), Heathcote (1970), Webb (1954), Gillier (1966)
	Cowpeas	Oxisol	Sierra Leone	Rhodes and Kpaka (1982)
	Maize	Acid red soils and low seed Mo content	Zimbabwe	Tanner (1976), Tanner and Grant (1977), Tanner (1982)
Zn	Maize	Most soils, Ferrallitic soil derived from gneiss	Nigeria, Madagascar	Kayode and Agboola (1983), Velly et al. (1974)
	Oil palm	Depleted sandy soil	Congo	Ferrand et al. (1951)
	Cacao	Isolated cases (high soil pH)	Ghana, Nigeria	Greenwood and Hayfron (1951), Egbe and Omotoso (1972)
	Banana	Induced by liming of soil	Guinea, Ivory Coast	Moity (1954), Charpentier and Martin-Prevel (1967)
	Maize	High rates of lime and P application, Sandy soils	Nigeria, Zimbabwe	Friesen et al. (1980), Osiname et al. (1973), Agboola et al. (1970), Tanner and Grant (1973), Tanner and Grant (1974)
	Rice	Vertisols	Nigeria, Sudan	Kang and Okoro (1976), Magar and Babikir (1965)

product values were put in Zone V. The concentration product values in Zones I and II represent low available concentrations of nutrient elements and those soil samples with a high probability of deficiencies, especially those in Zone I. Table 8.6 shows the percentage distribution of the concentration products for the soil and wheat sample analyses in the two lowest zones (I and II) for six micronutrients.

Considering the results in Table 8.6 on an element by element basis, it can be seen that significant percentages of low concentration products of B occur in Nigeria, Malawi and Zambia with a small percentage in Ethiopia. All of the African countries in the study (except Egypt) had high, or relatively high, percentages of potentially Cu-deficient soils. The soil samples from these countries were predominantly acid (<pH 5.6) and some had relatively high organic matter contents. Low Fe levels were found in Nigeria, Tanzania and Zambia. These countries differ from many others, where Fe deficiencies were induced by calcareous and/or high pH soils, in that their soils were predominantly acid. Egypt was the only participating country in Africa that had samples that were deficient in Mn. Ghana, Malawi, Nigeria, and especially Sierra Leone and Zambia, had high percentages of Mo deficient soils mainly due to acidity. Zinc deficiencies were not shown to be prevalent in the African countries that participated in the study. The only countries with samples in Zone I for Zn were Tanzania (3%) and Malawi (1% of samples).

A number of African countries were also involved in a follow-up project aimed at quantifying the effects of micronutrients on yields of economically important crops as well as determination of critical deficiency and toxicity levels and the development of guidelines for solving micronutrient problems (Sillanpää, 1990). These were Ethiopia, Malawi, Sierra Leone, Tanzania, Zaire, and Zambia. The results of the study were complicated by a number of factors and not all predicted deficiencies were confirmed in field trials. In contrast to the earlier study (Sillanpää, 1982), Zn exhibited the most frequent deficiency followed by B, Mo, Cu, Mn, and Fe. The fact that Zn exhibited the most frequent deficiency could be due to differing

Table 8.6 Percentage distribution of concentration products for micronutrients in Zones I and II of Africa (Sillanpää, 1982)

		Concentration products of elements in Zones I and II											
		B		Cu		Fe		Mn		Mo		Zn	
Country	Samples	I	II	I	II	I	II	I	II	I	II	I	II
Egypt	198	–	–	–	–	–	1	8	13	–	–	–	2
Ethiopia	125	4	6	21	6	–	1	–	–	–	4	–	1
Ghana	93	–	–	23	24	1	–	–	–	16	22	–	2
Malawi	97	9	14	7	6	–	4	–	–	7	24	1	1
Nigeria	153	11	14	12	20	5	12	1	3	20	12	–	1
Sierra Leone	48	–	2	68	21	–	4	–	2	81	10	–	–
Tanzania	163	–	1	14	9	6	7	–	–	6	7	3	2
Zambia	44	2	11	36	16	16	11	–	–	50	9	–	2

approaches and samples and crop species/cultivars used. Likewise, the predicted widespread occurrence of Cu deficiency was not borne out in the field experiments, with Cu being the fourth most common deficiency.

In a number of studies in Africa, attempts have been made to map micronutrient deficiencies. In the Lombin (1983) study for Nigeria, Zn extractability was low in about one-third of the soil samples collected but this was not widely reflected in appreciable cotton yield reductions. It was concluded though that the results indicated that Zn was an imminent fertility problem on Oxisols and Ultisols. Peters and Schulte (1994, 1996) collected soil and plant samples from 567 maize, millet, and groundnut fields scattered throughout The Gambia to assess the fertility status of common agricultural soils. The main finding regarding micronutrients was that Zn was the principal limiting nutrient for maize and that B deficiencies also appeared to be problematic in many areas. El-Fouly et al. (1984) reported the results of more than 10,000 soil and plant analyses conducted on samples from reclaimed sandy and calcareous desert soils, the Nile Valley and the Nile Delta in Egypt. The results indicated that Fe deficiency was common on calcareous soils, Zn deficiency occurred on sandy soils, and Mn deficiency occurred mainly on loamy alluvial soils. This data differs from that reported by Sillanpää (1982) for Egypt probably due to different approaches and sample sites.

A more recent study by Buri et al. (2000) indicated that mean topsoil available Zn levels were adequate for rice production (1.23 and 1.56 mg kg^{-1} for river flood plains and inland valley swamps, respectively) but that 66% of West African lowlands had available Zn levels below the critical soil level of 0.83 mg kg^{-1} necessary for rice production. Other micronutrient levels were moderate and Fe was high in some cases.

8.4 The Impact of Micronutrient Deficiencies in Different Farming Sectors with Specific Reference to South Africa

8.4.1 Commercial Farming Sector

Agricultural production in South Africa (and to a lesser extent the rest of Africa) can be divided into two sectors, a well-developed commercial farming sector and a resource-poor subsistence farming sector (Steyn and Herselman, 2005). In the first sector commercial agricultural practices dominate in that regular fertiliser inputs are made and outputs are in the form of produce that is sold on national and international markets. This sector dominates in terms of the monetary value of products.

In most of the commercial agricultural sectors, macronutrients are usually adequately supplied (and sometimes even over-supplied). Due to high yields, these areas experience high micronutrient removal rates and Fageria et al. (2002) noted this as one of the causes of such deficiencies in general. Being part of a commercial production system, micronutrient deficiencies are often identified through a number of diagnostic techniques (visible signs, plant and soil laboratory analysis) by extension officers, fertiliser company representatives, or subject field specialists. Deficiencies can be corrected in a number of ways with just as many (or more) different products from various companies. Even though the tools exist for efficient alleviation of micronutrient deficiencies, it is presumed that some deficiencies go unnoticed and are widespread.

Although expenditure on agricultural research in South Africa is lower than in most developed countries, it is safe to say that the commercial agriculture sector spends a significantly higher amount on soil and plant testing for optimal yields than the subsistence or resource-poor sector. The driving force behind this expenditure is mainly the maximizing of profit. Many fertiliser companies have testing programs that are conducted as a service for the farmer, the cost of which is included in the cost of fertiliser.

8.4.2 Subsistence Farming Sector

The subsistence agriculture sector differs in many ways from the commercial one discussed above. This sector is characterized by highly irregular or small fertiliser inputs and produce that is consumed within producer households or sold in small local markets. Due to the nature of the production, very little income is generated for households and this is usually spent on necessities and very seldom on fertilisers. In most cases, N fertilisers are the only types procured and then only in small quantities from suppliers other than the primary ones. Very few testing facilities cater for the needs of subsistence farmers with the exception being a number of facilities that subsidize the first sample of each client. In this way a subsistence farmer can submit one sample at a very low cost and obtain information, albeit restricted, on his soils.

Although small in terms of monetary value of produce when compared to commercial production, this sector dominates when the number of people dependent on it is considered. In this sector most macronutrients are in short supply and therefore limit yields. Laker (2004) quotes soil P data from four developing farming areas in the Northwest Province of South Africa which indicate that 87% of samples had a Bray-1 (soil test) P status of between 0 and 9 mg kg^{-1} with the median value being 2 mg kg^{-1}. Under such conditions it is to be expected that crop production would not be possible without significant P-fertiliser inputs. Without extensive analytical data, it can therefore only be postulated that micronutrient deficiencies are often masked by more pronounced deficiency symptoms of the macronutrients P (Fig. 8.3 in Colour Section) and K.

Yields in this sector are often far below potential as was found during investigations conducted in the Eastern Cape (Steyn and Herselman, 2005). Due to low yields, the micronutrient removal rate is expected to be low and the transfer to humans is therefore also limited. The geographical extent of subsistence production (per household) is usually restricted due to reliance on manual labour (Fig. 8.4 in Colour Section) and it is concentrated around homesteads due to the close proximity of manure sources such as kraals. Roberts, Adey and Manson (2003) provide soil

Fig. 8.3 Severe P-deficiencies in maize masking possible micronutrient deficiencies in a subsistence agriculture field in Mpumalanga province, South Africa (From J.H. van der Waals) (*See Colour Plates*)

Fig. 8.4 The limited geographical extent of a subsistence farmer's maize field in the Mpumalanga province, South Africa. Note the uneven colour through the field due to variable soil conditions and fertiliser application (From J.H. van der Waals) (*See Colour Plates*)

fertility data (a range of macronutrients, Mn and Zn, and soil acidity) for two resource-poor farming communities in the KwaZulu-Natal province of South Africa.

Community 1. The Obonjaneni community is in the Bergville district in a high rainfall area (1,050 mm $annum^{-1}$). An important characteristic of the soils is their high acidity, with average pH (KCl) 4.0 and acid saturation values over 50%. The "home fields", small garden plots near the homesteads, have a median adjusted AMBIC-2 extractable P level of 14 mg L^{-1} and 66% of the fields have adjusted P levels of higher than 12 mg L^{-1}, the recommended requirement for maize. This figure of 66% exceeds the 41% of samples from maize fields of commercial farmers in KwaZulu-Natal that had adjusted P levels of higher than 12 mg L^{-1}. The outfields had a median adjusted P value of only 4 mg L^{-1}. According to Roberts et al. (2003), the outfields would require between 4 and 13 t ha-1 lime, with a median of 9.3 t ha^{-1} to reduce the acid saturation to the acceptable level of 20%.

Community 2. The community in the Valley of a Thousand Hills is situated in an area with a mean annual rainfall of between 595 and 830 mm. The soils have no acidity problems with an average pH (KCl) of 5.3 and only 1% acid saturation. The median adjusted P value for the home fields is 11 mg L^{-1}, which means that more than 50% of them are P-deficient for maize. For the outfields the median adjusted P value is only 3 mg L^{-1}.

The emphasis of the study by Roberts et al. (2003) is on P and the fact that it is considered the most important constraint to food production. Although the levels of Mn and Zn are not discussed, their data indicate that median Mn levels are similar for all the fields and the median Zn levels are lower for the outfields. This indicates that, although some form of fertilisation is supplied to the home fields, outfields could exhibit deficiencies of macro- and micronutrients due to very low applications of manure and other fertilisers.

Steyn and Herselman (2005) provide information on high poverty levels in the Eastern Cape Province of South Africa and question the role of conventional trace element research in the face of this poverty. The fact that resource-poor farmers do not have access to diagnostic facilities or remediation measures for correcting micronutrient deficiencies does not bode well for the health and nutrition of entire communities in Africa. In this sense, micronutrient deficiencies of crops also paint a bleak picture due to the larger number of additional nutrients required by humans.

A number of studies have been conducted by personnel of the South African Agricultural Research Council's Institute for Soil, Climate and Water (ISCW) related to trace elements in South African soils as well as different aspects related to resource-poor farming communities.

- Steyn and Herselman (2005) cite studies conducted by the ISCW where Swiss chard was grown on a wide range of soils collected from different parts of South Africa and where Swiss chard was sampled from fruit and vegetable markets and shops in Pretoria. Also discussed are maize meal samples from shops in Pretoria. Large variations for the trace elements Cd, Co, Cu, Mn, Ni, Se, and Zn, from very low to very high, were found in both studies indicating that people consumed widely varying quantities of nutrients and potentially toxic elements (Table 8.7). This aspect raises a concern where it is known that resource-poor communities do not consume a wide variety of foodstuffs and that deficient nutrient levels are presumed to manifest in health problems. The problem looms even bigger when considering that the per capita arable land for Africa is predicted to decrease from 0.23 ha in 1995 to 0.08 ha in 2050 compared to 0.89 ha (1995) and 0.69 ha (2050) for North America (Lal, 2001).

- Background trace element concentrations for South African soils have indicated that large areas have low ammonium-EDTA-extractable Cu, Zn and Co levels (Steyn and Herselman, 2005). In commercial agriculture the low Zn levels have been addressed through the addition of Zn to fertiliser mixtures but subsistence, or resource-poor, farmers have not benefited, largely due to restricted access to fertilisers.
- Steyn and Herselman (2005) present data showing that a range of trace elements exhibited higher concentrations in irrigated commercial production soils than in soils of resource-poor farmers (Table 8.8 provides a summary of the micronutrients Cu, Fe, Mn, Mo, and Zn determined in the study). The reason for the difference is not clear and could be attributed to inherent soil differences

Table 8.7 Range of trace element concentrations as determined in Swiss chard grown in different soils and Swiss chard obtained from local markets (Steyn and Herselman, 2005)

	Range ($mg\ kg^{-1}$)		
Element	Swiss chard (Glasshouse)	Swiss chard (Markets and shops)	Maize meal (Shops)
Cd	not detected – 2.59	not detected – 0.7	not detected – 0.7
Co	0.018–51.158	1.4–14.67	0.06–1.91
Cu	2.72–152.12	9.4–111.71	1.97–14.57
Mn	0.07–1373	–	–
Ni	0.09–18.44	not detected – 35.75	0.90–19.82
Se	0.2–47.58	0.61–5.18	not detected – 0.18
Zn	11.92–623.77	34.10–815.99	36.19–99.43

Table 8.8 Median concentrations ($mg\ kg^{-1}$) of selected micronutrients per land use as extracted by NH_4EDTA (Steyn and Herselman, 2005)

	Land use[a]			
Element	Rangeland	Irrigation	Dry land	Resource poor
Cu	5.22[a]	10.87[b]	5.51[a]	3.13[c]
Fe	448.0[a]	195.8[b]	211.0[b]	102.7[c]
Mn	233.9[a]	381.2[b]	286.5[a,b]	102.4[c]
Mo	0.14[a]	0.13[a]	0.13[a]	0.11[a]
Zn	1.97[a]	11.32[b]	4.12[c]	3.65[c]

[a] Values in the row followed by the same symbol do not differ significantly at $p < 0.05$ according to Fisher's protected *t*–test least significant difference

between the location of the different socio-economic groups or to trace element depletion or enrichment factors. The required plant micronutrients Cu, Fe, Mn, Mo, and Zn indicated a risk of being deficient in many soils as was the case for the essential nutrients for humans Co, Cr, I, and Se.

- Studies on oesophageal cancer in the Transkei area and Mseleni joint disease in Northern KwaZulu-Natal province of South Africa have indicated that Se deficiencies could be related to the incidence of these diseases (Steyn and Herselman, 2005). Although many factors could be related to the incidence of the health problems, Steyn and Herselman (2005) state that it is surprising that very few studies have been conducted on micronutrients (in this case Se) in Africa, when compared to the body of knowledge available in Europe.

8.5 Strategies for the Alleviation of Micronutrient Deficiencies

Several possible strategies exist for the alleviation of micronutrient deficiencies in the different farming sectors of Africa. Firstly, the existing nutrient problems will have to be ascertained through surveys similar to but much larger in extent than those conducted by personnel of the ISCW. Government funding and support would be a prerequisite as these surveys tend to be costly and labour intensive.

Secondly, more intensive research should be conducted and extension services provided to those farming sectors where internally generated resources are not adequate to "buy" such services. In this sense, governmental institutions should play the leading role, but are generally hampered through lack of funds and manpower.

Thirdly, and related to the second point, resource-poor farmers should be aided and encouraged to expand and intensify their farming operations to become "commercial". In this sense there are many projects underway, governmental and non-governmental, to facilitate the process. Through the commercialization of farming practices, the need for and means to afford analytical services will arise.

The use of cattle manures is often touted as a sound strategy for the alleviation of nutrient deficiencies. Although it is a good source of micronutrients for crop production, levels in manure from smallholder farms were generally lower than those of experimental stations in Ethiopia (Lupwayi et al., 2000). Similar trends are found in manures in South Africa where the micronutrient contents tend to be higher in feedlot manures than manures from smallholder farmers (unpublished data). Cattle manure is limited though and it is insufficient to support crop production on a wide scale (Lupwayi et al., 2000).

Another approach that is often seen as a solution for low micronutrient levels in crops is the breeding of crops with higher micronutrient contents or improved capacities for micronutrient uptake. This approach makes sense for vitamins produced by the plant (such as vitamin A) but is probably not sustainable for nutrients that have to be extracted from already deficient soils.

Fourthly, due to the difficulty and time consuming nature of the first three points mentioned above, food fortification should be considered. In this sense the South

African Government has already promulgated regulations regarding the fortification of staple foods such as maize flour and bread. This approach is restricted in the instance where communities produce their own food necessitating one of the previously mentioned approaches.

8.6 Conclusions

Information on micronutrient deficiencies in Africa relates mainly to the commercial production sector (or cash crops) and is lacking in terms of staples consumed by the majority of the population. A number of studies have been conducted, but data from different workers (as well as the same ones!) often seem contradictory. The contradictions are ascribed to the "patchy" nature of the results and test sites as well as the limited extent of research on the continent. It is envisaged that these contradictions will disappear if more complete information is generated in future comprehensive studies.

Without adequate information it is considered that micronutrient deficiencies are widespread. Information is difficult to generate and the deficiencies difficult to quantify due the nature and dominance of resource-poor communities and subsistence farming practices. Unless governments step in to facilitate and fund research on the extent of these deficiencies and their impact on human health, it is envisaged that entire communities will suffer chronic and serious effects.

In the commercial agricultural sectors, it is to be expected that micronutrient deficiencies will increase with the intensification of production practices. The introduction of subsistence farmers into the commercial sector holds a number of threats of which the accelerated depletion of micronutrients is the most immediate. Adequate extension practices are required to bring across the need as well as to facilitate regular monitoring through soil and plant testing for nutrient replacement and fertiliser additions.

References

Agboola, A.A., Akinnifisi, F.K. (1993) Soil fertility management for the humid tropics of Africa. Proceedings Of the 2nd African Soil Science Society Conference, Cairo, 1991: 107–125.

Agboola, A.A., Corey, R.B., Obi, O. (1970) A survey of western state soils on the response of maize to fertilisers. Nig. Agric. J., 13: 150–167.

Agricultural Research Council – Institute for Soil Climate and Water. (2004) Overview of the status of the agricultural natural resources of South Africa. ARC-ISCW Report GW/A/2004/13. ARC-ISCW, Pretoria.

Alexander, K.E.F. (1967) A nutrient survey of cane on TMS soils in Natal. S.A.S.T.A. Proc., 41: 197–200.

Asomaning, E.J.A., Kwakwa, R.S. (1967) A note on boron deficiency and fruit malformations in cocoa (*Theobroma cacao* L.). Ghana J. Sci., 7: 126–129.

Bachy, A., Fehling, G. (1957) La fusariose du palmier a huile en Cote d'Ivoire. J. Agric. Trop. Bot. Appl., 4: 228–240.
Birot, Y. (1967) L'effet de la carence en bore sur l'Eucalyptus station de Dinderesso (Bobo-Dioulasso). Colloque sur le fertilite des sols Tropicaux. Tanarive, Malagasy, I: 1019–1023.
Blondel, D. (1970) Induction d'une chlorose ferrique en sol sableux (Dior) par des aux d'irrigation calco-magnesiennes. Agron. Trop., 25: 506–561.
Bornman, J.J., Ranwell, J.F., Venter, G.C.H., Vosloo, L.B. (Eds.) (1989). Fertiliser Society of South Africa Fertiliser Handbook. 3rd Edn. ISBN 0909071667, FSSA, Pretoria, p. 275.
Braud, M., Fitz, A., Megie, C.H., Quillon, P.J. (1969) Sur la deficience en bore du cottonier. Observations preliminaries. Cotton Fibres Trop., 24: 465–467.
Brown, P.H. (2004) Principles of Micronutrient Use. IFA International Symposium on Micronutrients. February 23–25, New Delhi, India.
Buerkert, A., Moser, M., Kumar, A.K., Fürst, P., Becker, K. (2001) Variation in grain quality of pearl millet from Sahelian West Africa. Field Crops Res., 69: 1–11.
Buri, M.M., Masunaga, T., Wakatsuki, T. (2000) Sulfur and zinc levels as limiting factors to rice production in West Africa lowlands. Geoderma, 94: 23–42.
Carpentier, J.M., Martin-Prevel, P. (1967) Les carences minerales du Bananier. Colloque sur le fertilite des sols Tropicaux. Tanarive, Malagasy, I: 286–293.
Cooper, G. (2005) Micronutrient and heavy metal excess and toxicity in Southern African soils. Combined Fertiliser Society of South Africa (FSSA) and South African Sugarcane Research Institute (SASRI) Symposium, Mount Edgecombe, South Africa, February 25, 2005.
Cottenie, A., Kang, B.T., Kiekens, L., Sajjapongse, A. (1981) Micronutrient status. In Greenland, D.J. (Ed.) Characterization of soils in relation to their classification and management. Examples from areas in the humid tropics. Oxford University Press, Oxford, pp. 149–163.
Dabin, B., Leneuf, N. (1960) Les sols de bananeraie en cote d'Ivoire. Fruits, 15: 112–127.
Deckers, J., Laker, M., Vanherreweghe, S., Vanclooster, M., Swennen, R., Cappuyns, V. (2000) State of the art on soil-related geo-medical issues in the world. In Låg, J. (Ed.) Geomedical Problems in Developing Countries. Norwegian Academy of Science and Letters, Oslo, pp. 23–42.
De Villiers, J.M. (1980) Highlights of soil chemistry research in southern Africa, 1953–1978. Proceedings of the 8th National Congress, Soil Science Society of South Africa. Technical Communication No. 165, Department of Agricultural Technical Services, ISBN 0621053104, pp. 71–82.
Dickmahns, E.C. (1957) Boron deficiency in sisal. E. Afric. For. J., 22: 197.
Du Preez, C.C., Steyn, J.T., Kotze, E. (2001) Long-term effects of wheat residue management on some fertility indicators of a semi-arid Plinthosol. Soil Tillage Res., 63: 25–33.
Du Toit, J.L. (1956) Copper deficiency discovered in the sugar belt. S. Afr. Sug. J., 40: 797.
Edwards, D.G., Kang, B.T. (1978) Tolerance of cassava (*Manihot Esculenta* (Crantz)) to high soil acidity. Field Crops Res., 1: 337–346.
Egbe, N.E., Omotso, T.I. (1972) Nutrient deficiencies of cocoa in Nigeria. Proc. 4th International Cocoa Research Conference, Trinidad.
El-Fouly, M.M., Fawzi, A.F.A., Firgany, A.H., El-Baz, F.K. (1984) Micronutrient status of crops in selected areas in Egypt. Commun. Soil Sci. Plant Anal., 15(10): 1175–1189.
Eswaran, H., Almaraz, R., van den Berg, E., Reich, P. (1997) An assessment of the soil resources of Africa in relation to productivity. Geoderma, 77: 1–18.
Fageria, N.K., Baligar, V.C., Clark, R.B. (2002) Micronutrients in crop production. Sparks D. L. (Ed.). Adv. in Agron., 77: 185–268.
FAO-UNESCO. (1977) Soil map of the World, Africa. Publ. FAO-UNESCO, Paris, p. 299.
Fenner, R.J., Goldring, O.L. (1968) The application of boron-containing fertilisers to maize. Rhod. Agric. J., 65: 112–116.
Ferrand, M., Bachy, A., Ollagnier, M. (1951) Les oligo elements dans la fumure du palmier a huile au moyen-Congo. Oleagineux, 11: 629–636.
Fewerda, J.D. (1954) Boron deficiency in oil palms in the Kasai region of the Belgian Congo. Nature (London), 173: 1097.

Friesen, D.K., Miller, M.H., Juo, A.S.R. (1980) Liming and lime-phosphorus-zinc interactions in two Nigerian Ultisols. II. Effects on maize root and shoot growth. Soil Sci. Soc. Am. J., 44: 1227–1232.

Gillier, P. (1966) Groundnut and molybdenum. C. r. Séanc. hebd. Acad. Agric. Fr. 52: 446–449. In Soils and Fertiliser, 29 (Abstr.): 4399.

Grant, P.M. (1980) Advances in soil fertility and crop nutrition research in southern Africa, 1953–1978. Proceedings of the 8th National Congress, Soil Science Society of South Africa. Technical Communication No. 165, Department of Agricultural Technical Services, ISBN 0621053104, pp. 83–87.

Greenwood, M., Hayfron. R.J. (1951) Iron and zinc deficiencies in cocoa in the Gold Coast. Emp. J. Exp. Agric., 19: 73–86.

Grobler, J.H., Hugo, P.H.J. (1963) Die sinkstatus van gronde van die Hoëveldstreek. (The zinc status of soils of the Highveld Region) Proc. Internal Symp. on Zinc, Agric. Res. Inst. of the Highveld Region, Potchefstroom, February 1963, pp. 2–8.

Haque, I., Kamara, C.S. (1978) Effect of inoculation and sodium molybdate on groundnuts. Trop. Grain Legume Bull., 11, 12: 35–37.

Haque, I., Odell, R.T., Walker, W.M. (1981a) Micronutrient cations status of some upland soils of Sierra Leone. Trop. Agric. (Trin.), 58: 89–95.

Haque, I., Odell, R.T., Walker, W.M., Kamara, C.S. (1981b) Micronutrient cations status of some lowland soils used for rice production in Sierra Leone. Trop. Agric., (Trin.), 58: 81–88.

Heathcote, R.G. (1970) Soil fertility under continuous cultivation in northern Nigeria. II. Response to fertilisers in the absence of organic manures. Expl. Agric., 6: 345–350.

Heathcote, R.G., Smithson, J.B. (1974) Boron deficiency in cotton in northern Nigeria I. Factors influencing occurrence and methods of correction. Expl. Agric., 10: 199–208.

Ibrahim, A.A. (1982) Mineralogy and micronutrient status of the major soils in the Gezira scheme (Sudan Africa). Ph.D. thesis. Texas A&M University, Texas.

Jones, M.J. (1979) Summaries of work on soil fertility at Swaziland experimental sites 9. Magnesium, sulphur, and micronutrients at Malkerns. Agric. Res. Div. (Swaziland) Research Report, 24: 18.

Kang, B.T. (1978) Effects of liming and nitrogen application on cowpea (*Vigna unguiculata* (L.) Walp.) grown on an Ultisol. Proc. 8th Int. Colloq. Plant Anal. Fert. Problems, Auckland, New Zealand: 231–237.

Kang, B.T., Balasubramanian, V. (1990) Long term fertiliser trials on Alfisols in West Africa. In Trans. 14th International Congress of Soil Science, Kyoto, Japan, August 1990, Volume IV, pp. 20–25.

Kang, B.T., Okoro, E.G. (1976) Response to flooded rice grown on Vertisol from north Nigeria to zinc source and methods of application. Plant Soil, 44: 15–25.

Kang, B.T., Osiname, O.A. (1985) Micronutrient problems in tropical Africa. Fertiliser Res., 7(1–3): 131–150.

Kang, B.T., Abifarin, A.O., Sajjapongse, A. (1976) Iron deficiency problems with upland rice in western Nigeria. Nig. Agric. J., 13: 3–15.

Kayode, G., Agboola, A.A. (1983) Investigations on the use of macro- and micronutrients to improve maize yield in South Western Nigeria. Fertiliser Res., 4: 211–221.

Laker, M.C. (1964) Die invloed van kalk- en fosfaattoedienings op die opname van sink en fosfaat deur plante. (The effects of lime and phosphate applications on the uptake of zinc and phosphate by plants) M.Sc.Agric. dissertation, University of Stellenbosch, South Africa.

Laker, M.C. (1967) Effect of previous applications of lime and zinc on the subsequent uptake of phosphorus and fertiliser zinc by rye plants in a pot experiment. S. Afr. J. Agric. Sci., 10: 11–18.

Laker, M.C. (2000) Soil resources: Distribution, utilization, and degradation. In Fox, R., Rowntree, K. (Eds.) The Geography of South Africa in a Changing World. Oxford University Press Southern Africa, Cape Town, South Africa, pp. 326–360.

Laker, M.C. (2003) Advances in the South African Soil Classification System. In Eswaran, H., Rice, T., Ahrens, R., Stewart, B.A. (Eds.) Soil Classification: A Global Desk Reference. CRC, Boca Raton, FL, pp. 199–218.

Laker, M.C. (2004) Challenges to soil fertility management in the "Third Major Soil Region of the World", with special reference to South Africa. 15th International Symposium of the International Scientific Center for Fertilisers (CIEC), Pretoria, South Africa, September 27–30, 2004.

Laker, M.C., Remmelzwaal, A. (1994) Arenosols: Concept, Classification and Economic Value. Trans. 15th World Congress of Soil Science, Acapulco, Vol. 6a, pp. 860–867.

Lal, R. (2001) Managing world soils for food security and environmental quality. Adv. Agron., 74: 155–192.

Le Mare, P.H. (1970) Soil research in Tanzania. In Soils of the Humid Tropics. National Academy of Science, Washington DC, pp. 210–211.

Lombin, G. (1983) Evaluating the micronutrient fertility of Nigeria's semiarid savannah soil: 2. Zinc. Soil Sci., 136(1): 42–47.

Lupwayi, N.Z., Girma, M., Haque, I. (2000) Plant nutrient content of cattle manures from small-scale farms and experimental stations in the Ethiopian highlands. Agric. Ecosyst. Environ., 78: 57–63.

Magar, B. (1965) Review of the work done on micronutrients in the clay plain of the Republic of Sudan. In Symposium on Maintenance and Improvement of Soil Fertility. Khartoum. OAU/STRC Publications Bureau, Watergate House, York Building, London.

Marchal, J., Pinon, A., Penel, J.P. (1980) Arriere – effets du chaulage en culture d'ananas (var Cayenne lisse) sur un sol ferralitique de cote d'Ivoire. Fruits, 35: 301–308.

Martin, G. (1969) Quelques symptomes de carence en bore chez le palmier a huile. Oleagineux, 24: 613–614.

Martin, G., Fourier, P. (1965) Les oligo-elements dans la culture de l'arachide du nord Senegal. Oleagineux, 20: 287–291.

Meyer, J.H. (2005) Overview of past research and trends in micronutrient deficiency in the South African sugar industry. Combined Fertiliser Society of South Africa (FSSA) and South African Sugarcane Research Institute (SASRI) Symposium, Mount Edgecombe, South Africa, February 25, 2005.

Meyer, J.H., Wood, R.A., Du Preez, P. (1971) A nutrient survey of sugarcane in the SA Sugar Industry with special reference to trace elements. Proc. S. Afr. Sug. Technol. Ass., 45: 196–204.

Moity, M. (1954) La carence en zinc sur bananier. Fruits, 9: 354.

Nyandat, N.N., Ochieng, P.N. (1976) Copper content and availability in soils. A survey of arable and range areas of Kenya. E. Afric. Agric. For. J., 42: 1–7.

Okoye, H.C. (1972) A preliminary assessment of the micronutrient status of the sandy soils of central eastern region of Nigeria. Nig. Agric. J., 9: 183–186.

Ollagnier, M., Valverde, G. (1968) Contribution a l'etud de la carence en bore du palmier a huile. Oleagineux, 6: 359–366.

Osiname, O.A., Kang, B.T., Schulte, E.E., Corey, R.B. (1973) Zinc response of maize (*Zea mays* L.) grown on sandy Inceptisols in western Nigeria. Agron. J., 65: 875–877.

Peters, J.B., Schulte, E.E. (1994) Soil test survey of The Gambia: An overview. Commun. Soil Sci. Plant Anal., 25: 1713–1733.

Peters, J.B., Schulte, E.E. (1996) Using plant analysis to guide research and extension programs in The Gambia. Commun. Soil Sci. Plant Anal., 27(5–8): 1311–1329.

Pinkerton, A. (1967) Copper deficiency on wheat in the Rift Valley. Kenya J. Soil Sci., 18: 18–26.

Pinkerton, A., Barret, M.W., Guthrie, E.J. (1965) A note on copper deficiency in the Njoro area, Kenya. E. Afric. Agric. For. J., 30: 257–258.

Prevot, P. (1959) Oleagineux tropicaux et oligo-elements. Oleagineux, 14: 371–375.

Rhodes, E.R., Kpaka, M. (1982) Effects of nitrogen, molybdenum, and cultivar on cowpea growth and yield on an Oxisol. Comm. Soil Sci. Plant Anal., 13: 279–283.

Roberts, V.G., Adey, S., Manson, A.D. (2003) An investigation into soil fertility in two resource-poor farming communities in KwaZulu-Natal (South Africa). S. Afr. J. Plant Soil, 20: 146–151.
Rothwell, A., Bryden, J.W, Knight, H., Coxe, B.J. (1967) Boron deficiency in cotton in Zambia. Cott. Grow. Rev., 44: 23–28.
Savory, B.M. (1962) Boron deficiency in Eucalyptus in northern Rhodesia. Comm. For. Rev., 41: 118–126.
Sika, M., Terrab, A., Swan, P.B., Hegarty, P.V.J. (1995) Composition of selected Moroccan cereals and legumes: Comparison with the FAO table for use in Africa. J. Food Compost. Anal., 8: 60–70.
Sillanpää, M. (1982) Micronutrients and the nutrient status of soils: A global study. FAO Soils Bulletin No. 48, FAO, Rome.
Sillanpää, M. (1990) Micronutrients assessment at the country level: an international study. FAO Soils Bulletin No. 63, FAO, Rome.
Smaling, E.M.A. (1991) Africa's soils are being mined. Ecol. Farm., 3: 19–20.
Smaling, E.M.A. (1993) The soil nutrient balance: And indicator of sustainable agriculture in sub-Saharan Africa. Peterborough, UK; Fertiliser Society Proceedings – Fertiliser Society No. 340, pp. 1–18.
Smaling, E.M.A., Braun, A.R. (1996) Soil fertility research in sub-Saharan Africa: New dimensions, new challenges. Commun. Soil Sci. Plant Anal., 27(3, 4): 365–386.
Smart, J. (1961) A guide to groundnut cultivation in northern Rhodesia. Rhod. Agric. J., 58: 99–102.
Smithson, J.B. (1972) Differential sensitivity to boron in cotton in the northern states of Nigeria. Cott. Grow. Rev., 49: 350–353.
Smithson, J.B., Heathcote, R.G. (1974) Boron deficiency in cotton in northern Nigeria II. The effect of variety. Expl. Agric., 10: 209–218.
Stanton, D.A. (1964) Studies on zinc in selected Orange Free State soils. D.Sc.Agric. thesis, University of the Orange Free State, Bloemfontein, South Africa.
Stephens, D. (1959) Field experiments with trace elements on annual food crops in Ghana. Emp. J. Expt. Agric., 27: 324–332.
Steyn, C.E., Herselman, J.E. (2005) Trace elements in developing countries using South Africa as a case study. Commun. Soil Sci. Plant Anal., 36: 155–168.
Tanner, P.D. (1976) Manganese toxicity and molybdenum deficiency effects on maize grown in an acid fersiallitic red sandy loam. Rhod. J. Agric. Res., 14: 95–99.
Tanner, P.D. (1982) The molybdenum requirements of maize in Zimbabwe. Zimbabwe Agric. J., 79: 61–64.
Tanner, P.D., Grant, P.M. (1973) Effectiveness of zincated fertilisers for young maize as influenced by fertiliser pH, and method of applying zinc. Rhod. J. Agric. Res., 11: 69–75.
Tanner, P.D., Grant, P.M. (1974) The effectiveness of N, P, and K fertiliser and lime as carriers of zinc for maize (*Zea mays* L.) under field conditions. Rhod. J. Agric. Res., 12: 163–168.
Tanner, P.D., Grant, P.M. (1977) Response of maize (*Zea mays* L.) to lime and molybdenum on acid red and yellow-brown clays and clay loams. Rhod. J. Agric. Res., 15: 143–150.
Tanner, P.D., Cooper, G.R.C., Madziva, T.J.T., Vere, I.P. (1981) Copper deficiency probable cause of the purpling symptom in irrigated winter wheat. Zimbabwe Agric. J., 78: 185–187.
Tinker, P.B.H., Ziboh, C.O. (1959) A study of some typical soils supporting oil palms in southern Nigeria. J. W. Afric. Inst. Oil Palm Res., 3: 16–51.
Vancompernolle, G., Stoops, G., Kayenga, J. (1965) Analyse quantitative de quelques elements traces dans les sols Congolais par fluoresence aux rayons-X. Pedologie, 15: 5–47.
van Heerden, S.M., Schönfeldt, H.C. (2004) The need for food composition tables for southern Africa. J Food Compost Anal, 17: 531–537.
Velly, J., Celton, J., Marquette, J. (1974) Experimentation sur les oligo-elements en culture seche et en riziere a Madagascar. Comm. Reunion d'Agron., IRAT, Paris, 26.
Vlek, P.L.G. (1995) The soil and its artisans in sub-saharan African agriculture. Geoderma, 67: 165–170.

Wapakala, W.W. (1973) The effect of nitrogen, phosphorus, and copper on grain yield of wheat on a copper deficient soil. E. Afric. Agric. For. J., 38: 229–239.
Webb, R.A. (1954) The mineral deficiencies of the Gambian soils. Proc. Second Int. African Soils Conf., I: 419–424.
White, J.G., Zasoski, R.J. (1999) Mapping soil micronutrients. Field Crops Res., 60: 11–26.

Chapter 9
Micronutrient Deficiency Problems in Agricultural Crops in Europe

Alex H. Sinclair and Anthony C. Edwards

Abstract This chapter reports on deficiencies of boron, copper, manganese and zinc in agricultural crops in Europe. Growers rely heavily upon the advice of suppliers of products and on results from plant and/or soil tests to guide decision-making on the need for, and the use of trace elements. The grower must have confidence that his decisions to use certain products on crops are based on correct prediction or diagnosis of deficiency and a satisfactory level of interpretation of likely response expected from specific element applications. Areas and yields of wheat and barley are compared for different countries and the link between yield and potential deficiency is discussed. The impact of changing legislation is also discussed.

9.1 Soils and Agriculture in Europe

The European Union (EU) comprises 27 states. These states are Luxembourg, Ireland, Denmark, Austria, the Netherlands, the United Kingdom, Belgium, Sweden, Finland, France, Germany, Italy, Spain, Cyprus, Greece, Slovenia, Malta, Portugal, Czech Republic, Hungary, Slovakia, Estonia, Lithuania, Poland, and Latvia. Bulgaria and Romania joined on 1 January 2007. States aspiring to be EU members are Croatia and Turkey. Norway, Switzerland and Iceland prefer to stay outside the EU (See Fig. 9.1).

The soil map of Western Europe shows the range of soil types (with the FAO-UNESCO classification) (Fig. 9.2 in Colour Plates). Cambisols are most widespread, e.g., SW England, NW France, Portugal, Spain, Italy and Germany. Highly leached Podzols occur in northern regions, e.g., N Scotland, Denmark and N Germany. Gleysols occur in high rainfall, poorly drained areas of Ireland and central England. Rendzina soils occur in S England, France and Italy.

B.J. Alloway (ed.), *Micronutrient Deficiencies in Global Crop Production.*

Fig. 9.1 Political map of the European Union (www.fco.gov.uk)

9.2 Occurrence of Micronutrient Deficiencies

9.2.1 Boron

Boron (B) may be a component of silicate minerals, adsorbed on clay particles or in organic matter (Shorrocks, 1990). The adsorbed B is probably the main source for plant uptake and the amount will vary with the clay content of the soil. Light sandy soils will contain less adsorbed B than will heavier textured soils. The degree of adsorption is affected by soil pH, being greatest at high pH. Boron deficiency is widespread throughout Europe in responsive dicotyledonous crops grown on calcareous soils (Calcisols), especially under semiarid conditions, e.g., eastern Austria where sugar beet is grown (M. Dachler, 2005, personal communication).

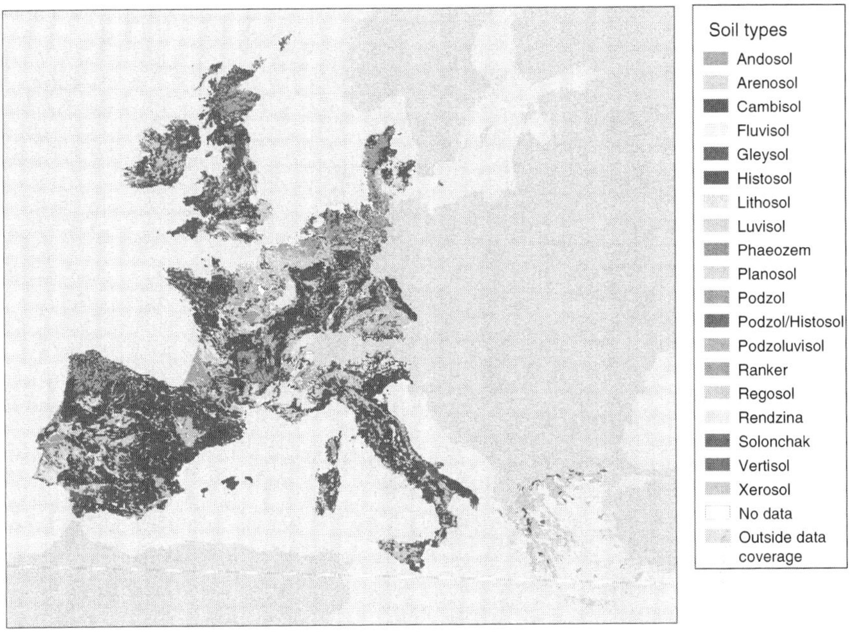

Fig. 9.2 The distribution of soil types in Western Europe (http://dataservice.eea.eu.int/atlas/viewdata/viewpub.asp?id=11) (*See Colour Plates*)

Boron is removed quite readily by leaching from light (sandy) soils with small adsorption capacities. Deficiency therefore tends to occur on light soils with pH values above 6.5. These are found in Poland, Finland, Germany, Greece, Sweden, Denmark, Latvia and in the United Kingdom. Liming will tend to reduce the availability of B. Soil organic matter is also an important reserve of available B for plant nutrition.

Boron fertilisers in common use are Borax, Solubor, liquid inorganics, and B in compound and blended fertilisers (see also Chap 1, Table 1.4). For some responsive crops, e.g., sugar beet, swedes and turnips, soil tests are ignored and B fertiliser use is standard practice, often added to the NPK fertiliser in areas with relatively high rainfall.

9.2.2 Copper

Sinclair and Withers (1995) summarised information on concentrations of total Cu in igneous (typically 10–100 mg kg^{-1}) and sedimentary (4–45 mg kg^{-1}) rocks, and concentrations of total and EDTA-extractable Cu in England and Wales. The total Cu content in mineral and organic agricultural soils in the UK ranges from 1 to about 100 mg kg^{-1}, although much larger concentrations can be found near Cu mining areas, or where there has been very long-term, frequent use of Cu-containing foliar sprays or of waste products with high Cu content. However, normal ranges of total Cu concentrations are 1–15 mg kg^{-1} in very sandy soils, and 25–60 mg kg^{-1} in loamy and clayey soils. More recently Alloway (2005) has reviewed the occurrence of Cu-deficient soils in Europe, and found more published information on Cu deficiency for France than for any other European country. Alloway estimated that the total area of soils likely to cause deficiency in Europe is around 1.2 million km^2.

Copper deficiency is much more likely in acid soils derived from highly siliceous parent material and alkaline soils derived from carbonate-rich sediments. Organic and peaty soils, reclaimed heathland sands and shallow chalk soils with moderate to high (6–12%) organic matter contents are therefore most commonly deficient in Cu. The most extensive areas of Cu-deficient soils in the UK are in south-west and south-east England (shallow chalks) and in East Anglia (peats and heathland soils). In Scotland, Cu deficiency occurs in soils derived from acid schists and granitic rocks, as well as peaty soils.

Copper is readily adsorbed by the clay and organic fractions in the soil, so only a part of the total Cu present in soils is readily available for plant uptake. Some of the Cu is also immobilised by micro-organisms. Copper does not leach easily through the soil, although mobility is slightly greater in sandy than in peaty or clayey soils, which leads to more severe deficiency in dry seasons. Mobility is, however, increased considerably in poorly drained soils. Most of the applied Cu remains in the cultivated topsoil layer of well-drained agricultural soils, often resulting in a sharp decrease in Cu content at the subsoil boundary.

Information supplied for Denmark (L. Knudsen, 2005, personal communication), Germany (Haneklaus, 2005), Finland (Alloway, 2005), Hungary (G. Fuleky, 2005, personal communication), England, Wales and Scotland (Sinclair and Withers, 1995), Portugal (J. C. Pinto, 2005, personal communication), Italy (P. I. Graziano, 2005, personal communication) and Austria (M. Dachler, 2005, personal communication) allowed Table 9.1 to be produced on the approximate extent of agricultural land that is at risk of copper deficiency in responsive crops in different countries.

Deficiency can be treated by soil or foliar application of Cu. Copper sulphate (25% Cu) and Cu oxychloride powder (52% Cu) are used for soil application and, depending on the amount applied and soil texture, the correcting effect could last up to 10 years. A typical application rate is 2.5–5.0 kg Cu ha^{-1}.

Foliar applications are usually of Cu oxychloride (about 25% Cu in liquid formulation) or of chelated Cu (typically around 9% Cu w/v). Application rate is

Table 9.1 Approximate extent of copper deficiency in different European countries (% of agricultural land)

Scotland	30%
Denmark, Germany	25%
Finland	20%
France, Hungary	Significant
England and Wales	5%
Portugal	Rare but widespread use of copper fungicides
Italy	Rare on cereals
Austria	No acute deficiency

between 200 and 500 g Cu ha^{-1} for Cu oxychloride, depending on soil analysis, and 70 g Cu ha^{-1} for chelated Cu. 'Cocktail' foliar feeds containing Cu generally supply much smaller amounts of Cu at recommended product rates than the specific inorganic or chelated Cu products. Copper oxysulphate (a mixture of CuO and $CuSO_4$, typically 20% Cu in powder form) and fritted Cu are also used. The recycling of sewage sludge bioproducts or other wastes, particularly pig slurry, FYM and distillery effluent, to agricultural land can supply substantial amounts of Cu.

9.2.3 Manganese

Manganese deficiency is the most widespread trace element problem in arable crops in the UK and is most commonly seen in cereals, with an estimated 15–20% of the crop area being treated with Mn annually. This deficiency has increased in importance over the years in the UK (Chalmers et al., 1999). Finck (1987) also reported an increased incidence of Mn deficiency in cereal crops in the Schleswig-Holstein area of Germany, which he associated with higher-yielding (8–10 t ha^{-1})) crops and their greater crop demand for Mn. The disorder appears to be most common in cool temperate regions.

Mild and transient deficiency is also commonly seen in cereal crops grown on poorly structured, fine-textured (clay loam) soils with a soil pH over 7.0. Leached sand and podzolic soils are particularly poor in Mn but in most other soils Mn is relatively abundant. Consequently, Mn deficiency is usually induced by low availability of soil Mn for crop uptake, rather than being due to an absolute shortage of soil Mn. The field conditions that can induce Mn deficiency are: high soil pH; high organic matter content; poor root development; poor root–soil contact (in under-consolidated (fluffy) seedbeds), low soil temperatures; and below average rainfall. The overall combination of these factors will dictate the severity of Mn deficiency in crops in any one season. The higher the organic matter content, the lower the soil pH needs to be to prevent deficiency occurring. A temporary shortage of Mn is also often induced under poor soil physical conditions, especially after periods of cold, dry weather that put a poorly rooted crop under stress. Bright sunny weather conditions can also accentuate Mn deficiency, compared with dull, humid conditions.

Over-liming has been responsible for Mn deficiency in diverse regions in the world and it was identified over 50 years ago as the main cause of Mn deficiency in some European countries (Steenbjerg, 1935).

9.2.4 Zinc

Total soil Zn content usually ranges from 10 to 300 mg kg^{-1} depending on the soil parent material. Soils originating from basic igneous rocks have relatively high Zn contents whereas siliceous parent materials have a low Zn content.

Concentrations in soils of England and Wales are given in the Soil Geochemical Atlas of England and Wales (McGrath and Loveland, 1992). The values given are for total Zn (*aqua regia*) and not for the extractable amounts often used in soil tests for deficiency. The median total Zn concentration was 82 mg Zn kg^{-1} dry soil. The concentration in 10% of samples was less than 38 mg Zn kg^{-1} and in 10% was greater than 147 mg kg^{-1}.

The availability of Zn in the soil is affected by pH. Availability is low in soils of high pH (>7.0) and especially when free calcium carbonate ($CaCO_3$) is present. Liming can reduce the availability of Zn. High soil phosphate (P) can be associated with reduced availability of Zn through the formation of insoluble Zn phosphate complexes (Agbenin, 1998). The concentration of soil P must be high for this effect to be significant but could occur where fertiliser P is placed next to the seed.

Zinc deficiency has been reported on calcareous soils (Calcisols) in a number of European countries including Austria, Bulgaria, Cyprus, France, Greece, Portugal, Spain and Turkey, but not, for example, on the calcareous soils of England and Italy. Zinc deficiency has been reported on sandy soils in France, Ireland, the Netherlands, Poland, Portugal, Sweden and Switzerland. The best documented examples of Zn deficiency in Europe and adjacent areas are in Turkey (Cakmak, this volume). However, there is a dearth of statistics on the extent of Zn deficiency in individual countries, in or near Europe, other than from Turkey.

The usual inorganic source is Zn sulphate (monohydrate $ZnSO_4.H_2O$ 35% Zn or septahydrate $ZnSO_4.7H_2O$ 22% Zn). This may be applied to the soil or as a foliar spray (usually around 10% w/w). Zinc oxide (ZnO 67 – 80% Zn) and Zn chloride ($ZnCl_2$ – 45% Zn) are sometimes used. Zinc frits and Zn oxysulphate can be used as slower release soil treatments. Chelated Zn is available in liquid formulations for foliar sprays.

9.3 Soil Analytical Methods

As there is no universally recognised technique for determining the concentration of plant-available trace elements in soil, it is inevitable that there are differences in approaches used by different laboratories within and between countries. Different analytical methods may be used to determine ‘extractable’ levels of a particular element. Consequently threshold levels for deficiency/acceptability may differ according to the proportion of the total element and content extracted by each method. This has been a continuing problem within the industry when attempts have been made to cross-compare results on ‘absolute’ values. Successful interpretation of laboratory results is dependent upon a full understanding of the analytical methods used, and the corresponding threshold levels for deficiency that apply to one particular method. This potential for misinterpretation is clearly illustrated when the threshold values for satisfactory soil trace element status are compared between different analytical laboratories within the UK (Table 9.2).

Table 9.2 Range of threshold levels (mg kg^{-1}) for 'extractable' soil trace elements considered by different laboratories within the UK to be satisfactory for crop growth (Adapted from Chalmers et al., 1999)

Boron	Copper	Manganese	Zinc
>0.5–1.2	>1.7 to >4.1	>2.5 to > 9.0[a]	>1.6 to 2.0

[a]Most laboratories indicated that Mn analysis from soil was very unreliable.

Table 9.3 Range of threshold levels (mg kg^{-1}) for hot water extractable boron considered by different countries within the EU to be deficient for satisfactory crop growth

Soil deficiency level (mg kg^{-1})	Country	Source
<0.3	Greece	
<0.4	Portugal	Calouro, 2005
<0.5–1.2	United Kingdom	Chalmers et al., 1999
<1.0	Ireland, Italy	P. I. Graziano, 2005, personal communication
<3.0	Denmark	L. Knudsen, 2005, personal communication

Table 9.4 Classification levels (mg B kg^{-1}) for acetic acid extractable boron (Dachler, 2005, Personal communication)

Class	Light soil	Medium/heavy soil
Low	<0.2	<0.3
Medium	0.2–2.0	0.3–2.5
High	>2.0	>2.5

The most widely accepted diagnostic test within Europe is hot water extraction of B (modified method of Berger and Truog, 1944). There is not universal agreement in interpretation of a soil deficiency level for crops sensitive to B deficiency (Table 9.3).

However, concentrations at which deficiency symptoms could appear can vary with soil texture. On heavy clay soils, deficiency is probable at less than 0.8 mg B kg^{-1}, on medium texture soils at less than 0.5 mg B kg^{-1} and on light sandy soils at less than 0.3 mg B kg^{-1} (Borax, 2001). In Austria, determination of 'plant available' B is carried out by acetic acid extraction, and classified by texture and extractable value (Table 9.4).

Soil tests used in European countries to predict supply of available Cu and Zn to crops are shown in Table 9.5.

Alloway (2005) has summarised critical concentrations of Cu used in the interpretation of soil and plant analysis in Europe, and estimated the extent of Cu deficiency in different countries. Most test methods contain the extractant EDTA or DTPA. Choice of test can depend on soil pH, but interpretation of test values is crucial. Some countries have different critical levels for deficiency depending on

Table 9.5 Soil tests to predict supply of available copper and zinc to crops in Europe

Soil test	Countries
EDTA (0.05 M; pH 7.0)	Austria, Denmark, France, Ireland, UK
HCl (0.1 M)	France
Ammonium oxalate (0.1 M)	France
Ammonium chloride (M) + EDTA(0.2 M)	Norway
Acid ammonium acetate + EDTA – pH 4.65 (Lakanen and Ervio, 1971)	Finland, Portugal
EDTA + KCl	Hungary
DTPA – pH 7.3 (Lindsay and Norvell, 1978)	Italy, Portugal

Table 9.6 The critical values used in Portugal (LQARS, 2000; Adapted from Sillanpää, 1982)

Trace element	Very low (mg kg^{-1})	Low (mg kg^{-1})
Cu	≤0.3	0.4–0.8
Mn	≤7	8–15
Zn	≤0.6	0.7–1.4

soil texture, e.g., Hungary (G. Fuleky, 2005, personal communication). Median concentrations (mg kg^{-1} dry weight) of soils of Flanders were Cu 9.6 and Zn 34.5. Clay and organic carbon contents were significant in predicting soil trace metal contents in unpolluted soils (Tack et al., 1997). France also includes organic carbon, but there is no consistency across Europe.

Although some soils are known to have low total Cu contents, Cu deficiencies in crops may be rare, e.g., in Portugal (F. Calouro, 2005, personal communication; J. C. Pinto, 2005, personal communication). The Lakanen and Erviö (1971) method is used in Portugal to determine availability of Cu, Mn and Zn in soil (Table 9.6).

Teixeira et al. (1980) and Sequeira (1981) studied the psammosoils (Arenosols FAO-UNESCO) of the Mio-Plio-Pleistocenic of southern Portugal, covering an area of more than 1 Mha, where total (HNO_3-$HClO_4$-HF) Cu ranges from <0.5 to 6.9 mg kg^{-1} (with an average of 2.0 ± 1.2 mg kg^{-1}) and with an average Cu EDTA-extractable level of 0.17 mg kg^{-1}. In Alentejo, in the south of Portugal, there are low Cu levels on some neutral to alkaline soils used for wheat production. However, Cu deficiencies in crops are almost never found even on these soils. The main reason for that is the widespread use of Cu fungicides in the most important crops grown in Portugal: vineyards, orchards, olive trees, and vegetable crops.

Sillanpää (1982) still appears to be the key study in attempting to bring consistency to the interpretation of trace element soil tests. Sillanpää developed statistical correlations between crop yield and the extraction value in combination with one other soil parameter from the six studied (pH, texture, organic carbon content, CEC, electrical conductivity and $CaCO_3$ equivalent). The soil factor that gave greatest improvement in predicting Cu deficiency was organic matter, while pH was best for Mn and Zn. There is a wide range in topsoil carbon across Europe (Fig. 9.3 and in Colour Plates). A Europe-wide evaluation of existing databases

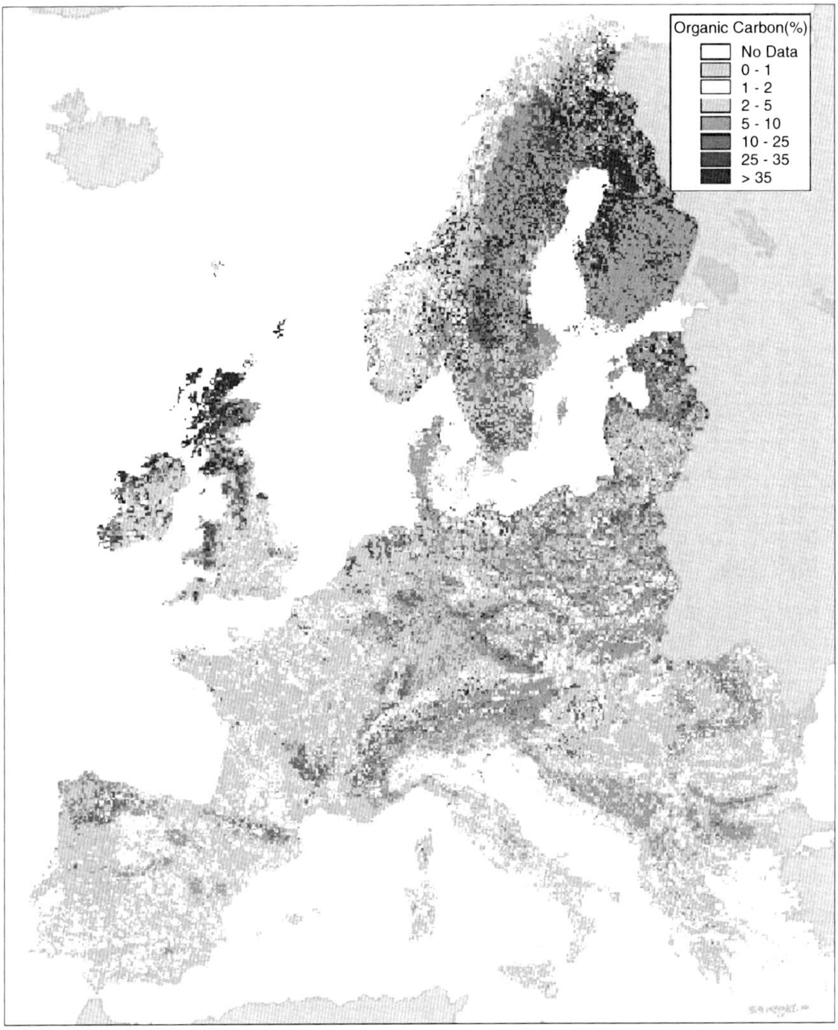

Fig. 9.3 Topsoil organic carbon concentrations in Europe (Jones et al., 2005) (*See Colour Plates*)

for trace elements and organic matter contents in soils started in 2000 (European Soil Bureau, 2001). Bussink and Temminghoff (2004) considered that the approach by Sillanpää should be regarded as a transition phase towards predicting micronutrient availability based on the extraction result, the solid phase characterisation and multi-surface models. These authors concluded

> 'For future development, no new extractants should be developed but the knowledge gained from soil chemistry should be embedded into making better predictions of micronutrient availability'.

9.4 Plant Analysis

As well as helping to diagnose observed deficiency symptoms, plant analysis may identify subclinical deficiencies and enable nutrient inputs for the current and possibly future crops to be correctly modified (Bergmann, 1992). As an alternative approach to using nutrient ranges in plants, the diagnostic and recommendation integrated system (DRIS), originally developed by Beaufils (1971), uses nutrient ratios together with soil and environmental parameters to assess plant nutrient status. However, Kadar et al. (1981) concluded that this system was no better than standard plant or soil testing techniques for determining fertiliser requirements, unless there is an extreme deficiency or excess of one or more nutrients. Bergmann (1992) also suggested that the use of nutrient ratios, to assess the nutritional status of the plant, provides no real advantage over a system based on ranges of nutrient concentrations, unless the nutrient levels in plants are extreme. Shuman et al. (1992), for example, found that the sufficiency range method gave almost identical results to the DRIS approach in predicting Mn deficiency or sufficiency in soya beans.

Some general guidelines on the use of plant analysis, and interpretation of results, are:

- The interpretation of plant analysis is not an exact science. The concentration of any specific nutrient depends on the plant species, age of the plant, part of plant analysed, variety and even the level of the other nutrients present. Plant analysis will indicate those nutrients that are clearly adequate or deficient. In those cases where the values are borderline or a number of nutrients appear deficient, expert advice must be taken.
- For most nutrients, values decline with maturity. Plant analysis ideally should be carried out early in the growing season when the plants are young. Interpretation of the results is easier and this also allows sufficient time for foliar nutrient sprays to be applied to overcome any deficiency.
- All samples submitted for analysis must be clean and free from dust or soil and ideally should not have been recently sprayed with any nutrient-containing spray. The presence of soil can lead to falsely elevated values, making interpretation impossible. Analysis of both affected (i.e., showing apparent deficiency symptoms) and, for comparison, unaffected plant samples is often helpful for interpreting the results.

Threshold guidelines used by ADAS (England and Wales) and SAC (Scottish Agricultural Colleges) for deficiencies of non-NPK nutrients in cereal plants are given in Table 9.7.

Table 9.7 ADAS and SAC guidelines on threshold concentrations for non-NPK nutrient deficiencies in susceptible crops

Nutrient	Deficiency threshold
Boron	Less than 20 mg kg^{-1}
Copper	Less than 3–4 mg kg^{-1} in leaf/ear
	Less than 2 mg kg^{-1} in grain
Manganese	Less than 20 mg kg^{-1}
Zinc	Less than 15 mg kg^{-1}

Plant analysis cannot replace soil analysis, but is a valuable tool in diagnosing trace element problems.

9.5 A Role for Trace Element Budgets

The lack of standardisation in analytical techniques and hence the disparity in guidelines on threshold levels for deficiency in soil or crop, do create problems in interpreting results. As an alternative approach to using nutrient ranges in plants, trace element budgets have been prepared. Very different balances can exist for individual elements.

Moolenaar (1999) suggest the aim of sustainable trace element management in agro-ecosystems is to ensure that the soil continues to fulfil its functions: in agricultural production, in environmental processes such as the cycling of elements, and as a habitat of numerous organisms. There has been increasing interest in the calculation and use of substance budgets for both major and trace elements. The approach makes it possible to calculate and therefore represent element flows over a range of geographical and temporal scales (see *European Journal of Agronomy* 2003 (Special issue Vol. 20 (1–2)) for examples). The complexity of individual budget calculations can vary considerably depending upon the range of information that they include. It is possible to represent substance flows at the individual field, farm, region or national scales. Of the four elements being considered here, most of the budget information present in the literature relates to Cu and Zn, less is available for Mn and very little for B. Table 9.8 shows a comprehensive assessment of the inputs and outputs of Cu and Zn for cultivated soils in Finland. In this example fertilizer and manure represented over 90% of the measured inputs of Cu and Zn. These calculations suggest that both elements would be expected to be accumulating in these soils. Regional scale averaged predictions for Switzerland indicated enrichment for Zn of between 76 and 525 g ha^{-1} for soils of arable and dairy farms,

Table 9.8 Copper and zinc balances of Finnish cultivated soils (g ha^{-1} $year^{-1}$) with the percentage of total inputs and outputs shown in brackets (Data from Mäkela-Kurtto, 1996 cited in Moolenaar et al., 1999)

	Cu	Zn
Deposition	5.9(3)	15.7(4)
Fertilisers	100.0(54)	91.0(22)
Manure	72.0(39)	278.0(68)
Lime	2.3(1)	11.6(3)
Sludge	6.0(3)	12.2(3)
Total in	186.2	408.5
Crops	18.7(56)	115.0(82)
Leaching	4.3(13)	8.0(6)
Runoff	10.5(31)	18.0(13)
Total out	33.5	141.0
Accumulation	152.7	267.5
Soil content (mg kg^{-1})	Cu 21	Zn 36

Table 9.9 Annual copper and zinc input (t) inventory to agricultural land in England and Wales for the year 2000 (Nicholson et al., 2003)

Source	Cu	Zn
Atmospheric deposition	631	2,457
Livestock manure	643	1,858
Sewage sludge	271	385
Industrial 'wastes'	13	45
Total inorganic fertilisers	53	266
Agrochemicals	8	21
Irrigation water	2	5
Composts	<1	<1
Total	1,621	5,038

respectively (Keller and Schulin, 2003). There appeared to be greater accumulation of Zn on grassland compared to arable land.

The total inputs of Cu and Zn for England and Wales are shown in Table 9.9 which demonstrates the relative significance of atmospheric deposition, especially for Zn where it represents approximately 50%.

Farm gate budgets for major nutrients, started in the Netherlands on dairy farms, and these have been extended to include trace elements (Tables 9.10–9.12). In Table 9.10, 'Total removal' includes metals in animal products (milk, meat, etc) as well as export of manure.

There is a need for a common methodology (Öborn et al., 2003) to be adopted. One particular issue related to calculating budgets is the quality of information that is actually available, because this can vary considerably for individual substances. Farm-level Zn budgets have been calculated as part of the Swedish Food 21 research programme (Food 21). Balances can differ greatly for individual metals. For example, Bengtsson et al. (2003) calculated the field balance for a Swedish organic and conventional dairy system. While Zn showed a positive balance, Cu showed a negative balance. Inputs from the atmosphere and losses through runoff were significant contributors to the balance where: Manure + Urine + Fertiliser + Lime + Deposition – Crop – Leachate – Run-off (Table 9.13).

This has allowed comparisons to be made between individual crop and farm types. For example, Gustafson et al. (2003) compare barn level flows and balances of Zn for a 50 cow Organic and Conventional farm. There were no differences in Zn concentrations of individual materials between farm systems (Table 9.14). At the barn level, there were greater fluxes of Zn out than could be accounted for in inputs, suggesting an 'internal' source of Zn. This could be due to galvanised cladding and would not have been picked up by farm-gate balances.

The example shown in Table 9.14 demonstrates that it is possible that individual sources of a particular substance, e.g., Zn from runoff from galvanized roofing or Cu from drinking water pipes can be identified. Inputs from farm buildings are likely to be spatially localised and therefore can introduce variability at the farm scale.

Trace element budgets are influenced by farming and manure management systems (Tables 9.15 and 9.16).

Table 9.10 Average farm gate metal balances (g ha^{-1} $year^{-1}$) for arable land on sandy soils (17 sites) and grassland with cattle production on peat (17 sites) in the Netherlands (After Groot et al., 1998)

	Arable land (sand)		Grassland (peat)	
Source	Cu	Zn	Cu	Zn
Fertiliser	22	76	7	30
Manure	323	566	44	66
Atmospheric deposition	6	23	9	40
Feedstuff	0	0	95	321
Other	23	88	4	10
Total supply	374	753	160	467
Total removal[a]	44	241	3	50
Leaching	56	163	44	97
Net surplus	274	349	113	320

[a]Includes metals in animal products (milk, meat, etc.) as well as export of manure.

Table 9.11 Average fluxes of Cu and Zn on all 84 study farms. Values in brackets give the range between 5% and 95% (90 percentile range)

	Metal flux (g ha^{-1} $year^{-1}$)	
Metal Source	Cu	Zn
Input	269 (63–608)	742 (190–1,511)
Output	8.6 (1.5–37)	82 (31–207)
Leaching plough layer	131 (84–224)	461 (163–1,003)
Accumulation plough layer	129 (−157–460)	199 (−780–1,081)
Leaching unsaturated zone	37 (7.1–81)	234 (34–741)
Accumulation unsaturated zone	223 (1.3–546)	426 (−221–1,279)

Table 9.12 Average fluxes of Cu and Zn for four farm types. Both leaching and accumulation refer to the plough layer (0–30 cm)

		Metal flux (g ha^{-1} $year^{-1}$)			
Metal	Type of farm	Input	Output	Leaching	Accumulation
Cu	Intensive cattle	398	4.1	104	289
	Cattle sand	206	2.4	110	94
	Cattle peat	160	2.2	178	−21
	Arable	362	31	155	176
Zn	Intensive cattle	1,170	63	340	767
	Cattle sand	663	50	435	177
	Cattle peat	467	46	633	−212
	Arable	710	199	464	47

9.6 Intensity of Use of Agricultural Land and Future Pressures

There is a large range in intensity of use of agricultural land in Europe (Fig. 9.4 and in Colour Plates). Southern England, the Netherlands, northern France through the central area to Poland is the most intensive. Crop yield also varies widely across

Table 9.13 Annual inputs and outputs of Cu and Zn (g ha^{-1} $year^{-1}$) for the arable fields in an organic (50 ha) and conventional (39 ha) farming system at the Öjebyn farm 1998, given as an average for the crops included in the crop rotation (From Bengtsson et al., 2003)

	Cu		Zn	
	Organic	Con.	Organic	Con.
Input				
Manure	50	64	369	487
Urine	7	16	18	42
Fertiliser	–	6	–	4
Lime	0	0	2	5
Deposition	3	3	82	82
Total input	60	89	471	620
Output				
Harvest	37	35	127	143
Drainage[a]	36	36	18	18
Surface run-off[a]	68	68	17	17
Total output	141	139	162	178
Balance (in–out)	−81	−50	+310	+442

[a]Water flow is an average value for organic and conventional fields with ley and is the same for all crops.

Table 9.14 Flows of Zn in organic and conventional dairy systems, where F – Harvested feed crops and purchased feed, B – bedding. M – milk, C – calves, U – urine and MA – solid manure. Barn balance between input and output (F + B) – (M + C + U + MA) (kg $year^{-1}$ per cow)

	F	B	M	C[a]	U + MA	Balance
Organic	0.36	0.004	0.03	0.002	0.46	−0.12
Conventional	0.42	0.004	0.04	0.002	0.42	−0.036

[a]Total pools of the elements of incoming heifers and cows for slaughter cancelled each other.

Table 9.15 Trace elements (mg kg^{-1}) in manure from different farming and manure management systems (Steineck et al., 1999)

	Cattle				Pig	
	Ecological		Conventional		Conventional	
	Solid	Slurry	Solid	Slurry	Solid	Slurry
Mn	208 ± 75	201 ± 47	230 ± 32	247 ±78	264 ± 74	302 ± 63
Zn	148 ± 66	156 ± 43	174 ± 32	190 ± 41	791 ± 294	635 ± 153
Cu	29 ± 14	32 ± 16	31 ± 8	49 ± 42	130 ± 50	178 ± 40

Table 9.16 Concentrations of trace elements calculated g t^{-1} (wet weight) stored solid manure and slurry (Steineck et al., 1999)

	Cattle				Pig	
	Ecological		Conventional		Conventional	
	Solid	Slurry	Solid	Slurry	Solid	Slurry
Mn	34	15	38	22	63	25
Zn	24	11	32	18	180	55
Cu	4.5	2.0	4.8	2.9	30.4	14.4

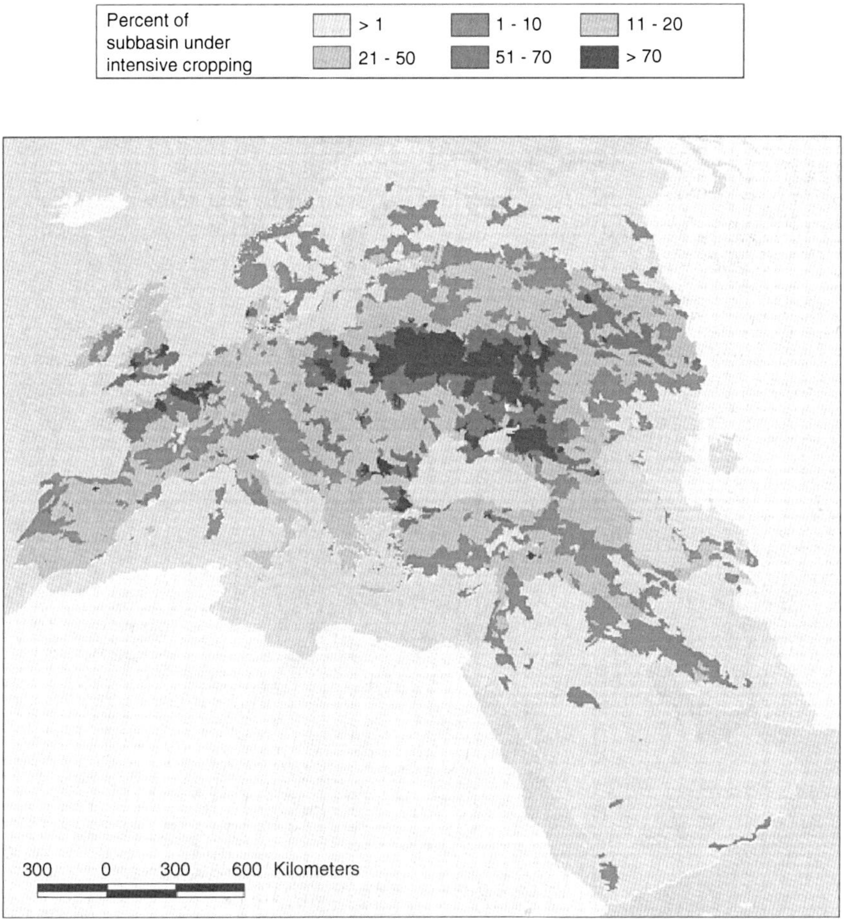

Fig. 9.4 Intensive agricultural land use by river sub-basin in Europe and the Middle East (World Resources Institute (Page, 2000)) (*See Colour Plates*)

Table 9.17 Areas and yields of wheat and barley for the 2003 harvest in Europe

Country	Wheat area (ha)	Wheat (t ha^{-1})	Barley area (ha)	Barley (t ha^{-1})
Finland	191,300	3.55	529,500	3.21
France	4,905,000	6.23	1,750,000	5.61
Germany	2,967,379	6.50	2,087,100	5.11
Greece	851,300	1.92	100,300	1.85
Italy	2,266,760	2.75	309,306	3.32
The Netherlands	134,700	9.12	56,200	6.61
Portugal	180,000	1.26	10,000	1.00
Spain	2,218,000	2.83	3,089,000	2.82
Sweden	411,740	5.55	367,580	4.20
United Kingdom	1,837,000	7.78	1,078,000	5.91

Europe. Areas and yields of wheat and barley for the 2003 harvest (taken from FAOSTAT) are shown in Table 9.17. Highest yields were obtained in the Netherlands and the lowest yields in Greece and Portugal. Very different situations exist between countries with negligible trace element application to cereals in Italy (P. I. Graziano, 2005, personal communication), whereas Mn application is widespread in the UK with a 65% yield loss if severe Mn deficiency is left untreated. However, differences in yield between countries are probably strongly correlated with annual precipitation (Fig. 9.5 and in Colour Plates). Annual precipitation ranges from less than 400 mm to more than 2,500 mm. There is a coastal influence with wet coastal areas and dry central areas.

With effect from January 2005, the granting of the majority of payments under the reformed Common Agricultural Policy (CAP) within the EU will be linked to environmental, food safety, animal welfare and plant health standards. A number of statutory requirements are specified which must be complied with in the context of

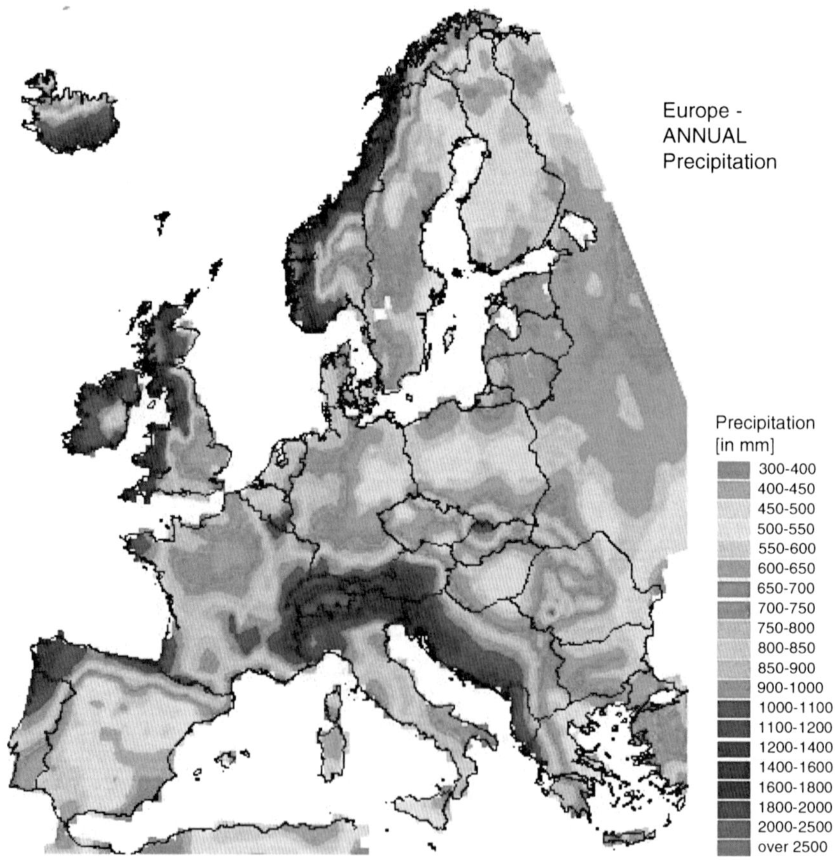

Fig. 9.5 Map of annual precipitation across Europe (http://www.iiasa.ac.at/Research/LUC/GIS/img/eur_prcy.jpg) (*See Colour Plates*)

cross-compliance and receipt of Single Farm Payments. In addition to this, Article 5 (1) of the Regulation establishes that all farmland must be kept in Good Agricultural and Environmental Condition (GAEC) (Scottish Executive, 2004). There is currently a debate on what the effect of this reform will have on the intensity of agriculture within the EU. One strong possibility is lower financial inputs in the less intensive areas. Other EU legislation will reduce N and P inputs.

The European Nitrates Directive 91/676/EEC requires legally binding rules to be put in place for Nitrate Vulnerable Zones (NVZs) to reduce nitrate loss from agricultural land when nitrate levels exceed, or are likely to exceed, the levels set in the Directive. These rules often result in lower inputs of N fertilisers. The EU Water Framework Directive sets targets for achieving good ecological status for water bodies and requires that the entire water environment is managed as a whole, from source to sea, considering impacts on rivers, lochs, estuaries, coastal and ground waters. It will extend environmental protection for point and diffuse sources of pollution and is expected to limit applications of P fertiliser. It is possible that these reforms will result in an increase in trace element deficiencies in some crops, but a reduction in others.

Higher cereal yields means that more Cu and other micronutrients are required, and removed from the soil in the harvested crop. This raises the question of whether soils can provide an adequate supply of micronutrients to meet the greater demands of a higher yielding crop. Field trials in north-east Scotland in 1985–1986 investigated whether increased levels of extractable soil Cu were needed for high-yielding cereal crops (Sinclair et al., 1990). Yield responses of 14–15% were obtained at two sites, one with the lowest soil Cu status (0.5 mg kg^{-1}) and the other at the lowest-yielding site, out of the five sites testing foliar applications of 0.5 kg ha^{-1} Cu oxychloride, applied as equal splits at Zadoks Growth Stage (GS) 22/30 and GS 31/32. Another site tested a foliar applied 'cocktail' of Cu + Mn + Zn, also applied as equal splits at GS 22/30 and GS 31/32, combined with three N rates (150, 200, 250 kg ha^{-1}) to create different yield levels. A significant ($P < 0.05$) yield response (+10%) to the Cu + Mn + Zn 'cocktail' was obtained at 150 kg ha^{-1} N, but micronutrient application had no effect on yield at the higher N rates, which produced site yields of 9.2–9.4 t ha^{-1}. Sinclair et al. (1990) suggested that non-pH dependent mobilisation of Cu increases as fertiliser N rate is increased, probably due to greater production of chelating agents in root exudates by high-yielding, vigorous cereal crops.

Two sites also tested seedbed applications of 10 kg ha^{-1} Cu oxychloride, combined with three rates of N (nil, 150, 200 kg ha^{-1}). Significant ($P < 0.05$) yield responses of 22–25% were obtained at nil N input, despite a moderate soil Cu status (1.6–2.1 mg kg^{-1}), but there was no response to Cu application at either 150 or 200 kg ha^{-1} N rates, despite doubling the plant off-take of Cu in those treatments which had nil Cu but received N fertiliser. Soil solution measurements at these two sites showed that mobilisation of Cu into solution in the root-zone soil, which occurred through the season in N-fertilised plots gave similar concentrations of Cu to those plots with nil N.

The results from these Scottish trials were consistent with the idea that high-yielding crops may be more efficient in obtaining Cu from the soil, given similar

Table 9.18 Mn and P concentration of wheat shoot dry matter after 116 days from sowing

	Soil treatment				
Element	P^0L^0	P^1L^0	P^0L^1	P^1L^1	S.E.
Mn (mg kg $^{-1}$)	59.1	40.6	24.3	12.8	0.78
P (mg g^{-1})	1.93	2.20	2.00	2.49	0.004

P^0 = no added PO_4; P^1 = added PO_4; L^0 = no added lime; L^1 = added lime
S.E. = Standard error of treatment means.

levels of extractable soil Cu prior to seed sowing, as far greater mobilisation of Cu occurs under a high-yielding crop. Thus, reducing N inputs may lead to an increase in Cu deficiency in crops.

The capacity of soil surfaces to absorb and retain transition element ions at a given pH is known to be enhanced by the adsorption of P onto oxide surfaces (Diaz-Barrientos et al., 1990). The increasing and continued use of P fertilisers in arable agriculture has resulted in a build-up of P in some soils. It is possible that this has resulted in increasing adsorption of Mn onto oxide surfaces in these soils, thus causing depressed soil solution concentrations and enhanced Mn deficiency in various crops.

A glasshouse experiment was designed to compare Mn concentrations in soil solutions and the Mn nutrition of wheat plants in soils which had received different amounts of both P and lime some 18 years previously (Neilsen et al., 1992). These treatments had significantly altered both P status and soil pH. The use of these historically treated soils avoided problems of interpretation resulting from non-equilibrium conditions that occur in recently adjusted soils. Throughout the growing season, the concentration of Mn in leaf tissue was lower in the limed treatments and in the P treatments, while leaf P concentrations were higher in plants from high P status soils and in those from limed soils compared with unlimed soils (Table 9.18).

The depressed Mn values for limed treatments were explained in terms of depressed soil solution Mn concentrations resulting from elevated pH. However, the results for high P soils could not be related to soil solution composition. Neilsen et al. (1992) suggested that high soil P resulted in elevated plant P that interfered in the uptake and/or translocation of Mn, rather than by a direct effect of soil chemistry on Mn availability.

9.7 Conclusions

The lack of standardisation in analytical techniques and hence the disparity in guidelines on threshold levels for deficiency in soil or crop, creates problems in interpreting results. As an alternative approach to using nutrient ranges in plants, trace element budgets have been prepared. Very different balances can exist for individual metals. There is a need for a common methodology to be adopted. One particular issue related to calculating budgets is the quality of information that is actually available because this can vary considerably for individual substances.

Recent EU legislation aimed at minimising nitrate loss to watercourses is expected to reduce inputs of N fertiliser, while future legislation is expected to result in lower P inputs. Scottish trials are consistent with the idea that high-yielding crops may be more efficient in obtaining Cu from the soil, as far greater mobilisation of Cu occurs, under a high-yielding crop. Thus, reducing N inputs may lead to an increase in Cu deficiency in crops. Conversely fewer high P soils may reduce the incidence of Mn deficiency.

References

Agbenin, J.O. (1998). Phosphate-induced zinc retention on a tropical semi-arid soil. European Journal of Soil Science, 49: 693–700.

Alloway, B.J. (2005). Copper-deficient soils in Europe. International Copper Association, New York, 129 pp.

Beaufils, E.R. (1971). Physiological diagnosis – a guide for improving maize production based on principles developed for rubber trees. Fertiliser Society of South Africa Journal, 1: 1–30.

Bengtsson, H., Öborn, I., Jonsson, S., Nilsson, I., Anderson, A. (2003). Field balances of some mineral nutrients and trace elements in organic and conventional dairy farming – a case study at Öjebyn, Sweden. European Journal of Agronomy, 20: 101 – 116.

Bergmann, W. (Ed.) (1992). Nutritional disorders of plants – visual and analytical diagnosis. Colour Atlas. Gustav Fischer Verlag Jena. Stuttgart. New York, 386 pp.

Borax (2001). Boron deficiency: It's Prevention and Cure, Borax Europe, Guildford, UK. Available at www.borax.com

Bussink, W., Temminghoff, E. (2004). Soil and tissue testing for micronutrient status. Proceedings No.548, International Fertiliser Society, York, UK, 1–41.

Chalmers, A.G., Sinclair, A.H., Carver, M. (1999). Nutrients other than nitrogen, phosphorus and potassium (NPK) for cereals. HGCA Research Review No.41, pp. 88–90.

Diaz-Barrientos, E., Madrid, L., Contreras, M.C., Morillo, E. (1990). Simultaneous adsorption of zinc and phosphate on synthetic lepidocrocite. Australian Journal of Soil Research, 28: 549–557.

European Soil Bureau (2001). Progress Report on 'Trace element and organic matter contents of European soils' – First results of the second phase of the 'Short Term Action', E.U. Ispra, Italy.

Finck, A. (1987). Improvement of manganese supply by fertilization in high-yielding cropping of northern Germany. Paper for Food and Agriculture Symposium, Geneva. January 1987.

Food 21 http://www-mat21.slu.se/eng/synthesis/synthesiswork.htm (accessed on the 12/11/05)

Groot, M.S.M., Bronswijk, J.J.B., Willems, W.J., de Haan, T., del Castilho, P. (1998). National Soil Quality Monitoring Network: Results of 1995, Report No. 714801024. National Institute for Public Health and the Environment (RIVM) Bilthoven, The Netherlands.

Gustafson, G.N., Salomon, E., Jonsson, S., Steineck, S. (2003). Fluxes of P, K and Zn in a conventional and an organic dairy farming. system through feed, animals, manure, and urine – a case study at Öjebyn, Sweden. European Journal of Agronomy, 20: 89–99.

Jones, R.J.A., Hiederer, R., Rusco, E., Montanarella, L. (2005). Estimating organic carbon in the soils of Europe for policy support. European Journal of Soil Science, 56: 655–671.

Kadar, I., Pusztai, A., Lasztity, B., Sarkadi, J., és Welisch, P. (1981). Diagnosis and recommendation system (DRIS): A review on a new possibility of estimating the fertilizer requirement. Agrokémia és Talajtan, 30: 465–486.

Keller, A., Schulin, R. (2003). Modelling regional-scale mass balances of phosphorus, cadmium and zinc fluxes on arable and dairy farms. European Journal of Agronomy, 20: 181–198.

Lakanen, E., Erviö, R. (1971). A comparison of eight extractants for the determination of plant available micronutrients in soils. Acta Agr. Fenn., 123: 223–232.

Lindsay,W.L., Norvell, W.A. (1978). Development of a DTPA soil test for zinc, iron, manganese and copper. Soil Science Society of America Proceedings, 42: 421–428.

LQARS (2000). Manual de fertilização das culturas. Lisboa: Laboratório Químico Agrícola Rebelo da Silva, 221 pp.

Mäkela-Kurtto, R. (1996). Heavy metal balances in Finnish cultivated soils (unpublished report of the Agricultural Research Centre of Finland).

McGrath, S.P., Loveland, P.J. (1992). The Soil Geochemical Atlas of England and Wales. Blackie, Glasgow, UK.

Moolenaar, S.W. (1999). Heavy-metal balances, Part II. Management of Cadmium, Copper, Lead, and Zinc in European Agro-Ecosystems. Journal of Industrial Ecology, 3(1): 41–53.

Neilsen, D., Neilsen, G.H., Sinclair, A.H., Linehan, D.J. (1992). Soil phosphorus status, pH and the manganese nutrition of wheat. Plant and Soil, 145: 45–50.

Nicholson, F.A., Smith, S.R., Alloway, B.J., Carlton-Smith, C., Chambers, B.J. (2003). An inventory of heavy metals inputs to agricultural soils in England and Wales. The Science of the Total Environment, 311: 205–219.

Obron, I., Edwards, A.C., Witter, E., Onenema, O., Ivarsson, K., Withers, P.J.A., Nilsson, S.I., Stinzing, A.R. (2003). Elements balances as a tool for sustainable nutrient management: a critical appraisal of their merits and limitations within an agronomic and environmental context. European Journal of Agronomy, 20 (1–2):211–225.

Sequeira, E.M. (1981). [Psammosoils of the Mio-Plio-Plistocenic of the Southern Portugal. V – Copper] Solos arenosos da região Mio-Plio-Plistocénica a Sul do Tejo. V – O cobre. Agronomia lusitana, 41(2): 149–164.

Shorrocks, V.M. (1990). Behaviour, Function and Significance of Boron in Agriculture. Report on an international workshop, Oxford, July 1990.

Shuman, L.M., Wilson, D.O., Hallmark, W.B. (1992). Evaluating SOY-DRIS for predicting manganese deficiency and sufficiency. Communications in Soil Science and Plant Analysis, 23: 1019–1029.

Sillanpää, M. (1982). Micronutrients and the nutrient status of soils: A global study. FAO Soils Bulletin 48: 444.

Sinclair, A.H., Linehan, D.J., Ross, J.A.M. (1990). Are higher-yielding cereal crops more prone to micronutrient deficiency. Proceedings of Crop Protection Conference in Northern Britain, Dundee, pp. 101–106.

Sinclair, A.H., Withers, P.J.A (1995). Copper deficiency in UK cereal crops: Occurrence, Significance and Treatment. HGCA Research Review No. 31, Home grown Cereals Authority, London.

Steineck, S., Gustafson, G., Andersson, A., Tersemeden, M., Bergström, J. (1999). Plant nutrients and trace elements in livestock wastes in Sweden. Report 5111, Swedish Environmental Protection Agency, Kundservice, S-106 48 Stockholm.

Steenbjerg, F. (1935). The exchangeable manganese in Danish soils and its relation to plant growth. Trans. 3rd Int.Congr. Soil Sci. Oxford, 1: 198–201.

Tack, F.M.G., Werloo, M.G., Vanmechelen, L., Van Ranst, E. (1997). Baseline concentration levels of trace elements as a function of clay and organic carbon contents in soils in Flanders (Belgium). Science of the Total Environment, 201: 113–123.

Teixeira, A.J.S., Sequeira, E.M., Lucas, M.D., Santos, M.J. (1980). Some Psammosoils of the Mio-Plio-Plistocenic of Southern Portugal. Soil properties. Total and extractable micronutrients] Solos arenosos da região Mio-Plio-Plistocénica a sul do Tejo. I – Características, micronutrientes totais e extraíveis. Agronomia lusitana, 40(1): 41–78.

Chapter 10
Micronutrient Deficiency Problems in South America

Nand K. Fageria and Luís F. Stone

Abstract South America has largest land area at global level to produce food and fibre crops. In addition, climatic conditions (temperature and water availability) are favourable, which further enhances the role of this continent in providing world food security. The Brazilian Cerrado, or savanna, a total area of about 205 Mha of acid soils is a good case in point. Similarly, Colombia, Bolivia, Venezuela, Peru and Ecuador also have large land areas, which can be utilized for crop production. However, the major soils of this continent are acidic and infertile. Hence, liming and fertiliser application are essential. Micronutrient deficiencies are an emerging limiting factor for annual crop production. In annual crops such as rice, corn, wheat, soybean and common bean, deficiencies of Zn, Cu, B, Mn and Fe have been reported. Adopting sound soil and crop management practices will not only enhance crop productivity but also help in reducing deforestation of tropical rainforests in South America. This strategy will permit less CO_2 release to the atmosphere, conservation of soil, water and global climatic change. In South America, micronutrient management issues require a great of deal of research for improvement.

10.1 Introduction

The essential micronutrients for crop plants are zinc (Zn), iron (Fe), manganese (Mn), copper (Cu), boron (B), molybdenum (Mo), chlorine (Cl) and nickel (Ni). In some publications, cobalt (Co) is also cited as an essential micronutrient; however, the essentiality of this element is not proved for crop plants (Fageria et al., 2002). Chlorine is classified as a micronutrient, even though its concentration in plant tissue is often equivalent to those of macronutrients. Micronutrients are required by plants in small amounts compared to macronutrients, but their influence is as important as the macronutrients in crop production. Based on physiological properties, the essential plant micronutrients are all metals except for B and Cl. For micronutrients, the deficient and toxic concentration range is very narrow compared with macronutrients. For macronutrients, the sufficiency range is very broad and toxicity

B.J. Alloway (ed.), *Micronutrient Deficiencies in Global Crop Production.*

rarely occurs. Among the micronutrients the deficiency and toxicity range is narrower for Mo and B than for any of the other micronutrients. Most of the micronutrients are immobile in plants, therefore, deficiency symptoms first appear in the younger leaves. The accumulation of micronutrients by plants generally follows the order of Cl > Fe > Mn > Zn > B > Cu > Mo (Fageria and Baligar, 1997; Marschner, 1995). This order may change among plant species and with growth conditions (e.g., flooded rice).

Increasing crop yields has become an essential requirement of modern society to keep pace with the increasing world population. High food production could be achieved by expanding the area under crops and by increasing crop productivity on the land under cultivation. South America offers both of these opportunities to increase world food supply. In South America, there is great potential to increase food production because of the availability of a large land area and favourable climatic conditions. The Brazilian Cerrado, or savanna, comprising a total area of about 205 Mha of acid soils is a good case in point. Similarly, Colombia, Bolivia, Venezuela, Peru and Ecuador also have large land areas, which can be utilized for crop production.

The Amazon Basin covers an area of some 700 Mha and the central part is almost entirely located within Brazilian territory (Cerri et al., 2003). This region has the highest rates of deforestation in the world. Between 1.5 and 2.4 Mha of forest were cleared every year over the last 2 decades and the total area deforested now exceeds 50 Mha (Cerri et al., 2003). Most of this cleared land is used for pasture or crop production. The substitution of native vegetation on such a large scale for pasture or crop production is likely to have an impact on nutrient and organic matter composition as well as a regional impact on hydrology and water quality. Furthermore, conversion of tropical forest to agricultural production has important implications for carbon storage in soils and global climate (Cerri et al., 2003). Therefore, proper management of South American soils is of both socio economic and ecological importance for the region and the world as a whole.

In the future, the strategy should be not to clear the forest land for agricultural management but increase crop productivity per unit area. To achieve this objective, it is important to understand the soil fertility of the region. The incidence of micronutrient deficiency in crops grown in Cerrado region of Brazil and other South American regions has increased markedly in recent years. This has happened due to intensive cropping, loss of top soil by erosion, losses of micronutrients through leaching, liming of acid soils, decreasing proportions of farmyard manure inputs to chemical fertilisers, increased purity of chemical fertilisers, high nutrient demands of modern crop cultivars and involvement of natural and anthropogenic factors that limit plant availability and create element imbalances. The majority of South American soils are highly weathered, subject to intensive leaching, and are vulnerable to soil erosion. Degradation of these soils poses a great challenge to sustainable agriculture in the regions. The objectives of this chapter are to discuss micronutrient problems in South America and suggest appropriate measures to correct deficiencies for annual crop production.

10.2 Major Soil Groups in South America and Micronutrient Deficiencies

The major soil groups in South America according to the USDA Soil Taxonomy classification system are: Oxisols, Ultisols, Alfisols, Mollisols, Entisols, Inceptisols, Vertisols, Aridisols, Andisols, Histosols, and Spodosols (Sanchez and Logan, 1992). However, Oxisols and Ultisols dominate in most of the South American countries (Table 10.1). Key profile characteristics of Oxisols are: highly weathered, uniform texture, high in iron (Fe) and aluminium (Al) oxides with kaolinite clay (low activity clay), low cation exchange capacity and small amounts of exchangeable bases. Similarly, Ultisols are characterized by an argillic horizon with a base saturation of <35% when measured at pH 8.2 and presence of Fe and Al oxides (Brady and Weil, 2002).

Micronutrient deficiencies in annual crops in South America are mainly Zn, Cu, B, Mn and Fe. Deficiencies of these micronutrients are generally observed early in the growth cycle of annual crops (35–50 days after sowing) (Fageria, 1989; Fageria et al., 2002). This may be associated with rapid growth of crop plants during early growth stages and an insufficient supply of micronutrients from the soil to keep pace with crop growth. The availability of micronutrients is mainly associated with the total contents of these elements in the soil and available forms in soil solution. The dominant forms of micronutrients that occur in soil solution are presented in Table 10.2.

The major factors which affect the concentration of micronutrients in soil include parent materials, pH, organic matter content, texture and mineralogy. The total concentration of micronutrients in the surface horizon is generally high and decreases with increasing soil depth (Table 10.3). Micronutrient deficiencies in the Oxisols and Ultisols of South America have been frequently reported (Fageria et al., 2002). For example, deficiency of Zn is very common in upland rice and corn grown on Oxisols of central Brazil (Fageria et al., 1997). Fageria and Baligar (1997) reported significant

Table 10.1 Distribution of Oxisols and Ultisols in South America (Cochrane, 1978)

Country	Area (Mha)	Total area of the country (%)
Brazil	572.71	68
Colombia	67.45	57
Peru	56.01	44
Venezuela	51.64	58
Bolivia	39.54	57
Guiana	12.25	62
Surinam	11.43	62
Paraguay	9.55	24
Equador	8.61	23
French Guiana	8.61	94
Chile	1.37	2
Argentina	1.28	0.4

Table 10.2 Dominant micronutrient forms in soil solution (Lindsay, 1979, Fageria et al., 1997, Brady and Weil, 2002)

Micronutrient	Forms in soil solution
Zinc	Zn^{2+}, $Zn(OH)^{+}$
Copper	Cu^{2+}, $Cu(OH)^{+}$
Iron	Fe^{3+}, Fe^{2+}, $Fe(OH)^{2+}$, $Fe(OH)_2$
Manganese	Mn^{2+}, Mn^{4+}
Boron	H_3BO_3, $H_2BO_3^-$
Molybdenum	MoO_4^{2-}, $HMoO_4^-$
Chlorine	Cl^-

Table 10.3 Micronutrient concentrations (mg kg^{-1}) at 0–10 and 10–20 cm soil depths in a Brazilian Inceptisol under different soil fertility treatments across three common bean crops (Compiled from Fageria and Baligar, 1996)

Soil fertility level	Cu	Zn	Mn	Fe
0–20 cm				
Low	6.9	2.8	52	1,020
Medium	7.1	3.9	83	1,026
High	7.2	4.9	47	885
Medium + green manure	6.2	3.2	54	898
20–40 cm				
Low	5.6	1.4	42	759
Medium	5.9	1.8	58	729
High	5.9	1.9	33	631
Medium + green manure	5.6	1.4	37	765

The fertility levels were low (without addition of fertilisers); medium (35 kg N ha^{-1}, 52 kg P ha^{-1}, 50 kg K ha^{-1}, 40 kg fritted glass material as a source of micronutrients) and high (all the nutrients were applied at the double the medium level). Pigeon pea (*Cajanus cajan* L.) was used as a green manure at the rate of 25.6 Mg ha^{-1} green matter.

increases in dry matter yields of common bean, upland rice and corn grown on Brazilian Oxisols with the application of Zn, Fe, Cu, B, Mo and Mn. Rosolem et al. (2001) reported responses of cotton, peanut and soybean to application of Zn, B, and Mn fertilisers in Brazilian Oxisols. Similarly, Galrão (1991) reported responses in soybean to Cu application in Oxisols of central Brazil. Barbosa Filho et al. (2001) reported Zn deficiency in rice, corn and wheat in Brazilian soils.

10.3 Crop Requirements and Management Strategies

Although micronutrients are required in small amounts, each micronutrient has certain specific roles to play in the plant. Their availability in adequate amounts is important for good plant growth and for completing its life cycle. The nutrient requirements of a crop depend on its productivity. Micronutrient requirements for annual crops are small compared with macronutrients (Table 10.4). However, their

Table 10.4 Uptake of macro and micronutrients by upland rice, common bean, corn and soybean grown on Brazilian Oxisols (Fageria, 2001a)

Nutrient	Straw	Grain	Total	Required to produce 1 t of grain
Upland rice				
Nitrogen (kg ha^{-1})	56	70	126	28
Phosphorus (kg ha^{-1})	3	9	12	3
Potassium (kg ha^{-1})	150	56	206	45
Calcium (kg ha^{-1})	23	4	27	6
Magnesium (kg ha^{-1})	14	5	19	4
Zinc (g ha^{-1})	161	138	299	65
Copper (g ha^{-1})	35	57	92	20
Iron (g ha^{-1})	654	117	771	169
Manganese (g ha^{-1})	1,319	284	1,603	351
Boron (g ha^{-1})	53	30	83	18
Common bean				
Nitrogen (kg ha^{-1})	19	68	87	45
Phosphorus (kg ha^{-1})	1	6	7	4
Potassium (kg ha^{-1})	25	36	61	32
Calcium (kg ha^{-1})	16	6	22	11
Magnesium (kg ha^{-1})	7	4	11	6
Zinc (g ha^{-1})	29	74	103	54
Copper (g ha^{-1})	8	22	30	16
Iron (g ha^{-1})	268	144	412	215
Manganese (g ha^{-1})	73	27	100	52
Boron (g ha^{-1})	20	14	34	18
Corn				
Nitrogen (kg ha^{-1})	72	127	199	24
Phosphorus (kg ha^{-1})	4	17	21	3
Potassium (kg ha^{-1})	153	34	187	23
Calcium (kg ha^{-1})	33	8	41	5
Magnesium (kg ha^{-1})	20	9	29	4
Zinc (g ha^{-1})	184	192	376	46
Copper (g ha^{-1})	53	14	67	8
Iron (g ha^{-1})	2,048	206	2,254	274
Manganese (g ha^{-1})	452	82	534	65
Boron (g ha^{-1})	103	43	146	18
Soybean				
Nitrogen (kg ha^{-1})	33	91	124	86
Phosphorus (kg ha^{-1})	4	8	12	9
Potassium (kg ha^{-1})	30	28	58	40
Calcium (kg ha^{-1})	33	6	39	27
Magnesium (kg ha^{-1})	14	10	24	16
Zinc (g ha^{-1})	43	78	121	84
Copper (g ha^{-1})	53	31	84	58
Iron (g ha^{-1})	778	190	968	671
Manganese (g ha^{-1})	193	32	225	156
Boron (g ha^{-1})	22	21	43	30

role in crop production is as important as those of macronutrients. Overall, Fe and Mn were having minimum utilization efficiency (grain yield per unit of nutrient accumulation in straw and grain), whereas Cu and B were having maximum utilization efficiency for annual crop production (Table 10.4). Zinc ranks intermediate among micronutrients in grain production efficiency for annual crops. Among micronutrients, Cl is supplied through rainwater and Mo deficiency is corrected by liming in acid soils. Therefore, this review will discuss the requirements and management strategies of Zn, Cu, B, Mn and Fe.

Understanding nutrient interactions is vital for improving uptake and use efficiency of micronutrients in crop production. Two or more growth factors are said to interact when their influence individually is modified by the presence of one or more of the others (Sumner and Farina, 1986). Interactions may be positive or negative depending on the growth response. If the growth response is greater with two combined factors as compared to the sum of their individual effects, it is a positive (or synergistic) interaction, and when the combined effects are less, the interaction is negative (Sumner and Farina, 1986).

Furthermore, soil and plant tissue testing are important nutrient management strategies for maximizing crop yields and similar for all essential nutrients. Hence, before discussing requirements and management of individual micronutrients, soil and plant tissue subject is discussed here to cover five micronutrients.

10.3.1 Soil and Plant Tissue Testing

Soil and plant tissue testing are important micronutrient deficiency/toxicity diagnostic techniques. In South American countries, soil testing is more commonly used to identify micronutrient disorders in crop plants than plant analysis. The reason for this is that soil tests are cheaper and farmers' need for fertiliser recommendations. Soil testing is the term applied to chemical or physical measurements that are made on a soil. The main objective of soil chemical testing is to measure soil nutrient status and lime requirements in order to make fertiliser and lime application recommendations for profitable farming. The use of soil analysis as a fertiliser recommendation method is based on the existence of a functional relationship between the amount of nutrient extracted from the soil by chemical methods and crop yield. When a soil analysis test shows a low level of a particular nutrient in a given soil, application of that nutrient would be expected to increase crop yield. The best method so far reported for Fe, Mn, Cu and Zn extraction seems to be the DTPA (diethylene triamine pentacetic acid) test developed by Lindsay and Norvell (1978), which involves 1:2 soil/solution extraction with 0.005 DTPA, 0.01 M $CaCl_2$ and 0.1 M triethanolamine, adjusted to pH 7.3. Adequate and toxic soil levels of Zn, B, Cu and Mn for upland rice, dry bean, corn, soybean and wheat grown on Brazilian Oxisols are shown in Table 10.5. Adequate concentrations of Fe in plant tissue reported for some important crops are presented in Table 10.6.

Plant analysis is based on the concept that the concentration of an essential element in a plant or part of the plant indicates the soil's ability to supply that nutrient. For annual crops, the primary objective of plant analysis is to identify nutritional

Table 10.5 Adequate and toxic micronutrient levels in soil and plant tissue of major field crops grown on a Brazilian Oxisol (Fageria, 2000b, c, 2001b)

	DTPA-extractable soil level ($mg\ kg^{-1}$)[b]		Plant tissue level ($mg\ kg^{-1}$)	
Micro nutrient	Adequate[a]	Toxic[a]	Adequate[a]	Toxic[a]
Upland rice				
Zinc[c]	4.0	35.0	67	673
Boron[d]	0.4	2.3	10	20
Copper[e]	1.0	28.0	15	26
Manganese[f]	4.0	80.0	520	4,560
Dry bean				
Zinc	0.3	25.0	18	133
Boron	0.9	2.8	24	135
Copper	0.5	18.0	6	10
Manganese	6.0	88.0	400	1,640
Corn				
Zinc	1.0	60.0	27	427
Boron	1.3	5.7	20	68
Copper	1.5	32.0	7	11
Manganese	4.0	336.0	60	2,480
Soybean				
Zinc	0.3	33.0	20	187
Boron	2.6	5.2	75	155
Copper	0.5	6.0	7	10
Manganese	4.0	56.0	67	720
Wheat				
Zinc	0.3	34.0	19	100
Boron	0.4	4.3	13	144
Copper	8.5	28.0	14	17
Manganese	3.0	40.0	173	720

[a]Adequate level in soil and plant tissue was determined on the basis of 90% of relative yield of dry matter and toxic level was determined on the basis of 10% reduction in relative yield. [b]Boron was extracted by hot water. [c], [d], [e], [f]Plants were harvested at 6, 4 and 4 weeks after sowing and at physiological maturity, respectively

Table 10.6 Iron sufficiency levels for different crops (Compiled from various sources by Fageria et al., 1990)

Crop	Plant part analyzed	Growth stage	Sufficiency range (mg Fe kg^{-1} dry matter)
Rice	Whole top	Tillering	70–300
Wheat	Whole tops	Heading	50–150
Barley	Whole tops	Heading	50–150
Corn	Ear leaf	At silk	50–200
Sorghum	Third leaf below head	At bloom	65–100
Common bean	Full developed trifoliate	Flowering	100–450
Peanut	Upper stem and leaves	Flowering	50–300
Soybean	Full developed trifoliate	Prior to pod set	51–350
Sugarcane	Blade	Not given	60–140
Cotton	Mature leaves	Early bloom	30–300
Sugarbeet	Blade	Not given	60–140

Table 10.7 Concentration of Fe in the shoot of lowland rice at different growth stages. Values in parenthesis represent plant age in days after sowing (Compiled from Fageria, 2004)

Plant growth stage	Fe concentration (mg kg^{-1})
Initiation of tillering (22)	418
Active tillering (35)	373
Panicle initiation (71)	175
Booting (97)	155
Flowering (112)	179
Physiological maturity(140)	261

problems or to determine or monitor the nutrient status during the growing season. If a deficiency of element(s) is identified early in the growth stage of a crop, it is possible to correct such a deficiency during the current season. Otherwise, steps should be taken to correct such deficiency on or before the next cropping season. Nutrient concentrations in plant tissues generally decreased with the advancement of plant age (Table 10.7). Adequate and toxic levels of plant Zn, Cu, B and Mn for upland rice, common bean, corn, soybean and wheat grown on Brazilian Oxisols are presented in Table 10.5.

10.4 Zinc

Zinc deficiency has been reported in various parts of the world for a large number of annual crops (Cakmak et al., 1998; Fageria et al., 2002). Additionally, about 50% of the soils used worldwide for cereal production contain low levels of plant-available Zn (Graham et al., 1992; Welch, 1993). Zinc deficiency is widely reported in South American soils and it is mainly associated with low total contents of this element in the highly weathered Oxisols and Ultisols (Fageria et al., 2002; Fageria et al., 2003). Deficiency of Zn in South American soils is further aggravated by high soil pH due to excessive lime application. Soil pH is more important than any other single property for controlling Zn mobility in soils. Increasing soil pH generally decreased Zn availability to plants and such decreases were usually due to higher adsorption of Zn (Table 10.8). As soil pH increases above pH 5.5, Zn is adsorbed on hydrous oxides of Al, Fe, and Mn (Moraghan and Mascagni, 1991).

10.4.1 Zinc: Functions and Deficiency Symptoms

Zinc is a cofactor for several enzymes that are involved with N metabolism (e.g., glutamate dehydrogenase) and anaerobic metabolism (e.g., alcohol dehydrogenase) (Fageria et al., 2003). Zinc increases root growth in annual crops grown on Brazilian Oxisols (Fageria, 2000b). It also promotes growth hormones, starch formation, and seed maturation (Fageria et al., 2002). Zinc deficiency symptoms in crop plants are presented in Table 10.9.

Table 10.8 Influence of soil pH on acquisition of Zn, Cu, Fe, and Mn by upland rice grown in an Oxisol of Brazil (Fageria, 2000a)

Soil pH	Zn (μg $plant^{-1}$)	Cu (μg $plant^{-1}$)	Fe (μg $plant^{-1}$)	Mn (μg $plant^{-1}$)
4.6	1,090	75	4,540	11,160
5.7	300	105	1,860	5,010
6.2	242	78	1,980	4,310
6.4	262	64	1,630	3,610
6.6	163	61	1,660	2,760
6.8	142	51	1,570	2,360
R^2	0.98**	0.89*	0.97**	0.99**

*, **Significant at 5% and 1% probability levels, respectively.

Table 10.9 Micronutrient deficiency symptoms in crop plants (Bennett, 1993, Baligar et al., 1998, Fageria et al., 1997, 2002)

Micronutrient	Symptoms
Zinc	Deep yellowing of whorl leaves (cereals). Dwarfing (rosette) and yellowing of growing points of leaves and roots (dicots). Rusting in strip on older leaves with yellowing in mature leaves. Leaf size reduced. Main vein of leaf or vascular bundle tissue becomes silver-white, and marked stripes appear in middle of leaf. Root length in cereals as well as in legumes reduced with zinc deficiency.
Copper	Yellowing of younger leaves. Rolling and dieback of leaf tips. Leaves are small. Tillering is retarded. Growth of tops and roots is stunted.
Iron	Interveinal yellowing of younger leaves with distinct green veins. Entire leaves become white with severe deficiency and leaf border turn brown and die. Root growth is reduced in iron-deficient plants.
Manganese	Interveinal tissue becomes light green with veins and surrounding tissue remaining green on dicots and long interveinal leaf streaks on cereals. Develop necrosis in advanced stages. Root growth is reduced in cereals as well as legumes.
Boron	Death of growing points of shoot and root. Failure of flower buds to develop. Blackening and death of tissues, especially inner tissue of brassica plants.

10.4.2 Zinc: Interactions with Other Nutrients

Understanding interactions between micronutrients and various mineral nutrients is important for balancing nutrient supplies to plants, improving growth and yields of plants, and eliminating deficiencies and toxicities imposed on plants (Fageria et al., 2002). Zinc interactions with P and other micronutrients such as Fe, Cu and Mn are important. The most widely reported interaction with Zn is that of P. High P applied to low Zn soils enhanced the plant accumulation of P thereby increasing the internal plant Zn requirement because of Zn precipitation (Robinson and Pitman, 1983). Fageria and Zimmermann (1979) reported that liming and P reduced Zn uptake in upland rice grown on Brazilian Oxisols and induced Zn deficiency. Phosphorus deficiency is very common in South American Oxisols and Ultisols and requires large amounts of P for achieving good crop yields. High levels of P have also resulted in increased absorption and retention of Zn in roots and decreased translocation to leaves (Iorio et al., 1996). This may lead to Zn deficiency in crops grown on these highly weathered Oxisols and Ultisols low in available Zn (Fageria, 1989). Higher levels of Fe, Cu and Mn decrease uptake of Zn and vice versa (Fageria et al., 2002; Fageria, 2002).

10.4.3 Zinc: Management Practices

Use of adequate sources of Zn, rate and methods are vital to correct Zn deficiency in crop plants. Important compounds used as micronutrient fertilisers, including sources of Zn are listed in Chap. 1 (Table 1.4-1.10). Similarly, Zn application rates to soil and in foliar spray solutions are presented in Table 10.10.

In Brazilian Oxisols, application of $ZnSO_4$ (23% Zn) at the rate of 30–50 kg ha^{-1} as a top dressing can correct Zn deficiency in upland and lowland rice, and maize (Fageria et al., 1989). Application of Zn to soil as $ZnSO_4$ can increase its total content significantly (Table 10.11). Use of Zn-efficient genotypes is a very attractive strategy for correcting Zn deficiency in crop plants. Results obtained by the authors at the National Rice and Bean Research Center of EMBRAPA, Brazil, showed that upland rice genotypes differed significantly in relation to shoot dry weight, grain yield, panicle number, grain harvest index and Zn harvest index, when tested under two Zn levels (0 and 10 mg Zn kg^{-1} soil) (Table 10.12). Similarly, genotypic variations in Zn uptake and utilization in Zn-deficient soils have been reported (Fageria et al., 2002; Fageria and Boligar, 2005; Graham and Rengel, 1993).

10.5 Copper

Copper deficiency in annual crops grown on Brazilian Oxisols has been reported in recent years (Galrão, 1999; Fageria, 2001b). This deficiency is related to intensive cropping as well as use of liming material to improve the pH of these acid soils. In addition, crop recovery values for micronutrients generally range from only 5% to 10% (Mortvedt, 2000). Some of the reasons for the low efficiency of micronutrient fertilisers are poor distribution of the low rates applied to soils, fertiliser reactions with soils to form unavailable reaction products, and low mobility in the soil to reach plant root (Martens and Lindsay, 1990; Fageria et al., 1997).

Table 10.10 Zinc: Methods of correcting micronutrient deficiencies (Martens and Westermann, 1991, Fageria et al., 1997, 2002)

Micronutrient	Soil application	Foliar application
Zinc	1–10 kg Zn ha^{-1}	0.1–0.5% solution of $ZnSO_4.7H_2O$ and required about 400 L water per hectare of field crop
Copper	1–2 kg Cu ha^{-1}	0.1–0.2% solution of $CuSO_4.5H_2O$ and required about 400 L water per hectare of field crop
Iron	Soil application is not recommended in South American limed Oxisols and Ultisols due to high rate requirements which is uneconomical	1–2% solution of $FeSO_4.7H_2O$ and required about 400 L water hectare of field crop
Manganese	5–50 kg Mn ha^{-1}	0.1–0.2% solution of $MnSO_4.H_2O$ and required about 400 L of water per hectare of field crop
Boron	1–6 kg B ha^{-1}	0.1–0.25% solution of boric acid and required about 400 L of water per hectare of field crop

Table 10.11 Influence of broadcast application of zinc sulfate and copper sulfate fertilisers on DTPA-extractable soil Zn and Cu in Brazilian Oxisols determined after harvest of three bean and two upland rice crops

Zn rate (kg ha^{-1})	Zn in soil (mg kg^{-1})	Cu rate (kg ha^{-1})	Cu in soil (mg kg^{-1})
0	2.0	0	1.5
5	2.6	2.5	1.7
10	3.0	5	2.2
20	4.4	10	2.9
40	6.4	20	4.8
80	13.1	40	8.9
R^2	0.98*	R^2	0.99*

*Significant at the 1% probability level.

10.5.1 Copper: Functions and Deficiency Symptoms

Copper plays an important role in many biochemical functions in plants such as chlorophyll formation, a precursor of several enzymes, participation in both protein and carbohydrates metabolism, and is required in symbiotic N_2 fixation (Fageria and Gheyi, 1999; Mengel et al., 2001). It is essential for chloroplast development and photosynthesis. Impaired lignification of cell walls is the most typical anatomical change induced by Cu deficiency in higher plants. This gives rise to the characteristic distortion of young leaves, bending and twisting of stems and increase in lodging susceptibility (Marschner, 1995). Copper is immobile in phloem in plants, hence deficiency of this element first appears on younger leaves. Copper deficiency symptoms in crop plants are presented in Table 10.9.

Table 10.12 Shoot dry weight, grain yield, number of panicles, grain harvest index (GHI), (across two Zn levels) and Zn harvest index (ZnHI) of 10 upland rice genotypes

Genotype	Shoot dry wt. ($g\ pot^{-1}$)	Grain yield ($g\ pot^{-1}$)	Number of panicles pot^{-1}	GHI	ZnHI Zn_0	ZnHI Zn_{10}
Bonança	64.25ab	52.93c	21.00bc	0.45ab	0.60abc	0.25ab
Caipó	89.52de	55.60bc	19.17c	0.38e	0.62ab	0.25ab
Canastra	79.02c	60.47abc	24.00abc	0.43ab	0.64ab	0.35a
Carajas	62.07ab	60.67abc	24.50ab	0.50cd	0.76a	0.32ab
Carisma	70.38a	61.82abcd	26.17ab	0.47ab	0.60abc	0.31ab
CNA8540	66.85ab	68.78d	25.83ab	0.53cd	0.68a	0.37a
CNA8557	79.83c	71.33d	26.83a	0.47ab	0.46bc	0.30ab
Guarani	58.43b	64.92abcd	25.67ab	0.53cd	0.70a	0.22b
Maravilha	83.98ce	60.45abc	27.33a	0.42ab	0.57abc	0.26ab
IR42	96.26d	68.15d	35.83d	0.41ab	0.40c	0.21b
Average	75.06	62.51	25.63	0.46	0.60	0.28
F test						
Zn level	**	*	NS	NS	**	
Genotype	**	**	**	**	**	
Zn X G	NS	NS	NS	NS	**	
CV (%)	9	12	15	8	12	

*, **, NS Significant at the 5% and 1% probability levels and non-significant, respectively.
Means in the same column followed by the same letter are not statistically different at the 5% probability level by the Tukey's test.

10.5.2 Copper: Interaction with Other Nutrients

Nutrient interactions are generally measured in terms of growth responses and changes in mineral nutrient concentrations in plants. Applications of relatively high levels of N and P fertilisers have induced Cu deficiency in plants grown in low Cu soils (Fageria et al., 2002). Copper uptake is inhibited by other divalent cations, especially Zn^{2+} (Table 10.11). Since Zn and Cu are absorbed into the root by the same carrier, each of these nutrients competitively inhibits uptake of the other (Fageria et al., 2002).

10.5.3 Copper: Management Practices

Copper deficiency can be corrected by banding or broadcasting Cu to soil or as foliar sprays. Band application is generally one-third of broadcast application. The rate applied depends on soil type, concentration in the soil and crop species. However, in general 1–2 kg Cu as a soil application and a 0.1–0.2% of $CuSO_4$ solution foliar spray can correct Cu deficiency in most crop plants (Table 10.10). Table 10.11

shows that Cu applied as $CuSO_4$ can significantly increase the total Cu concentration in the soil. Use of Cu-efficient crop genotypes is an important strategy to overcome Cu deficiency problems in South American soils, but this information is not available for these soils.

10.6 Manganese

Manganese deficiency in South American soils is mainly due to soil erosion and the liming of acid soils to raise pH or base saturation. Data in Table 10.14 show that available Mn contents of Brazilian Oxisols are significantly reduced with increasing base saturation or pH. In the Cerrado region of Brazil, the average pH of unfertilized and unlimed Oxisols is about 5.0 in water. However, for most annual crops maximum yields are normally obtained with a pH of 6.0–6.5 or base saturation of about 60% for these soils (Fageria et al., 1997). Therefore, liming is an essential and effective practice to improve crop yields on South American acid soils. Increasing soil pH decreases Mn uptake by upland rice grown on Brazilian Oxisols (Table 10.8).

10.6.1 Manganese: Functions and Deficiency Symptoms

The role of manganese in many biochemical functions in plants such as the formation of chlorophyll, as a precursor of respiratory enzymes, activating several important metabolic reactions, accelerating germination and maturity, and increasing the availability of P and Ca is well-documented (Fageria and Gheyi, 1999). It is involved in the O_2-evolving system of photosynthesis and is a component of the enzyme arginase and phosphotransferase (Fageria et al., 1997). Manganese is immobile in the plant tissues and hence its deficiency first appears in the younger leaves. Deficiency symptoms of Mn are given in Table 10.9.

10.6.2 Manganese: Interaction with Other Nutrients

Uptake of Mn in crop plants is inhibited by Ca, Mg, Zn, Cu and Fe (Fageria et al., 2002). Increasing rates of Zn significantly reduce uptake of Mn in upland rice grown on Brazilian Oxisols (Table 10.13). In lowland rice Mn is applied to correct iron toxicity under reduced soil conditions (Fageria et al., 1990; Fageria et al., 1997). Similarly, increasing concentration of Fe in the growth medium may also decrease Mn toxicity (Marschner, 1995). Problems associated with Fe–Mn interactions have been related mainly to chemical interactions at the root–soil interface (Kochian, 1991).

Table 10.13 Influence of Zn on uptake of Cu and Mn in upland rice plants grown on an Oxisol of Brazil (Fageria, 2002)

Zn rate (mg kg^{-1})	Cu (mg kg^{-1})	Mn (mg kg^{-1})
0	13	1,075
5	14	740
10	12	663
20	12	643
40	12	785
80	11	593
120	10	440
R^2	0.80**	0.55*

*, **Significant at 5% and 1% probability levels, respectively.

Table 10.14 Influence of base saturation and pH on DTPA-extractable soil Mn determined after harvest of common bean crop in Brazilian Oxisol. Values are averages of 0–10 and 10–20 cm soil depths and 3 years field experimentation

Base saturation (%)	pH in water	Extractable Mn (mg kg^{-1})
23	5.4	8.1
54	6.4	6.4
68	6.8	5.5

Base saturation = Σ(Ca, Mg, K)/(CEC) × 100, CEC = cation exchange capacity.

Table 10.15 Influence of broadcast application of Mn and Fe sulfate on DTPA-extractable soil Mn and Fe in Brazilian Oxisols determined after harvest of two upland rice and three common bean crops

Mn rate (kg ha^{-1})	Soil Mn (mg kg^{-1})	Fe rate (kg ha^{-1})	Soil Fe (mg kg^{-1})
0	3.31	0	23.18
10	4.28	50	25.27
20	5.13	100	24.16
40	6.35	150	24.76
80	8.22	200	25.08
160	12.72	400	27.34
R^2	0.91*	R^2	0.58*

*Significant at 1% probability level.

10.6.3 Manganese: Management Practices

Soil or foliar application of $MnSO_4$ can correct Mn deficiency. The quantity of Mn required varies from soil to soil, crop to crop and other management practices. Soil application rates may vary from 5–50 kg Mn ha^{-1}. Higher rates are applied as broadcast and lower rates as band applications. Data in Table 10.15 show that Mn in soil increased significantly with the addition of $MnSO_4$. Foliar sprays of 0.1–0.2% solution of $MnSO_4$ can correct Mn deficiency. Sometimes more than one application is required to achieve satisfactory results. Use of Mn-efficient crop genotypes is a very promising strategy to correct Mn deficiency in crop plants. Variations in uptake and use efficiency of Mn among crop species and genotypes of the same species have been reported (Marschner, 1995; Fageria et al., 2002).

10.7 Iron

Iron deficiency is a worldwide problem and occurs in numerous crops (Fageria et al., 1990; Marschner, 1995). In Brazil, Fe deficiency has been observed in upland rice, soya bean and sorghum grown on Oxisols. Iron deficiency in these crops occurs not because of Fe scarcity in soil but because of various soil and plant factors that affect Fe availability and inhibit its absorption or impair its metabolic utilization (Marschner, 1995; Fageria et al., 2002). Availability of Fe is reduced by high soil pH (>6.0), high P, high levels of Zn, Cu and Mn, low and high temperatures, high levels of nitrate N, high organic matter content, poor aeration, unbalanced cation ratios and root infection by nematodes (Fageria et al., 1990). Hansen et al. (2003) reported that Fe chlorosis in soya bean was associated with a greater soil moisture content, high concentrations of soluble salts, high carbonate contents and had lower concentrations of DTPA-extractable Fe and Mn compared with nonchlorotic areas.

Soil reaction (pH) is one of the most important factors affecting Fe availability to plants. Soil pH is known to influence the solubility of Fe in soil (Fageria et al., 1990). As pH increases, Fe is converted to less soluble forms, principally to the oxide Fe_2O_3. The reaction responsible for the reduced solubility of Fe with increasing pH is well understood. It results in the precipitation of $Fe(OH)_3$ as the concentration of OH^- ions is increased as indicated by the following reaction:

$$Fe^{3+} + 3OH^- = Fe(OH)_3 \tag{1}$$

The $Fe(OH)_3$ is chemically equivalent to the hydrated oxide, $Fe_2O_3.3H_2O$. Acidification shifts the equilibrium, causing a greater release of Fe^{3+} as a solution ion. Liming is an essential management practice for crop production in South American acid soils. However, when the pH values of Cerrado soils in Brazil are increased from 5 to 6, Fe deficiency has been observed in upland rice, maize, sorghum and soybean, even when soil analysis showed a high total Fe concentration (Fageria et al., 1990). When Fe was applied as $FeSO_4$ fertiliser, DTPA-extractable Fe did increase, but the magnitude of the increase was much less than other micronutrients such as B, Zn, Cu and Mn (Table 10.15). Hence, conversion of Fe to plant-unavailable forms was much higher compared with other micronutrients.

10.7.1 Iron: Functions and Deficiency Symptoms

Iron is one of the most important elements for plant growth, especially for chloroplast development, photosynthesis, respiration and deoxyribonucleic acid (DNA) synthesis (Marschner, 1995). It is also an essential component of many heme and nonheme Fe enzymes and carriers, including the cytochromes (respiratory electron carriers) and the ferredoxins (Fageria et al., 1997; Marschner, 1995). Iron-deficient plants have smaller root systems. Iron is an immobile nutrient in plant tissues, hence its deficiency first starts in younger leaves in all crop plants. The deficiency is exhibited as a chlorosis developing interveinally in the new leaves. In the beginning, Fe deficiency is similar to Mn-deficiency but at an advanced stage, Fe-deficient leaves are bleached while Mn-deficient leaves develop interveinal necrosis resulting in dead brown tissue. Most crop plants are more susceptible to Fe-deficiency in the early stage of growth due to the small root system. Iron deficiency symptoms are summarized in Table 10.9.

10.7.2 Iron: Interaction with Other Nutrients

A number of nutrients appear capable of reducing the availability, absorption and utilization of Fe in crop plants. These so called negative interactions may result from interactions that occur either outside the root or within the root. Those interactions taking place in the external root environment are usually precipitation or similar reactions that reduce the chemical availability of the nutrient. Those that influence absorption or utilization processes alter the effectiveness of a nutrient by reducing its physiological availability (Fageria et al., 1990).

Iron deficiency induced by heavy application of P is widely reported (Sumner and Farina, 1986; Fageria et al., 1990). Phosphorus may precipitate Fe externally on the roots, but the interaction of Fe and P leading to Fe chlorosis appears to be caused by an internal immobilization of Fe probably due to the formation of Fe phosphate (Fageria et al., 1990). Iron deficiency is caused by the accumulation of excess Mn and Cu in the soil (Olsen and Watanabe, 1979). The interactions of Fe with Mn and Cu appear to be of a physiological nature. They may reflect the joint participation of these nutrients in some of the same biochemical systems, the proper functioning of which depends on the relative proportions of each of the nutrient present. High Mn in soils or plants may oxidize Fe to an inactive state. Competitive effects on Fe uptake have also been observed with excess Ca^{2+}, Mg^{2+}, K^{+}, Zn^{2+} and Mo (Fageria et al., 1990).

10.7.3 Iron: Management Practices

Iron deficiency of plants is difficult to correct by application of inorganic salts (Olsen, 1972). Some synthetic chelates such as FeEDDHA (ethylenediaminedi-o-hydroxyphenylacetic acid) and FeEDTA (ethylene diamine tetraacetic acid) effectively supply Fe to plants, but they are expensive. Hence, foliar application of Fe salts is recommended for correcting Fe deficiency in crops (Table 10.10). Another alternative for correcting Fe-deficiency is use of Fe-efficient crop species or genotypes of the same species. There are genotypic differences in Fe-uptake and utilization in annual crops (Fageria et al., 1990). Hansen et al. (2003) reported that there is a wide variation in susceptibility to Fe deficiency among soybean varieties, and variety selection is the most important management practice for producers with chlorosis-prone soils.

10.8 Boron

Among micronutrients, B deficiency is widespread and it has been reported in at least 80 countries on 132 crop species (Shorrocks, 1997). Plant species vary in B requirement with dicotyledons generally requiring 3–4 times more B than monocotyledons (Bennett, 1993). The authors (Fageria and Stone) conducted a field experiment with upland rice and common bean grown in rotation on a Brazilian Oxisol. Upland rice did not respond to B applications (data not presented) but the yield of common bean was significantly increased with B fertilization (Table 10.16). The relationship between grain yield (Y) and B application rate (X) was quadratic ($Y = 3184.66 + 15.20X - 1.20\,X^2$, $R^2 = 0.53^*$). Based on this quadratic response, the maximum bean yield was obtained with the application of 6 kg B ha^{-1} as a broadcast application. A number of soil properties influence B availability to plants and are reviewed by Fageria et al. (2002). Coarse textured, low organic matter soils located in humid regions are the most prone to B deficiency. (*significant at the 5% probability level)

10.8.1 Boron: Functions and Deficiency Symptoms

Boron plays an important role in cell development and elongation, protein synthesis, carbohydrate metabolism, pollen tube formation, and pollen viability (Bennett, 1993; Marschner, 1995). Boron deficiency of rice may be expressed solely in the form of reduced grain yields from floret sterility and may be mistakenly blamed on

Table 10.16 Response of common bean to B fertilization and hot water extractable soil B. Boron was applied with borax (11% B) as broadcast. Values are averages of 3 years field experimentation. Boron was determined after harvest of three bean and two upland rice crops

B rate (kg ha^{-1})	Grain yield (kg ha^{-1})	Soil B (mg kg^{-1})
0	3,111	0.79
2	3,202	0.98
4	3,353	1.17
8	3,232	1.59
16	3,041	2.21
24	2,888	3.11
R^2	0.53^*	0.99^*

*Significant at the 1% probability level.

poor environmental conditions during anthesis (Fageria et al., 2003). Fageria (2000c) reported significant increases in root weight of upland rice, common bean, corn and soybean with B fertilization in a Brazilian Oxisol. Boron, like other micronutrients, is also immobile in the plant tissues and its deficiency symptoms first appear on the younger plant tissues. Boron deficiency symptoms are summarized in Table 10.9.

10.8.2 Boron: Interaction with Other Nutrients

High N, Ca and Mg in the soil may reduce B uptake in plants. In low B soil, high N induced B deficiency in plants (Gupta, 1993). However, the effects of P, K, and S on uptake of B are not clear, and these minerals had positive, negative, and/or no effects on B uptake (Gupta, 1993). Boron became toxic to corn when grown under P deficiency conditions, and P applications alleviated B toxicity (Gunes and Alpaslan, 2000). Zinc fertilisation reduced B accumulation and toxicity on plants grown in soils containing adequate B (Moraghan and Mascagni, 1991).

10.8.3 Boron: Management Practices

The measures for correcting B deficiency are summarized in Table 10.10. Boron is usually applied at 1–6 kg ha^{-1} and higher rates are required for broadcast than band application or foliar sprays (Mortvedt and Woodruff, 1993). Borax or other soluble borates are usually applied to soil before planting. Boron fertiliser should not be placed in contact with seeds or at levels that may be toxic to crops. Boron toxicity was observed in common bean in a Brazilian Oxisol, when plots received 16 and 24 kg B ha^{-1} as a broadcast of borax having 11% B. The hot water extractable B in this soil was increased from 0.79 to 3.11 mg kg^{-1} soil after harvesting three bean and two upland rice crops (Table 10.16). Genotypic differences in B uptake and utilisation have been reported for field crops (Fageria et al., 2002). Hence, planting B-efficient genotypes is a very attractive strategy for correcting deficiency of this element in annual crops.

10.9 Conclusions

South America offers the largest land area on a global basis (Fig. 10.1). Furthermore, favorable climatic conditions (water and temperature) are another asset for crop production in most of the South American countries. For example, Brazil alone has 12% of the world's sweet (fresh) water availability. However, most of the soils of this continent are acidic and infertile. Hence, liming and fertiliser application are

Fig. 10.1 Geographic map of South America

essential. Micronutrient deficiencies are an emerging limiting factor for annual crop production. In annual crops such as rice, corn, wheat, soybean and common bean, deficiencies of Zn, Cu, B, Mn and Fe have been reported.

The main reasons for micronutrient deficiencies in South American soils are leached acid soils with low total contents of some micronutrients, intensive cultivation and use of modern high-yielding cultivars. Liming acid soils is also an important factor (at high pH level, availability of all the micronutrients decreases, except for Mo and Cl), with other nutrients and soil erosion. Furthermore, micronutrients are not included in the fertiliser recommendations for annual crops. The supply of micronutrients in a balanced form is pivotal for maximizing crop yields. Overall, B and Cu have the maximum utilization efficiency and Mn and Fe have the minimum utilization efficiency in crop production.

Climatic, soil and plant factors and their interactions influence the availability of micronutrients. Appropriate use of micronutrients in crop production is important for economic and environmental purposes. There are four methods by which micronutrient disorders in crop plants can be identified. These methods are: (1) visual symptoms, (2) soil testing, (3) plant analysis and (4) crop growth response. Use of chemical fertilisers (soil and foliar application), organic manures, liming acid soils and planting nutrient efficient genotypes are sound agronomic practices for increasing availability of micronutrients and improving crop production on South American soils. Adopting sound soil and crop management practices will not only enhance crop productivity but also help in reducing deforestation of the tropical rainforests in South America. This strategy will permit less CO_2 release to the atmosphere, conservation of soil, water and global climatic change. In South America, micronutrient management issues require a great of deal of research for improvement.

References

Baligar, V. C., Fageria, N. K., Elrashidi, M. A. (1998) Toxicity and nutrient constraints on root growth. HortScience, 33: 960–965.

Barbosa Filho, M. P., Cantarella, H., Wietholter, S. (2001) Rice corn and wheat. In: Ferreira, M. E., Cruz, M. C. P., Raij, B. V., Abreu, C. A. (Eds.) Micronutrients and toxic elements in agriculture. Jaboticabal, São Paulo, Brazil. pp. 285–318.

Bennett, W. F. (1993) Plant nutrient utilization and diagnostic plant symptoms. In: Bennett, W. F. (Ed.) Nutrient deficiencies and toxicities in crop plants. American Phytopathological Society Press, St. Paul, MN. pp. 1–7.

Brady, N. C., Weil, R. R. (2002) The nature and properties of soils (13th edition). Prentice-Hall, Upper Saddle River, NJ. 960 p.

Cakmak, I., Torun, B., Erenoglu, B., Ozturk, L., Marschner, M., Kalayci, M., Ekiz, H., Yilmaz, A. (1998) Morphological and physiological differences in the response of cereals to zinc deficiency. Euphytica, 100: 349–357.

Cerri, C. E. P., Coleman, K., Jenkinson, D. S., Bernoux, M., Victoria, R., Cerri, C. C. (2003) Modeling soil carbon from forest and pasture ecosystems of Amazon, Brazil. Soil Sci. Soc. Am. J., 67: 1879–1887.

Cochrane, T. T. (1978) An ongoing appraisal of the savanna ecosystems of tropical America for beef cattle production. In: Sanchez, P. A., Teggas, L. E. (Eds.) Pasture production on acid soils of the tropics, CIAT, Cali, Colombia. pp. 1–12.

Fageria, N. K. (1989) Tropical soils and physiological aspect of crops. EMBRAPA, Department of Technology Transfer, Brasilia. 425 p.

Fageria, N. K (2000a) Upland rice response to soil acidity in cerrado soil. Pesq. Agropec. Bras., 35: 2303–2307.

Fageria, N. K. (2000b) Adequate and toxic levels of zinc for rice, common bean, corn, soybean and wheat production in cerrado soil. Rev. Bras. Eng. Agric. Ambiental, 4: 390–395.

Fageria, N. K. (2000c) Adequate and toxic levels of boron for rice, common bean, corn, soybean and wheat production in Cerrado soil. Rev. Bras. Eng. Agric. Ambiental., 4: 57–62.

Fageria, N. K. (2001a) Response of upland rice, dry bean, corn and soybean to base saturation in cerrdo soil. Rev. Bras. Eng. Agric. Ambiental., 5: 416–424.

Fageria, N. K. (2001b) Adequate and toxic levels of copper and manganese in upland rice, common bean, corn, soybean, and wheat grown on an Oxisol. Commun. Soil Sci. Plant Anal., 32: 1659–1676.

Fageria, N. K. (2002) Influence of micronutrients on dry matter yield and interaction with other nutrients in annual crops. Pesq. Agropec. Bras., 37: 1765–1772.

Fageria, N. K. (2004) Dry matter yield and nutrient uptake by lowland rice at different growth stages. J. Plant Nutr., 27: 947–958.

Fageria, N. K., Baligar, V. C. (1996) Response of lowland rice and common bean grown in rotation to soil fertility levels on a varzea soil. Fertiliser Res., 45: 13–20.

Fageria, N. K., Baligar, V. C. (1997) Response of common bean, upland rice, corn, wheat, and soybean to soil fertility of an Oxisol. J. Plant Nutr., 20: 1279–1289.

Fageria, N. K., Baligar,V. C. (2005) Growth components and zinc recovery efficiency of upland rice genotypes. Pesq. Agropec. Bras., 40: 1211–1215.

Fageria, N. K., Gheyi, H. R. (1999) Efficient crop production. Federal University of Paraiba, Campina Grande, Brazil. 547 p.

Fageria, N. K., Zimmermann, F. J. P. (1979) Interaction among P, Zn, and lime in upland rice. R. Bras. Ci. Solo., 3: 88–92.

Fageria, N. K., Baligar, V. C., Wright, R. J. (1990) Iron nutrition of plants: An overview on the chemistry and physiology of its deficiency and toxicity. Pesq. Agropec. Bras., 25: 553–570.

Fageria, N. K., Baligar, V. C., Jones, C. A. (1997) Growth and mineral nutrition of field crops (2nd edition). Marcel Dekker, New York. 624 p.

Fageria, N. K., Baligar, V. C., Clark, R. B. (2002) Micronutrients in crop production. Adv. Agron., 77: 185–268

Fageria, N. K., Slaton, S. A., Baligar, V. C. (2003) Nutrient management for improving lowland rice productivity and sustainability. Adv. Agron., 80: 63–152.

Galrão, E. Z. (1991) Micronutrients and cobalt on soybean yield in cerrado soil. R. Bras. Ci. Solo., 15: 117–120.

Galrão, E. Z. (1999) Methods of copper application and evaluation of its availability for soybean grown on a cerrado red yellow latosol. R. Bras. Ci. Solo., 23: 265–272.

Graham, R. D., Rengel, Z. (1993) Genotypic variation in zinc uptake and utilization by plants. In: Robson, A. D. (Ed.). Zinc in Soils and Plants. Kluwer, Dordrecht, The Netherlands. pp. 107–118.

Graham, R. D., Ascher, J. S., Hynes, S. C. (1992) Selecting zinc efficient cereal genotypes for soils of low zinc status. Plant Soil, 146: 241–250.

Gunes, A., Alpaslan, M. (2000) Boron uptake and toxicity in maize genotypes in relation to boron and phosphorus supply. J. Plant Nutr., 21: 541–550.

Gupta, U. C. (1993) Factors affecting boron uptake by plants. In: Gupta, U. C. (Ed.) Boron and its role in crop production. CRC, Boca Raton, FL. pp. 87–123.

Hansen, N. C., Schmitt, M. A., Anderson, J. E., Strock, J. S. (2003) Iron deficiency of soybean in the upper Midwest and associated soil properties. Agron. J., 95: 1595–1601.

Iorio, A. F., Gorgoschide, L., Rendins, A., Barros, M. J. (1996) Effect of phosphorus, copper, and zinc addition on the phosphorus copper and phosphorus/zinc interaction in lettuce. J. Plant Nutr., 19: 481–491.

Kochian, L. V. (1991) Mechanisms of micronutrients uptake and translocation in plants. In: Mortvedt, J. J., Cox, F. R., Shuman, L. M., Welch, R. M. (Eds.) Micronutrients in agriculture (2nd edition). Soil Science Society of America, Madison, WI. pp. 229–296.

Lindsay, W. L. (1979) Chemical equilibria in soils. Wiley, New York. 449 p.

Lindsay, W. L., Norvell, W. A. (1978) Development of a DTPA soil test for zinc, iron, manganese, and copper. Soil Sci. Soc. Am. J., 42: 421–428.

Marschner, H. (1995) Mineral nutrition of higher plants (2nd edition). Academic, San Diego, CA. 889 p.

Martens, D. C., Lindsay, W. L. (1990) Testing soils for copper, iron, manganese, and zinc. In: Westermann, R. L. (Ed.) Soil testing and plant analysis (3rd edition). Soil Science Society of America, Madison, WI. pp.229–264.

Martens, D. C., Westermann, D. T. (1991) Fertiliser applications for correcting micronutrient deficiency. In: Mortvedt, J. J., Cox, F. R., Shuman, L. M., Welch, R. M. (Eds.) Micronutrients in agriculture (2nd edition). Soil Science Society of America., Madison, WI. pp. 549–592.

Mengel, K., Kirkby, E. A., Kosegarten, H., Appel, T. (2001) Principles of plant nutrition (5th edition). Kluwer, Dordrecht. 849p.

Moraghan, J. T., Mascagni, Jr. H. J. (1991) Environmental and soil factors affecting micronutrient deficiencies and toxicities. In: Mortvedt, J. J., Cox, F. R.,Shuman, L. M., Welch, R. M. (Eds.) Micronutrients in agriculture (2nd edition). Soil Science Society of America, Madison, WI. pp. 371–425.

Mortvedt, J. J. (2000) Bioavailability of micronutrients. In: Sumner, M. E. (Ed.) Handbook of Soil Science. CRC, Boca Raton, Fl. pp. D71–D88.

Mortvedt, J. J., Woodruff, J. R. (1993) Technology and application of boron for crops. In: Gupta, U. C. (Ed.) Boron and its role in crop production. CRC, Boca Raton, FL. pp. 157–176.

Olsen, S. R. (1972) Micronutrients interactions. In: Mortvedt, J. J., Giordano, P. M., Lindsay, W. L. (Eds.) Micronutrients in agriculture. Science Society of America, Madison, WI. pp. 243–261.

Olsen, S. R., Watanabe, F. S. (1979) Interaction of added gypsum in alkaline soils with uptake of iron, molybdenum, manganese, and zinc by sorghum. Soil Sci. Soc. Am. J., 43: 125–130.

Robinson, A. D., Pitman, M. G. (1983) Interactions between nutrients in higher plants. In: Lauchli, A., Bieleski, R. L. (Eds.) Inorganic plant nutrition. Encyclopedia of Plant Physiology, Vol. 1. Springer, New York. pp. 147–180.

Rosolem, C. A., Quaggio, J. A., Silva, N. M. (2001) Cotton, peanut and soybean. In: Ferreira, M. E., Cruz, M. C. P., Raij, B. V., Abreu, C. A. (Eds.) Micronutrients and toxic elements in agriculture. Jaboticabal, São Paulo, Brazil. pp. 319–354.

Sanchez, P. A., Logan T. J. (1992) Myths and science about the chemistry and fertility of soils in the tropics. In: Lal, R., Sanchez, P. A. (Eds.) Myth and science of soils of the tropics. Soil Science Society of America and American Society of Agronomy Special Publication number 29, Madison, WI. pp. 35–46.

Shorrocks, V. M. 1997. The occurrence and correction of boron deficiency. In: Dell, B., Brown, P. H., Bell, R. W. (Eds.) Boron in Soils and Plants: Reviews. Plant and Soil, 193. pp. 121–148.

Sumner, M. E., Farina, M. P. W. (1986) Phosphorus interactions with other nutrients and lime in field cropping systems. Adv. Soil Sci., 5: 201–236.

Welch, R. M. (1993) Zinc concentrations and forms in plants for humans and animals. In: Robson, A. D. (Ed.) Zinc in soil and plants. Kluwer, Dordrecht, The Netherlands. pp. 183–195.

Chapter 11
Micronutrient Use in Agriculture in the United States of America

Current Practices, Trends and Constraints

Patrick H. Brown

Abstract The use of micronutrient fertilisers to correct deficiencies in the USA is determined by a complex of factors including soil type, crop system, local environmental conditions and economic factors. These factors, and hence the patterns of micronutrient deficiency and use, vary dramatically across the US agricultural landscape. While soil mapping, crop analysis, and more recent modeling procedures have been used to identify broad regions and cropping systems in which the occurrence of a deficiency can be expected, this information generally lacks sufficient resolution to inform management decisions. Micronutrient management in the USA is highly localized and has historically been determined by the recommendations of local university or government scientists and extension agents.

Data on the consumption of micronutrients in the USA can be derived from records of commercial fertiliser sales and from information derived from private industry. This data, however, is scattered and incomplete and frequently of only local relevance. Notwithstanding this paucity of statistically rigorous data, it is clear that the patterns of use of micronutrients in the USA are changing rapidly, and over the last 5 years there has been a significant growth in micronutrient application across US agriculture and important changes in the manner in which fertilisers are used. Important driving forces for this change include an increasing demand for high quality and uniform product, changing agricultural practices and increasing intensification including precision agriculture, low-till management, and the use of genetically modified crops. There is a growing recognition of the environmental impact of agriculture, increasing interest in sustainable production systems and adoption of stricter environmental regulations. As the monetary, environmental and societal cost of synthetic fungicides and pesticides increases, there is also a growing interest in the concept of optimised plant health through balanced plant nutrition which will inevitably increase the use of micronutrients in US agricultural production.

The current growth in micronutrient use in the USA coincides with a significant downsizing of publicly funded, applied-research and extension and a greatly diminished applied-research effort by fertiliser manufacturers and distributors. Many of the recent changes in fertiliser practice have occurred, therefore, without the research base needed to implement those changes wisely and have resulted in a

B.J. Alloway (ed.), *Micronutrient Deficiencies in Global Crop Production.*

growing gap between historic public-sector derived recommendations and current practice. In the absence of renewed investment in micronutrient research and adaptation, the goal of increasing yields and crop quality while reducing environmental impact will be very difficult to achieve.

11.1 Introduction

Determining the occurrence of micronutrient deficiencies and defining trends in the use of micronutrient fertilisers in an area as large and diverse as the USA is a formidable task and one that is inevitably complicated by local conditions, crops and history of management. To help understand the trends in micronutrient use in the USA, information can be drawn from records of micronutrient fertiliser consumption, a review of current and historic data and micronutrient practices and an analysis of trends in cropping patterns and changes in the agricultural industry across the USA.

Data on the consumption of micronutrient fertilisers represents an important basis on which to define the patterns of micronutrient deficiency and demand for fertilisers. In US agriculture, micronutrient fertiliser consumption data is nominally available from a variety of state and federal agencies and through data compilations provided by government and private organisations including the US Geological Survey (Mineral Commodity Summaries), the Fertilizer Institute (TFI) (www.tfi.org) and the American Association of Plant Food Control Officers (AAPFCO) (www.aapfco.org). In all cases however, this data is inadequate since it only records sales of either raw minerals (much of which will ultimately have non-agricultural uses) or reports only limited classes of micronutrient fertiliser sales but fails to report the vast majority of micronutrients that are consumed in blends with macronutrients, or many of the micronutrients present in specialty products and complexes. Given the inherent differences in the efficacy of different micronutrient formulations on diverse crops, even an entirely accurate accounting of micronutrient consumption data would be very difficult to interpret. The difficulty in defining trends in micronutrient use is further complicated by the growing use of multi-element blends and a movement away from broadcast bulk blended fertilisers to the use of specialty blends, foliar fertilisers and specialty micronutrient formulations. This trend to specialty and higher-price micronutrient products, clearly reflects an increase in interest, demand and expenditure on micronutrients fertilisers, however, as product cost and effectiveness increase, total consumption may fall as low solubility materials are displaced from the market.

Literature review, historic records and maps of soil and plant analysis can all provide some additional insight into the occurrence of micronutrient deficiencies. These records however, are of highly divergent quality and rarely are of sufficient scale to be useful for management decisions. Given this diversity in the quality of information describing the occurrence of, and approach to, micronutrient deficiencies across the USA, a review of the geographic and crop-specific occurrence of deficiencies is neither possible nor of any real academic or practical value as a primary descriptive tool.

The complexities of the US farming industry, the diversity of crops and environmental conditions and the paucity of data suggest that this information alone cannot be utilized to determine either the extent of micronutrient deficiencies or the patterns of micronutrient use in the USA. Rather, an integrated analysis of the current and historic fertiliser recommendations and practices, data on consumption and ongoing changes in agricultural practices and composition of the agricultural industry are needed to provide insight into the use of micronutrients in crops in the USA. This chapter, therefore, we will focus on the practices that have guided micronutrient use in the past, the changes in knowledge, practice and consumption patterns that have occurred over the last decade and the challenges and opportunities that will be faced in the next decade. The following represents the views of the author as informed by extensive literature review and interviews with university, government and industry leaders.

11.2 Micronutrient Use in the USA

11.2.1 History

Early attempts to define micronutrient usage in the USA relied upon the use of maps generated from extensive plant and soil analyses to determine the regions of the USA, predicted to be prone to deficiencies (Kubota, 1980; Welch et al., 1991 and references therein). While of some value to guide fertiliser industry development and research investment, this approach was of insufficient resolution to influence farm management practice and has not been of significant utility for several decades.

Over the period from the 1930s until the 1990s, decisions on fertiliser use and management practices for a given crop and environment were based almost exclusively upon the research, standards and guidelines developed by public sector (university and USDA) scientists and extension advisors. These guidelines for micronutrient fertiliser usage have strongly determined the pattern of micronutrient in the USA in the last century. The role of the public sector as a primary determinant of micronutrient use is diminishing as funding for micronutrient research and extension decreases and growers increasingly derive their information from other sources.

The use of micronutrients in the USA has always been strongly influenced by the value of the crop being grown, thus horticultural crops use substantially greater amounts of micronutrients on a per acre basis than irrigated agronomic or rain-fed systems. Changes in the size of the average US farm, changes in crop choice, intensity of production, the use of genetically modified crops and the adoption of precision farming are all having a significant influence on the use of micronutrients. Micronutrient use is also being impacted by environmental legislation that increasingly demands that agricultural practices are environmentally sustainable. There is also a growing recognition that maintaining optimum plant and soil 'health' can minimize the demand for synthetic pesticides while maximizing the nutritional value of the resulting crop.

The principle determinants of micronutrient use in the USA are changing rapidly and new challenges and consumption patterns are emerging. The following analysis provides an examination of the historic determinants of micronutrient use and the emerging factors that will determine how micronutrients will be used in the future.

11.2.2 Occurrence of Micronutrient Deficiencies in the USA

The occurrence of boron (B), copper (Cu), zinc (Zn), manganese (Mn) iron (Fe) and molybdenum (Mo) deficient soils and responsive crops has been reviewed by Kubota (1980, 1987) and Welch et al. (1991 and references therein) who based their conclusions on a combination of extensive plant and soil analysis as well as consideration of soil type and geologic origin. Kubota (1987) generated a series of maps illustrating the occurrence of soils with soil micronutrient supply capacity. While these maps do not provide information of sufficient scale to assist in farm-level decision-making, they are of some value in the identification of potentially problematic regions and generally raising awareness of the potential impact. With the growing capability of geographic information systems (GIS), information management, precision agriculture, variable rate fertilisation technologies and improved micronutrient formulations, the prospect of effectively modeling and prescriptively defining fertiliser programs at an individual farm scale is rapidly becoming a reality. Thus, mapping approaches to micronutrient management may once again become more important. In the following analysis of micronutrient deficient sites, regions are defined by the classifications provided by AAPFCO as outlined in Fig. 11.2.

Recent advances in our understanding of the biology of B, and nickel (Ni) also provide some new insight that can help better define the occurrence of deficiencies of these elements, while changes in cropping systems will have a profound effect on the use of Mn and other micronutrients.

11.2.2.1 Boron

Boron deficiency occurs in scattered regions throughout the USA without any apparent relationship to a particular soil type. Boron is unique among the micronutrients in that B in the water used for irrigation is often the primary determinant of B status in plants. Species vary significantly in their sensitivity to B deficiency and toxicity and crop choice is often the most significant factor determining the occurrence of B deficiency (Goldbach et al., 2002). Tree fruit and nut species including pistachio (*Pistacia vera*), almond (*Prunus amygdalus*), Prunus species (Plum, Cherry, Apricot, Nectarine), walnut (*Juglans regia*); cruciferous crops (cabbage *Brassica oleraceae* var capitata), cauliflower (*Brassica oleraceae* var botrytis), cotton (*Gossypium* spp.), sugar beet (*beta vulgaris*), turnip (*Brassica rapa*) and canola (Brassica napus) are amongst the most important crops sensitive to B deficiency.

Two recent advances in our understanding of the function and transport of B in plants have had a significant effect on B use in the USA (for a review see Brown et al., 2002). The first major advance was the demonstration that B is unique among the plant nutrients in that species differ dramatically in their capacity for phloem B transport and their sensitivity to B deficiency (Brown and Shelp, 1997; Brown et al., 1999). In the majority of plants, B has restricted phloem mobility, while in species that produce polyols (sorbitol, mannitol, and dulcitol, etc.); B is readily translocated in the phloem to satisfy the demands for B in growth (Brown and Shelp, 1997). Species that transport polyols and exhibit phloem B mobility include *Prunus* spp. (peach/plum), *Malus* spp. (apple), *Pyrus* spp. (pear), *Olea* spp. (olive), *Daucus* spp. (carrot), and other vegetable and spices in the family Umbeliferae.

The second major advance was the demonstration in 2001 that B plays an essential function in the formation of the plant cell wall (O'Neill et al., 1996) and may function in the plant membrane (Brown et al., 2002). The role of B in cell wall formation, and the restrictions in within-plant transport in many species explains, in part, why B is critical for reproductive growth.

This new information on B function and transport has had important practical implications for the management of B in agriculture and has altered field practice in the USA for several crops, particularly tree species (Nyomora et al., 2000; Perica et al., 2001). While severe B deficiency results in the characteristic shoot dieback and deformation of leaves, it was only after the recognition of the differences in phloem B transport, that B toxicity symptoms were shown to be fundamentally different among these plants. Non-polyol producing plants show burning of the tip and margin of old leaves under B toxicity, while in the polyol-producing species these symptoms are absent and toxicity is expressed as meristematic dieback (Fig. 11.1 and in Colour Plates) (Brown et al., 1999). The differences in expression are a consequence of the site of B accumulation that occurs in the oldest tissues of phloem-immobile species and in the youngest tissues of species in which B is phloem-mobile. The ability to remobilize B from old to young leaves, in polyol producing species, reduces the occurrence of transient B deficiencies and provides remarkable tolerance of short-term B deprivation. Species that do not transport B in the phloem are very sensitive to short-term B deprivation.

As a consequence of these findings and numerous reports that reproductive growth is especially sensitive to B deficiencies (Huang et al., 2001; Brown et al., 2002), many producers of B sensitive species (see above) have adopted programs of prophylactic B application. Knowledge of the dynamics of B *in-planta* has also resulted in changes in fertilisation strategies and fertiliser formulations. Over the

11.1a: Pistachio (*Pistacia vera*)

11.1b: Peach (*Prunus dulcis*)

Fig. 11.1 Characteristic symptoms of B toxicity in species in which B is phloem-immobile (a-Pistachio) or phloem mobile (b-Almond). Boron toxicity in phloem immobile species is expressed as marginal leaf scorch on the oldest leaves (1a), in species exhibiting phloem B mobility toxicity is expressed as death of youngest growing tips, necrosis of vascular cambium and deformation of fruits. For a full list of species in each symptom category see Brown et al. (1999) (*See Colour Plates*) last 7 years there has been a very significant increase in fertiliser B consumption and product diversity especially in the regions growing high value, B-sensitive species (US Borax and Monterey Chemicals, 2005, personal communication).

11.2.2.2 Zinc

Zinc deficiency typically occurs in high-pH soils and in coarse textured leached soils. The distribution of Zn deficiency in the USA is not consistently associated with any particular gross climatic or edaphic factor and is most commonly seen in crops with a high Zn requirement. Zinc deficiency in the USA is observed more frequently in the predominantly calcareous soils of the west north central region and mountain region of the USA where >50% of total US consumption occurs (Figs. 11.2 and 11.6), the acid coarse textured soils of the east south central region of the USA and in Florida (Fig. 11.2).

11.2.2.3 Manganese

Manganese deficiency occurs throughout the USA on calcareous soils scattered in the mountain and west north central regions, the peat and muck soils (Histosols) of the east north central region and the coarse textured, poorly

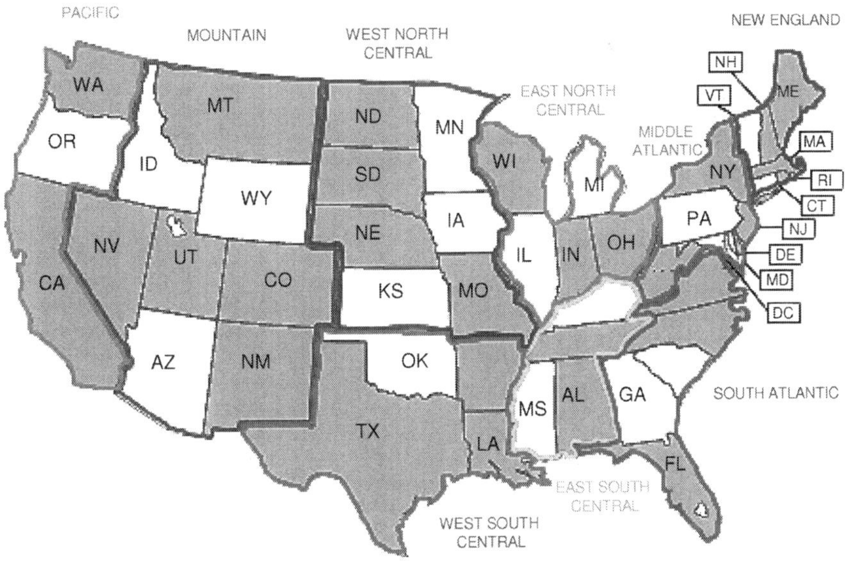

Fig. 11.2 Map of the USA showing generalized geographic regions utilised for fertiliser sales compilation (Based on AAPFCO classifications)

drained soils of the soils of the South Atlantic region. Manganese is used most heavily on the sensitive legume species including soybean, however, until quite recently the cost and relatively inefficient response to applied Mn limited application of Mn on large crop areas. Changes in cropping practice including the adoption of herbicide resistant crops have increased both the demand and opportunity to apply Mn fertilisers.

11.2.2.4 Iron

Iron deficiency occurs primarily in high-pH and calcareous soils, as well as in the acid or calcareous coarse textured soils of Florida and California. The occurrence of Fe deficiency is highly variable and crop-dependent and is most agriculturally relevant in sensitive crops (soybean) grown in the mountain and west north central regions. The relative sensitivity of the species and the intensity of the production system is a primary determinant of Fe use, thus California with a high density of production of sensitive high value species including citrus spp., grape (*Vitis vinifera*) and peach (*Prunus persica*), is a major market for Fe fertilisers.

11.2.2.5 Copper

Copper deficiency in the USA is primarily seen in the calcareous mountain region, the poorly drained soils of the South Atlantic region and in some horticultural species grown on high organic matter soils (Histosols) including the California delta, portions of Florida, Michigan and New England. Application is most common on higher value horticultural species with only very limited application in field crops.

11.2.2.6 Nickel

Nickel was first identified as an essential element by Brown and co-workers in 1987 (Brown et al., 1987) but not identified as occurring in agricultural situations until 2004 (Wood et al., 2004), when it was observed in Pecan (*Carya illinoinensis*). Nickel deficiency in pecan is associated with a physiological disorder 'Mouse-Ear' which occurs sporadically, but with increasing frequency, throughout the southeastern USA (portions of South Atlantic region) where it represents a substantial economic impact. Though the function of Ni in plants is not fully understood, it is known to be essential for the function of the enzyme urease (EC 3.5.1.5, urea amidohydrolase) and to be required in greatest amounts in plants that transport ureides as a prominent N-form. Examples of ureide transporting crop genera are *Annona, Carya, Diospyros, Juglans* and *Vitis* (Schubert and Boland, 1990). Additionally, tropical legumes (e.g., soybean, Phaseolus beans, mungbean, and cowpeas) are also relatively sensitive to Ni deficiency.

Fig. 11.3 Branches of Ni sufficient (*left*) and deficient (*right*) pecan (*Carya illinoinensis*). Symptoms include delayed and decreased leaf expansion, poor bud-break, leaf bronzing and chlorosis, rosetting, leaf tip necrosis (Photo courtesy of B. Wood) (*See Colour Plates*)

Nickel deficiency is currently known to occur only on the coarse textured, poorly drained soils of the South Atlantic region and is exacerbated by excessive soil Zn that occurs as a consequence of over-fertilisation. Symptoms resemble those of Zn deficiency or non-specific vigour loss (Fig. 11.3).

11.3 Existing Standards and Guidelines for Micronutrient Use in Crops in the USA

From the founding of the agricultural universities in the 1860s until today, decisions on micronutrient use in US agriculture have been driven primarily by university-based researchers and extension agents who have developed fertiliser use recommendations for the local cropping systems based upon the results of many years of regional field trials. Historically, research on micronutrients was substantially less than that devoted to the macronutrients, N, P and K. The US model of integrated and applied research for the benefit of rural society is by all measures a remarkable success and has been replicated in many countries (http://www.csrees.usda.gov/qlinks/extension.html). Historically US agricultural research has been conducted through a strong collaboration between researchers, crop organisations and commodity boards, the local agricultural extension service and the fertiliser industry with funding derived from either public or private sources. Governmental and industry funded organizations have also played an important direct role in fertiliser research and development. The most notable of these was the Tennessee Valley Authority – National Fertilizer Development Center, which was established in 1933 (http://www.tva.gov/heritage/bloom/).

The results of more than 80 years of research and field experimentation have been reported in tens of thousands of scientific journal articles, experiment station reports, videos, Internet articles and books. The most important source of information to guide fertilisation decisions at a field level has historically been the extension bulletins and workshops provided by local agricultural extension services, university researchers the US Department of Agriculture (USDA). An example of a publicly provided fertiliser recommendation is the 'Tri-State Fertilization Recommendations for Corn, Soybean and Wheat' (Bulletin E-2567, http://ohioline.osu.edu/e2567/). Most current public crop fertilisation recommendations are now available online and with 72% of all mid and large size farms in the USA having Internet access (National Agricultural Statistics Service, Farm Computer Usage and Ownership, July 2005), the role of the Internet in providing fertiliser recommendations is growing substantially.

11.3.1 The Historically Conservative Approach to Micronutrient Use

Whereas the fertiliser recommendations for the macronutrients N, P and K are based upon many years of trials and experimentation, micronutrient recommendations are frequently based on a much more limited experimental basis, especially in horticultural crops for which yield-based fertilisation trials are exceedingly difficult to conduct. Macronutrient fertiliser recommendations for the majority of crops in the USA are increasingly being based upon a conceptual model that strives to replace nutrients removed in the crop and to maintain the soil nutrient concentrations at or just above the critical level required for optimum yield (Fig. 11.4). This approach to fertilisation is termed the 'soil nutrient maintenance approach' and is favoured by most private crop consultants, many soil testing laboratories and the majority of the more progressive and leading farmers for whom maintaining optimum yield and avoiding the depletion of native soil fertility is a priority.

Significantly, many extension services in the USA utilise a more conservative approach to micronutrient fertilisation than is practiced for N, P and K. The so-called 'Nutrient Sufficiency Approach' advocates fertilisation only when a soil is demonstrated to have insufficient nutrients to maintain optimum yield (Hochmuth, 2003). The 'Nutrient Sufficiency' approach to fertilisation allows for the depletion of native soil nutrients to a level at which a response to supplemental fertilisation can be expected. While studies have suggested that the sufficiency approach is economically superior in the short term by virtue of its use of the 'free' nutrients

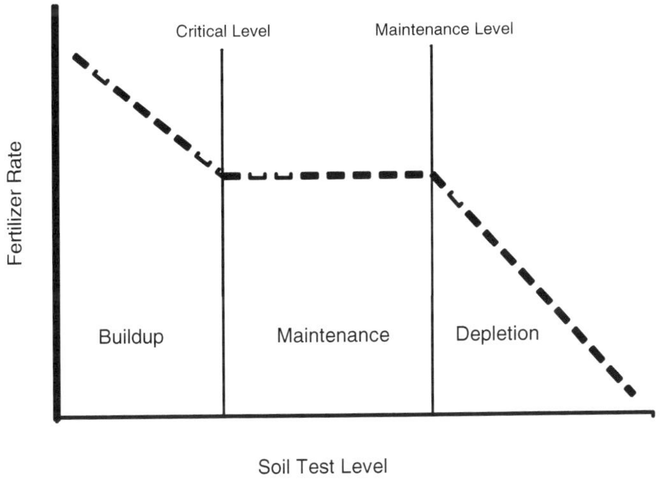

Fig. 11.4 The soil test and maintenance nutrition approach to fertiliser management as recommended in the major corn, wheat and soybean growing regions of the USA (Bulletin E2567). In this approach soil tests are used to ensure that soil fertility remains at or above a critical level at which no response to supplemental fertilisation is expected. This approach strives to maintain optimum fertility and to avoid the depletion of native soil fertility

present in the soil (Olson et al., 1982), considerable debate exists as to the ecological and economic sustainability of this practice in the longer term.

While the 'maintenance' approach to fertilisation for the macronutrients has been gaining favour, the use of micronutrients in the major field crops in the USA is still generally conservative with applications of micronutrients being restricted to situations where clear expression of deficiency symptoms are observed (Hoeft, 1997). The reason that this conservative practice has prevailed is likely to be threefold; first, the history of intensive commercial agriculture in North America is short and, as such, native soil fertility has not been depleted to the extent it has in Asia and Europe. Second, fertiliser trials focused on micronutrient responses are difficult to conduct, frequently variable in response and more local in relevance than trials involving the macronutrients and third, growers of crops with a small profit margin are reticent to invest in preventative fertilisation for which a clear yield benefit cannot be presumed. In many crops, difficulties in application, variability in effectiveness and high cost of micronutrient applications are additional significant factors that have limited the routine adoption of micronutrient fertilisation programs.

The prevalence of the 'nutrient sufficiency' philosophy of micronutrient use coupled with the historic desire of public extension agents to minimize input costs, has resulted in a highly cautious approach to micronutrient use in much of the USA.

Through much of the history of the fertiliser industry in the USA there has also been a degree of suspicion between the public sector representatives (scientists, extension agents, students, etc.) and the private fertiliser sector. This suspicion is manifested in the form of highly cautious statements that are often provided to balance the claims of fertilisers industry.

> Much research has been conducted with cotton and micronutrients in Alabama. These data have shown that cotton does not respond to any of the micronutrients except boron (B). (ANR-619, 1991)

> In the panhandle region of Oklahoma, Mn, B, Cl, Mo, Cu are "never deficient" in any crop and Fe and Zn are "seldom deficient. (Johnson, 2000)

Significantly, these statements do not reflect field practice, which shows substantial sales of B, Zn and Fe-containing micronutrient fertilisers in these same regions (Fig. 11.6 below).

11.3.2 Changing Sources of Fertiliser Recommendations in the USA

Since the mid 1980s, publicly funded and conducted fertiliser research has diminished substantially. Direct fertiliser research by private industry (either collaborative or independent) that was common in the 1950s through to the mid-1980s, has also diminished dramatically since that time. This decrease has occurred during a period of substantial change in the US agricultural sector, with a significant decrease in the number of small farmers and a coincident increase in farm size and management intensity. As funding for both public and private research and extension into fertiliser use has diminished, there has been a notable increase in the role of private consultants and industry sales persons as the primary source for micronutrient fertiliser practice throughout the USA. Increasing environmental constraints and the introduction of new agricultural practices including precision agriculture, genetically modified crops, minimum tillage and sustainable production systems has also introduced a new set of considerations for which the research base is inadequate.

Increasingly, growers are relying upon micronutrient recommendations provided directly from private consultants, fertiliser manufacturers and distributors. The information provided from these new private sources is generally based upon earlier public research with adaptation for local conditions, prevailing fertiliser formulations and practices. Recommendations made by private sources are frequently less conservative than public recommendations. With the significant changes that are occurring in the US agricultural system and the decrease in new research programs to address these changes there is a very significant and growing knowledge gap. Concern has already been expressed by all sectors that the lack of adequate adaptive research and extension may compromise the ability to develop truly sustainable production systems and result in sales of fertiliser products of untested value.

11.4 Trends in Usage

Micronutrients usage in the USA is recorded on the basis of fertiliser sales reported to state fertiliser control officers and tabulated by the American Association of Plant Food Control Officers (www.aapfco.org). Additionally, data can be obtained from sales records of private manufacturers and distributors. In both instances however, this data is incomplete and difficult to interpret. Data provided by AAPFCO, for example, reports only limited classes of micronutrient fertiliser sales and does not identify the vast majority of micronutrients that are consumed in blends with macronutrients, or the micronutrients present in specialty products and complexes. Nevertheless, it is reasonable to presume that trends in the classes of micronutrients reported reflect trends seen in micronutrients as a whole (David Terry, AAPFCO regional coordinator, 2004, personal communication). Data on elemental analysis of the fertiliser materials is not provided and there is no attempt to interpret relative nutrient solubility or efficiency. Thus, highly insoluble metal oxides or highly soluble metal-chelates contribute equally to consumption data.

Micronutrient sales data derived from private fertiliser manufacturers and distributors can provide additional information on trends in use patterns, and providing the data is representative of a substantial percentage of the micronutrient market, these trends can be quite informative. Data from private sources must also be considered with caution as apparent shifts in consumption patterns are frequently driven by industry changes due to mergers, competition, relocation of distribution plants and purchases destined for subsequent redistribution rather than real changes in local consumption. Currently, there are a vast number of micronutrient formulations including blends of several micronutrients in a cocktail, with or without an equally diverse range of adjuvants, 'carriers' and chelates. As with the AAPFCO data, information on elemental analysis and relative effectiveness of the products is not included in data derived from private sources.

The difficulty in defining trends in micronutrient use is further complicated by the growing use of multi-element blends and a movement away from broadcast bulk blended fertilisers to the use of specialty blends, foliar fertilisers and specialty micronutrient formulations. This trend to specialty and higher-price micronutrient products can result in a decrease in consumption of lower-cost oxides and thus an overall decrease in total micronutrient consumption reported. This drop in consumption however, reflects an increase in interest, demand and expenditure on micronutrient fertilisers.

Total micronutrient consumption across the entire USA has remained fairly steady over the past, frequently exhibiting a 2–3-year cycle of increase and decrease in consumption (Fig. 11.5). Averaged across the entire USA, Zn accounts for approximately 45% of all micronutrient use, Fe – 25%, B – 20%, Mn – 8% and Cu – 2% (*www.tfi.org*).

The consumption of micronutrients across the USA is not uniform and varies by cropping system and soil characteristics (Fig. 11.6). By far the greatest percentage

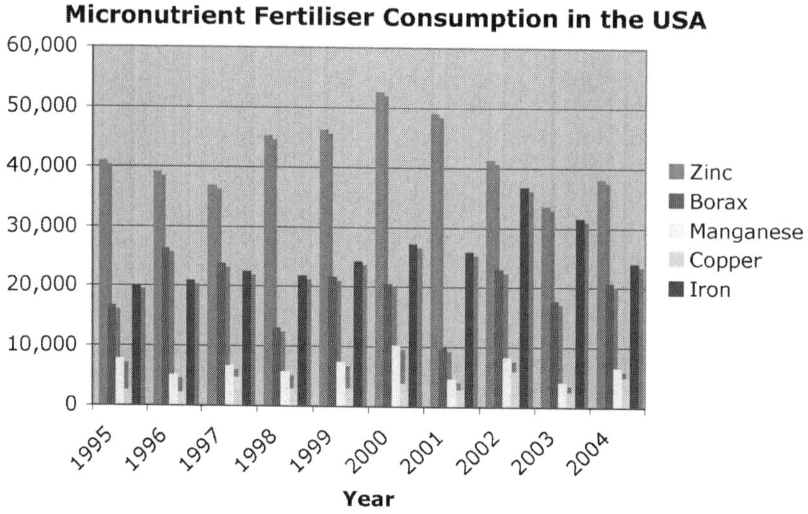

Fig. 11.5 Total Micronutrient consumption for select micronutrients over the period 1995–2004. This data is sourced from fertiliser tonnage sales reported to respective state agencies and Plant Food Control Officers in each state. Values shown are gross total nutrient consumed and do not consider form of the element or respective nutrient solubility. Micronutrients incorporated into bulk blended, finished fertilisers and other specialty fertilisers are not included in these data (Data courtesy of the American Association of Plant Food Control Agents (AAPFCO))

of Zn used in the USA is used on corn (*Zea mays*) and soybean (*Glycine max*) crops in the West North Central states of Minnesota, Iowa, Missouri, North Dakota, South Dakota, Nebraska and Kansas. Though cropping areas are roughly equivalent between the West North Central and East North Central regions, the soils in the former are typically more alkaline and deficiencies of Zn, B, Cu and Mn are more prevalent. The most significant consumption of Fe, B and Cu occurs in the Pacific region both as a consequence of a prevalence of alkaline soils in this region and the large acreages devoted to high value horticultural products in which micronutrients are used more frequently.

One of the most significant trends occurring in US agricultural use of micronutrients is the dramatic increase in use of Mn, Fe, Zn and to a lesser extent B that has occurred over the last 5 years in the primary soybean/corn growing states of Illinois, Indiana, Ohio (Fig. 11.7) and Iowa (not shown). This increase in use of almost 300% over a 4-year period coincides with the widespread adoption of genetically modified, herbicide and insect-resistant cultivars in these growing regions. The adoption of herbicide-resistant crops has stimulated micronutrient applications by providing the higher returns required for micronutrient purchase and by increasing the number of times growers enter the field with application equipment thereby reducing the marginal cost of micronutrient applications. Similarly, the spread of

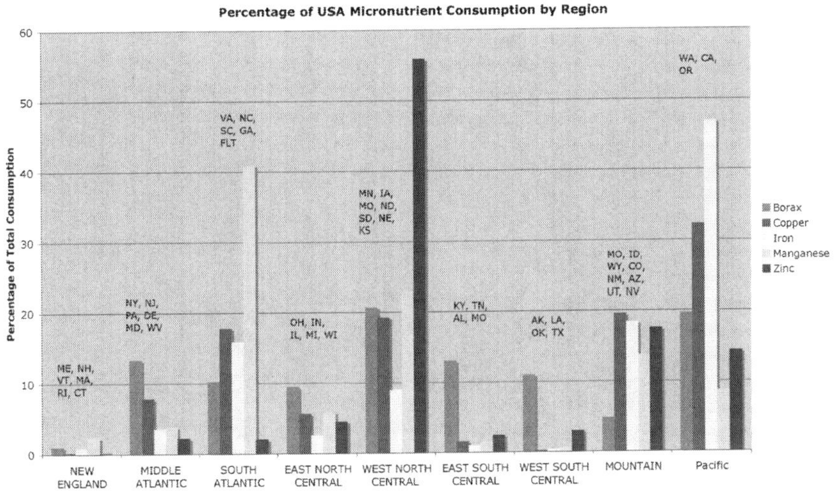

Fig. 11.6 Distribution of micronutrient sales as a percentage of US total consumption by region

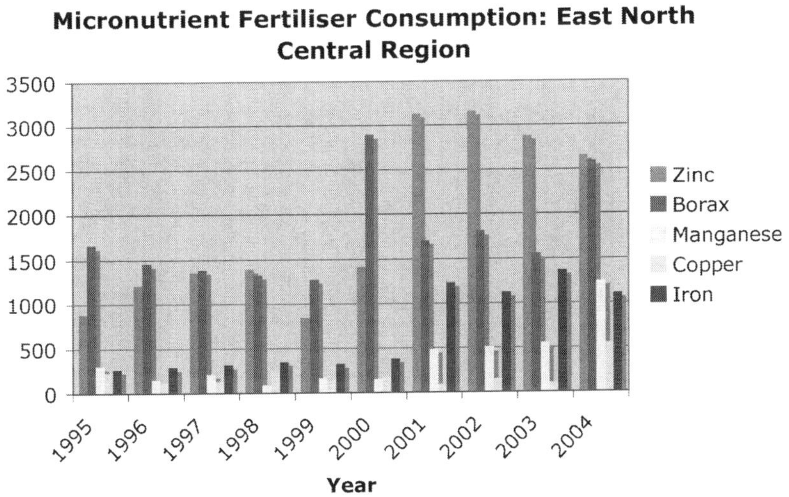

Fig. 11.7 Changes in micronutrient consumption patterns in primary corn and soybean production areas of East North Central USA

soybean rust (*Phakopsora pachyrhizi*) through the USA, and the potential for increased use of foliar fungicides is predicted to provide increased opportunities and incentive for micronutrient applications that would be uneconomical as standalone applications.

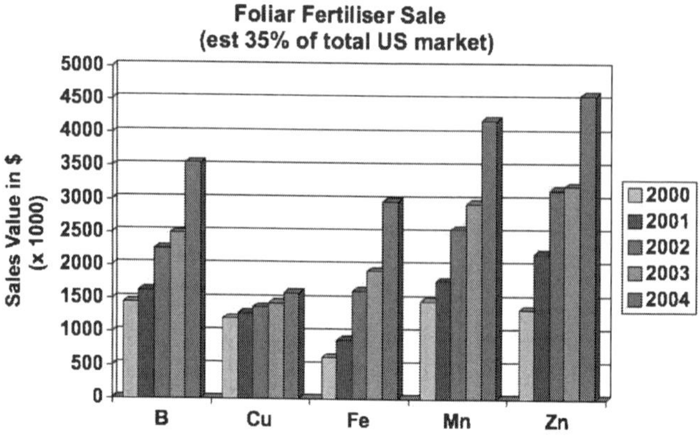

Fig. 11.8 Sales of foliar micronutrients by Brandt Chemicals, Illinois (2000–2004), (Vatren Jurin, 2005, Personal communication)

With the threat of soybean rust looming and in the absence of highly effective fungicides or resistant cultivars, there has been a renewed interest in the optimisation of plant nutrition to minimize disease impact. Manganese has long been recognized to play a role in plant defense to fungal diseases (Brown et al., 1984; Graham and Webb, 1991; Huber and Wilhelm, 1988), and recent evidence suggests that glyphosate use increases the prevalence of Mn deficiency in glyphosate-resistant soybeans (Huber et al., 2004). The potential for micronutrients in general and Mn in particular to play a role in minimizing the threat of soybean rust and correcting the antagonism that may occur with use of glyphosate resistant crops, has undoubtedly contributed to the growth in Mn use seen in the past few years and has been a focus of several applied extension bulletins (see Huber, 2000).

Trends in total micronutrient data reported by AAPFCO are also reflected in the sales of a leading producer of foliar micronutrients to the corn and soy growing industry (Brandt Chemicals, Illinois) who have seen a dramatic increase in sales of all micronutrients over the preceding 4 years. These trends are indicative of the foliar micronutrient industry as a whole.

11.5 Changing Cropping Patterns and Demands for Micronutrients

As the case of herbicide-resistant corn and soybean illustrates, changes in crop management can have a profound effect on micronutrient use patterns by increasing profit margins and increasing management options. The extent of adoption of genetically modified crops has been more rapid in the USA than in any other country and genetically modified cultivars now represent an estimated 90%, 50% and 85% of total acreage of soybean, corn and cotton, respectively (Monsanto, 2004). In general, any intensification of cropping systems also results in an increase micronutrient utilisation. This principle is illustrated in the example of the Pacific region, where a disproportionate percentage of the total consumption of Fe, Cu, B and Zn is found (Fig. 11.7). Agriculture in the Pacific region is dominated by irrigated agricultural systems and the intensive horticultural production systems of California which consume as much micronutrient fertiliser as the entire West North Central region that has more than 12 times as much total crop land (USDA-NASS, Agricultural Statistics, 2003). Changes in cropping systems, increased use of irrigation, adoption of high value crops and intensification of production can all be predicted to result in increased micronutrient use.

11.5.1 High Value Crops and Intensification of Cropping Systems

Over the last 10 years, there has been a substantial growth in agricultural land devoted to high value crops including fruits, nuts, vegetable, ornamental crops (Table 11.1) turf farms, and golf courses (data not shown). Over this period, horticultural production has grown from by 15–45% while agronomic crops in this same period have remained unchanged or have declined. While the overall acreage of land devoted to agronomic crops is approximately 50 times greater than that devoted to horticultural crops, the high level of micronutrient usage in horticulture represents a significant proportion of total US usage. All of these sectors utilise substantially greater quantities of micronutrients on a per hectare basis than are used in agronomic or forestry crops. The high investment in micronutrients in horticulture is a consequence of the relatively small cost of micronutrients as a proportion of total production costs and a willingness of producers to invest in inputs that enhance quality and minimise the chance of crop loss. There is also growing recognition that optimised micronutrient management can help reduce disease incidence and minimise applications of fungicides and pesticides to control diseases. Unlike agronomic crops, micronutrient fertilisation in horticultural crops is frequently conducted even when a deficiency is not apparent. Examples include the widespread use of B in tree and fruit species to ensure optimal flowering and fruiting (Nyomora et al., 2000; Perica et al., 2001), the application of preventative Fe applications in many ornamental and turf crops and some vegetable crops and the application of Zn routinely in tree nut species.

Horticultural crops also typically utilise a greater proportion of higher-cost foliar, chelated and otherwise enhanced micronutrient formulations than is used in other sectors of agriculture. The desire to use these enhanced materials is a consequence of their perceived benefits in reducing application costs, optimising compatibility with application equipment and enhanced effectiveness.

Coincident with the increase in horticultural acreage occurring in the USA is an increase in intensification and integration of farming activity on broad acre crops including soybean, corn and cotton. This intensification and integration now provides

Table 11.1 Trends in production areas for major agronomic and horticultural crops in the USA in 1994 and 2003 (USDA-NASS Agricultural Statistics 1994, 2003)

	1994	2003
	ha (×1,000)	
Principal field crops		
Corn	30,757	31,877
Wheat	31,161	24,980
Soybean	29,543	30,352
Cotton	4,904	5,459
Horticultural production		
Fruits (excl. citrus)	738	834
Citrus	383	419
Nut crops	247	353
Vegetable production (incl. melons)	1,173	1,331
Greenhouse/ornamental wholesale value (US$ × billion)	2.7	4.4

the impetus and the means to apply micronutrients where they have not previously been used. Several factors are contributing to this change: (1) an increased awareness and occurrence of micronutrient problems, (2) an interest in optimizing plant health to minimise disease impact, (3) an ability through precision agriculture and yield monitoring to identify areas of poor performance that may be suffering from micronutrient stress and to apply micronutrients to those areas through variable rate technology, (4) an opportunity to apply micronutrients in combination with fungicide or herbicide applications thereby reducing application costs, and (5) the development of prescription bulk blending and granule-coating techniques that allow a grower to specify fertiliser mixes for use on a farm or field basis.

11.6 Summary and Conclusions

The occurrence of micronutrient deficiencies in the USA does not follow any discrete geographic distribution but is more directly determined by the crop grown, as well as historic and contemporary management practices. In much of the USA, and specifically in land devoted to agronomic crops, the historic use micronutrients is as much a consequence of the prevailing and frequently conservative management philosophy, as it is a consequence of true demand. Thus, micronutrient use has been historically restricted to agronomic crops that show clear deficiency symptoms. This pattern of usage is, however, changing, and there has been a remarkable growth in micronutrient usage in both corn and soybean over the last 5 years. This growth is primarily a consequence of increased management intensity by virtue of the use of genetically modified crops and precision agriculture techniques that provide both the incentive and opportunity to supply micronutrients as normal practice. The industry as a whole is seeing greater commitment to the prevention, rather than the correction of micronutrient deficiencies.

In the horticultural crops, micronutrients have always been more widely used than they have in the agronomic crops. This is a consequence of the relatively low cost of these treatments (as a proportion of total production costs), the ability to apply micronutrients in combination with other field treatments or in irrigation, thereby reducing application costs, and a belief that optimised nutrition can contribute to product quality and reduced pesticide use. As horticultural crops expand in the USA, the use of micronutrients is expected to expand also.

The major changes in the patterns of micronutrient use that are occurring in the USA are occurring at a time of diminishing public investment in research and extension by universities and the USDA. During the same period, direct industry expenditure on micronutrient research has also declined. The coincidence of a changing market and increasing demand with reduced research and expenditure represents a very significant challenge to US agriculture and a potential threat to long-term sustainability.

Though the science of plant nutrition has provided some significant new advances in the understanding of the molecular basis of micronutrient nutrition, our practical ability to apply this information to optimise the micronutrient management of crop plants is still remarkably poor. Substantial challenges exist in the area of uptake efficiency, micronutrient translocation and efficiency of utilisation of micronutrients especially Mn, Zn, Fe and Cu.

An intelligent and thoughtfully implemented program that supports fundamental research to specifically address the issues of micronutrient nutrition in cropping systems is clearly needed.

References

ANR-619, New, December (1991). Mitchell, C.C., Burmester, C.H., Edmisten, K.L. Extension Agronomists. Cotton Fertilization Management in Alabama. Alabama Cooperative Extension System. Available at http://www.aces.edu/pubs/

Brown, P.H., Shelp, B.J. (1997). Boron mobility in plants. Plant and Soil, 193(1–2): 85–101.

Brown, P.H., Graham, R.D., Nicholas, D.J.D. (1984). The effects of manganese and nitrate supply on the levels of phenolics and lignin in young wheat plants. Plant and Soil, 81: 437–440.

Brown, P.H., Welch, R.M., Cary, E.E. (1987). Nickel: A micronutrient essential for higher plants. Plant Physiology, 85: 801–803.

Brown, P.H., Hu, H., Roberts, W.G. (1999). Occurrence of sugar alcohols determines boron toxicity symptoms of ornamental species. Journal of the American Society For Horticultural Science, 124(4): 347–352.

Brown, P.H., Bellaloui, N., Wimmer, M.A., Bassil, E.S., Ruiz, J., Hu, H., Pfeffer, H., Dannel, F., Romheld, V. (2002). Boron in plant biology. Plant Biology, 4(2): 205–223.

Bulletin E-2567 (2001). Tri-State Fertilization Recommendations for Corn, Soybean and Wheat (Bulletin E-2567) Retrieved September 2005 from http://ohioline.osu.edu/e2567/

Goldbach, H.E., Rerkasem, B., Wimmer, M.A., Brown, P.H., Theiller, M., Bell, R.W. (2002) (Eds.). Boron in Plant and Animal Nutrition, Kluwer/Plenum, Dordrecht, The Netherlands, Boston, MA, and London.

Graham, R.D., Webb, M.J. (1991). Micronutrients and disease resistance and tolerance in plants. In: Mortvedt, J.J., Cox, F.R., Shuman, L.M., Welch, R.M. (Eds.) Micronutrients in Agriculture (2nd edition). Soil Science Society of America, Madison, WI. pp. 329–370.

Hochmuth, G.J. (2003). Progress in mineral nutrition and nutrient management for vegetable crops in the last 25 years. HortScience, 38: 999–1003

Hoeft, R. (1997). Micronutrient Needs in Illinois. Cooperative Extension Service, University of Illinois. Issue No. 5, April 25, 1997.

Huang, L., Bell, R.W., Dell, B. (2001). Recent advances on boron in wheat reproduction: A mini-review. In: Horst, W.J. et al. (Eds.) Plant Nutrition, Food Security and Sustainability of Agro-Ecosystems, Kluwer. Dordrecht, The Netherlands, Boston, MA, and London.

Huber, D.M. (2000). 'Hidden Hunger' threatens many crops. Ag Answers 09/19/2000. Ohio State Extension and Purdue Extension Partnership. Retrieved July 2005 from http://www.agriculture.purdue.edu/AgAnswers/story.asp?storyID=2347

Huber, D.M., Wilhelm, N.S. (1988). The role of manganese in resistance of plant diseases. In: Graham, R.D., Hannam, R.J., Uren, N.C. (Eds.) Manganese in Soils and Plants. Kluwer, Dordrecht, The Netherlands. pp. 155–173.

Huber, D.M., Leuck, J.D., Smith, W.P., Christmas, E.P. (2004). Induced manganese deficiency in GM soybeans. Proceedings of the North Central Extension-Industry Soil Fertility Conference, 20: 80–83. Potash and Phosphate Institute, Brookings, SD.

Johnson, G. (2000). Secondary and Micronutrient Needs for Oklahoma Panhandle Crops. Web posted Powerpoint presentation. Retrieved January 2005 from www.dasnr.okstate.edu/NPK/ Panhandle%20Micronutrients%2003%2000.ppt

Kubota, J. (1980). Regional distribution of trace elcment problems in North America. In: Davies, B.E. (Ed.) Applied Soil Trace Elements. Wiley, London. pp. 443–466.

Monsanto (2004). Monsanto biotechnology trait acreage. Monsanto Financial Reports. Retrieved October, 2005 from http://www.monsanto.com/monsanto/content/ investor/financial/reports/2004/ Q42004Acreage.pdf

National Agricultural Statistics Service (2005). Farm Computer Usage and Ownership, July 2005. Retrieved September, 2005 from http://usda.mannlib.cornell.edu/reports/nassr/other/computer/

Nyomora, A.M.S., Brown, P.H., Pinney, K., Polito, V.S. (2000). Foliar application of boron to almond trees affects pollen quality. Journal of American Society for Horticultural Science, 125(2): 265–270.

Olson, R.A., Frank, K.D., Grabouski, P.H., Relm, G.W. (1982). Economic and agronomic impacts of varied philosophies of soil testing. Agronomy Journal, 74: 492–499.

O'Neill, M.A., Warrenfeltz, D., Kates, K., Doco, T., Darvill, A.G., Albersheim, P. (1996). Rhamnogalacturonan-II, a pectic polysaccharide in the walls of growing plant cell, forms a dimmer that is covalently crosslinked by a borate ester. J. boil. Chem., 271:22923–22930.

Perica, S., Brown, P.H., Connell, J.H., Nyomora, A.M.S., Dordas, C., Hu, H., Stangoulis, J. (2001). Foliar boron application improves flower fertility and fruit set of olive. HortScience, 36(4): 714–716.

Schubert, K.R, Boland, M.J. (1990). The ureides. In: Miflin, B.J., Lea, P.J. (Eds.) The Biochemistry of Plants, Vol. 16. Academic, San Diego, CA. pp. 197–283.

USDA-NASS (1993–2004). Retrieved September, 2005 from National Agricultural Statistics Service. Available at http://www.usda.gov/nass/

Welch, R.M., Allaway, W.H., House, W.A., Kubota, J. (1991). Geographic distribution of trace element problems. In: Mortvedt, J.J., Cox, F.R., Shuman, L.M. and Welch, R.M. (Eds.) Micronutrients in Agriculture (2nd edition). SSSA Book Series No.4. Soil Science Society of America, Madison, WI.

Wood, B.W., Reilly, C.C., Nyczepir, A.P. (2004). Mouse-ear of pecan:I. Symptomology and occurrence. HortScience, 29, 38:87–94.

Chapter 12
Linkages Between Trace Elements in Food Crops and Human Health

Ross M. Welch

Abstract Malnutrition accounts for more than 30 million deaths a year in mostly resource-poor families in the developing world. Much of this malnutrition is the result of insufficient intakes of available trace elements in the diets of the poor. Dysfunctional food systems are responsible for this global crisis in human health. Importantly, agricultural systems are the foundation upon which all nutrients enter the human food chain. Thus, agriculture must be contributing to dysfunctional food systems and the resulting malnutrition. Only through linking agricultural systems to human nutrition can sustainable solutions to malnutrition be forthcoming. This review focuses on many agricultural practices that can be used to change agricultural systems in ways that will help supply enough essential trace elements to the poor to meet their needs for healthy and productive lives.

12.1 Introduction

Incredibly, micronutrient deficiencies (i.e., micronutrient elements and vitamins) afflict over three billion people worldwide, mostly women, infants and children in resource-poor families. These micronutrients include a number of essential trace elements (e.g., iron (Fe), zinc (Zn), selenium (Se), iodine (I) and cobalt (Co)). Among these, Fe, Zn and I are known to be limiting in the diets of vast numbers of people (Kennedy et al., 2003; WHO, 1999, 2002). The consequences to human health, felicity, livelihoods, and national development are staggering, resulting in increased mortality and morbidity rates, decreased worker productivity, poverty and diminished cognitive ability in children with lower educational potential born to deficient mothers. The social and economic costs to societies around the world from these deficiencies are staggering (Darnton-Hill et al., 2005; Bouis, 1999). Dr. Bro Harlem Brundtland (Director General, World Health Organization, United Nations), declared at the World Economic Forum in 2000 that:

> Nutrition is a key element to any strategy to reduce the global burden of disease. Hunger, malnutrition, obesity and unsafe food all cause disease, and better nutrition will translate into large improvements in health among all of us, irrespective of our wealth and home country.

B.J. Alloway (ed.), *Micronutrient Deficiencies in Global Crop Production.*

Furthermore, the World Health Organization's 2002 "World Health Report" states that inadequate food and malnutrition leads to a downward spiral of increased susceptibility to illness, sickness and loss of livelihood ending in death (WHO, 2002). Current trends in micronutrient deficiencies continue to be increasing in many nations. For example, the global burden of Fe deficiency has risen from about 35% of the world's population in 1960 to over 50% by 2000, and Fe deficiency among poor women is increasing at an alarming rate in many developing countries (Broek, 2003). Current intervention programs (i.e., food fortification and supplementation programs) to alleviate Fe deficiency have not proven to be effective or sustainable in many countries (Underwood, 2003). Additionally, these programs have been limited to addressing only Fe, I and vitamin A deficiencies, with no programs operating for the rest of the limiting essential trace elements and vitamins in the diets of the poor (Gibson, 2003; Subbulakshmi and Naik, 1999; Underwood and Smitasiri, 1999; Yip, 1997).

Dysfunctional food systems are the cause of malnutrition and diet-related diseases (Graham et al., 2001). Importantly, these systems are dependent on agricultural systems that produce most of the foods that supply the nutritional needs of people everywhere. Historically, the linkages that exist between agriculture, nutrition, human health and well-being have never been widely viewed as important determinants in the alleviation of malnutrition and diet related diseases. Currently, intervention programs have not focused on approaches that address the basic underlying causes of these diseases and are, therefore, not sustainable in many nations (Underwood and Smitasiri, 1999; Tontisirin et al., 2002). To find sustainable solutions to malnutrition, agricultural systems must be used as a primary intervention tool (Welch and Graham, 1999). This chapter focuses on essential trace elements and how agricultural systems could be changed to meet the trace element needs of malnourished people worldwide.

Importantly, if agricultural technologies are directed at improving the nutritional quality of food crops, they must encompass a holistic food system perspective to assure that the planned interventions will be sustainable, and adopted by farmers and consumers. Furthermore, the agriculture sector must adopt a specific goal of improving human nutrition and health, and the nutrition and health sectors must adopt agricultural interventions as primary tools to fight this growing crisis in human health (Welch and Graham, 1999; Combs and Welch, 1998).

Humans require at least 51 known nutrients (i.e., both macronutrients, and micronutrients including various minerals and vitamins), in adequate amounts, consistently, to live healthy and productive lives (see Table 12.1). These nutrients include 17 micronutrient elements (i.e., trace elements) reported to be essential for human life. Unfortunately, global food systems are failing to provide adequate quantities of all of these micronutrients to vast numbers of people. Advances in crop production, achieved during the "Green Revolution", were dependent mostly on improvements in cereal cropping systems (i.e., rice, wheat and maize) and resulted in greatly increased food supplies for the world preventing massive starvation and famine in many regions. However, most cereals (as normally eaten after

Table 12.1 The known essential nutrients for human life[a]

Air, water and energy	Protein (amino acids)	Lipids–fat (fatty acids)	Macro-minerals	Essential trace elements	Vitamins
Oxygen	Histidine	Linoleic acid	Sodium – Na	Iron – Fe	A
Water	Isoleucine	Linolenic acid	Potassium – K	Zinc – Zn	D
Carbohydrates	Leucine		Calcium – Ca	Copper – Cu	E
	Lysine		Magnesium – Mg	Manganese – Mn	K
	Methionine		Sulphur – S	Iodine – I	C (Ascorbic acid)
	Phenylalanine		Phosphorus – P	Fluorine – F	B_1 (Thiamin)
	Threonine		Chlorine – Cl	Selenium – Se	B_2 (Riboflavin)
	Tryptophan			Molybdenum – Mo	B_3 (Niacin)
	Valine			Cobalt – Co (in B_{12})	B_5 (Pantothenic acid)
				Boron – B	B_6 (Pyroxidine)
				Níckel – Ni	B_7 (Biotin)
				Chromium – Cr	B_9 (Folic acid, folacin)
				Vanadium – V	B_{12} (Cobalamin)
				Silicon – Si	
				Arsenic – As	
				Lithium – Li	
				Tin – Sn	

[a]Numerous other beneficial substances in foods are also known to contribute to good health. See Nielsen, 1997, for a discussion of the essentiality of trace elements listed in this table.

milling), only supply needed carbohydrates for energy, a small amount of protein but few other nutrients in required amounts. This change in agricultural production to more monoculture cereal systems and away from more varied cropping systems appears to be contributing to micronutrient deficiencies by limiting food-crop diversity. This has had the unforeseen consequences of reducing available micronutrient supplies to the poor, formerly dependent on more diverse cropping systems which provided more traditional micronutrient-rich food crops (e.g., pulses, fruits, and certain vegetables) that are now in low supply and no longer affordable to this sector of societies (Welch et al., 1997).

Nutritional transitions are also causing increased rates of chronic diseases (e.g., cancer, heart disease, diabetes, obesity, osteoporosis, etc.) in many rapidly developing nations where people are switching from traditional diets to more calorie-rich diets derived from adopting developed nation's food systems (WHO, 2002, 2003). Clearly, there is an urgent need to tightly link the agricultural sector to human health to find ways to reduce the burden of diet-related diseases in the world (Sobal, 1999; WHO, 2002; Clugston and Smith, 2002; Fresco, 2000; Welch, 2001).

In 2004 the WHO at the 57th World Health Assembly acknowledged that malnutrition, including undernutrition and nutritional deficiencies, is still a major cause of death and disease worldwide. Non-communicable diseases were viewed as occurring in crisis proportions in developed countries and rapidly increasing in developing nations. In 2001, chronic diseases (many diet-related) accounted for almost 60% of the 56 million deaths annually and 47% of the global burden of disease. The Assembly stated that national food and agricultural policies should be consistent with the protection and promotion of public health. Agricultural policy and production have great effects on national diets and governments can influence agricultural production through many policy measures. The Assembly recommended that, as emphasis on health increases and consumption patterns change, Member States need to take healthy nutrition into account in their agricultural policies (WHO, 2004).

In 2003, the National Academy of Sciences, USA held a Workshop on "Exploring a vision: Integrating knowledge for food and health" (Rouse and Davis, 2004). At this Workshop, Charles Muscoplat was quoted as saying:

> It is time for the United States to shift to a new agricultural paradigm – one based on both what is good for the consumer and profitable for farmers.

Further, Dr. Van Hubbard (Director of the Division of Nutrition and Research at the National Institutes of Health, USA) stated:

> The time is ripe for an integrated approach to nutrition, health, and disease prevention.

The participants at this Workshop developed a diagram explicitly linking agriculture to human health (see Fig. 12.1). Clearly, one of the most prestigious scientific

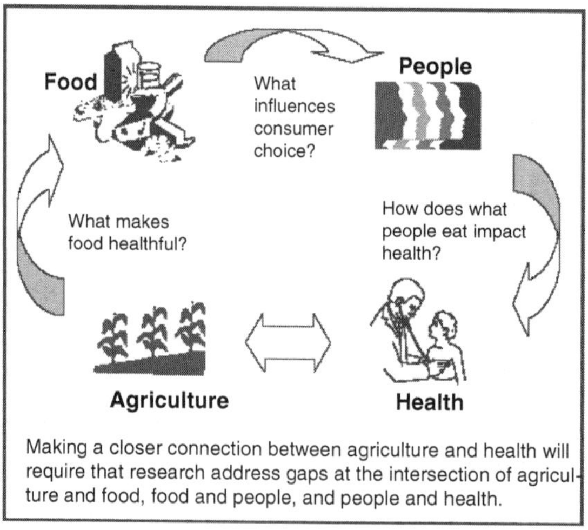

Fig. 12.1 Diagram showing the close link between agriculture and human health (From Rause and Davis, 2004)

organizations in the world now recognizes the important role that agriculture can play in enhancing human nutrition and health.

In Copenhagen, Denmark, an international meeting (the "Copenhagen Consensus Project") was held in 2004, where ten of the world's leading economists were invited to discuss what priorities should be given to the most critical challenges facing the world today (Copenhagen Consensus, 2004). The outcome of that meeting was a prioritized list of the ten most important challenges that should be addressed by world leaders and nations. Importantly, the second priority challenge was finding ways to provide adequate micronutrients to meet human needs, and the fifth was the development of new agricultural technologies to reduce malnutrition. Thus, from an economic perspective, investing in programs to reduce trace element malnutrition (possibly through agricultural strategies) has been given a very high priority by some of the world's most influential economists and is included among the most important challenges facing the world.

12.2 Global Magnitude of Trace Element Deficiencies in Humans

Malnutrition is a major health burden worldwide (WHO, 2002). Currently, undernourishment (protein-energy malnutrition) affects about 800 million people globally. Even more troublesome is that over three billion people (half the world population) are known to be afflicted with one or more trace element deficiencies (Kennedy et al., 2003; WHO, 1999). Malnutrition (both minerals and vitamins) accounts for 53% of all deaths in children under the age of five. About 800 million people are deficient in I, including 300 million with goiter and 20 million with permanent brain damage from maternal I deficiency in utero during fetal development. Around two billion people are thought to be Zn deficient (Hotz and Brown, 2004); about one billion people suffer from the worst form of Fe deficiency, Fe deficiency anemia (Müller and Krawinkel, 2005; Mason et al., 2001). Selenium deficiency (Keshan disease) is known to occur in numerous people especially in some regions of China (e.g., the Enshi District) (Fordyce et al., 2000; Ge and Yang, 1993). Cobalt (vitamin B_{12}/cobalamin) deficiency (pernicious anemia) has also been reported in many developing regions of the world, and there is a high prevalence of Co deficiency among vegetarians globally (Stabler and Allen, 2004). Figure 12.2 displays the global prevalence of Fe, vitamin A and I deficiencies and Fig. 12.3 shows the worldwide prevalence of inadequate Zn intakes. Clearly, micronutrient deficiencies are more concentrated in many developing nations although significant numbers of people are also affected in developed nations.

The global magnitude of other potential trace element deficiencies (e.g., Cu, Mo and B) in humans is not known with any certainty because no data are available to make such determinations. However, trace element deficiencies rarely occur in isolation because these deficiencies are usually associated with unbalanced diets lacking

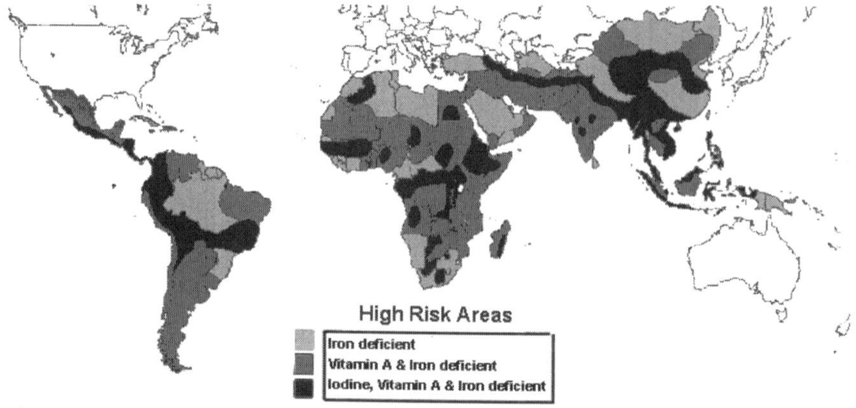

Fig. 12.2 Global prevalence of micronutrient deficiencies (Fe, I and vitamin A) (Sanghvi, 1996)

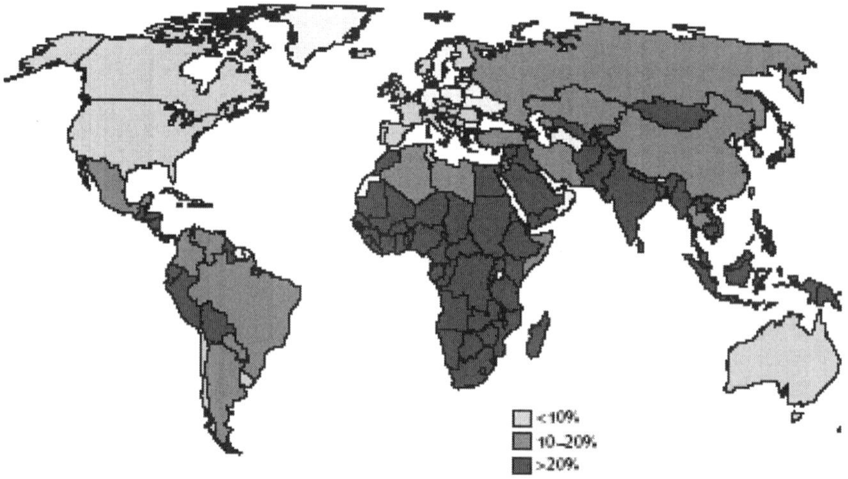

Fig. 12.3 Global percentage of people with inadequate Zn intakes (Hotz and Brown, 2004)

nutrient diversity, being dependent on mostly staple food crops that are low in bioavailable trace elements as eaten. Indeed, in Venezuela, cobalamin deficiency prevalence has been reported to be greater than 60% in infants, children, adolescents and pregnant women (Garcia-Casal et al., 2005). Therefore, it is probable that vast numbers of people also suffer from deficiencies of other trace elements as well.

12.3 Agricultural Tools to Meet Trace Element Requirements

How can agriculture be changed in ways that will result in sufficient essential trace element output of farming systems to assure adequate nutrition for all? Numerous approaches can be used by agriculture for improving the nutritional quality of food crops and the nutrient output of agricultural systems. However, this requires that the agriculture, nutrition and health sectors understand: (1) the importance of such action to human health and society, (2) how they might contribute, and (3) what nutrients are of greatest concern to the nutrition and health communities. Furthermore, government policies must reflect support for such action, consumers must understand the importance of a diverse and balanced diet for their health, productivity, and well-being, and farmers must be shown that participation would be profitable. Increasing consumer knowledge about the impact of poor nutrition on livelihood, educational attainment, employment opportunities, and health should provide a stimulus to increase the demand for better nutritional quality and diversity of foods (McInerney, 2002). Increased consumer demand for improved products and for more diversity of products available in the marketplace would motivate farmers to produce more nutritious and diverse agricultural products.

Some of the ways that agriculture can contribute to reducing malnutrition globally are discussed below. What is currently lacking, however, is the resolve of the agricultural community, the nutrition community, public health officials, private industry, and government policy makers to use agriculture as a primary tool to alleviate malnutrition. Hopefully, through communications, such as this, it will become abundantly clear to the world's leaders that agriculture holds the paramount means by which sustainable solutions to malnutrition can be found. Unless profitable ways are established that will allow agriculture to provide enough food for healthy diets consistently to all, and unless consumers are informed of the consequences of poor diets on their health and livelihoods, developing nations will continue to be plagued with trace element deficiencies with all of their unacceptable ramifications, and developed nations will continue to be burdened with increasing chronic disease rates.

12.3.1 Farming Practices

Today, current agricultural practices are almost always directed at maximizing production while minimizing costs. Recently, preserving the environment has become a more important objective of agriculture (i.e., "sustainable" agricultural goals) worldwide (Cakmak, 2002; Tilman et al., 2002). Unfortunately, maximizing nutrient output of farming systems has never been an objective of either agriculture or of public policy. Yet, scientific knowledge is available that could greatly improve the nutrient output of farming systems, and the available nutrient content of the food crops produced.

The following discussion briefly presents some examples of how some cultural and agronomic practices could be used to enhance the nutrient output garnered from farms.

12.3.1.1 Fertilisers and Soil Amendments

Both macronutrient fertilisers containing N, P, K, and S, and certain micronutrient fertilisers (e.g., Zn, Ni, I, Co, Mo and Se) can have significant effects on the accumulation of nutrients in edible plant products (Grunes and Allaway, 1985; Allaway, 1986). Other micronutrient fertilisers have very little if any effect on the amount of the micronutrient accumulated in edible seeds and grains when they are applied to soils or when used as foliar sprays (Welch, 1986). This is especially true for those micronutrient elements with limited phloem sap mobility such as Fe, B, V and Cr. Some examples of the effects of fertiliser practices on the micronutrient concentrations in edible plant parts are given below. For more detailed information concerning the effects of fertilization practices on micronutrient accumulation in plant foods refer to Salunkhe and Desai (1988), Welch (2001) and Rengel et al. (1999).

For certain essential micronutrient elements (e.g., Zn, Ni, I, and Se), increasing their supply to food crops can result in significant increases in their concentrations in edible plant products. For example, increasing the supply of Zn to pea plants (*Pisum sativum* L.) at levels in excess of that required for maximum yield has been shown to increase the concentration of bioavailable Zn in pea seeds (see Fig. 12.4) (Peck et al., 1980; Welch et al., 1974). Furthermore, increasing the supply of Zn and Se to wheat (*Triticum aestivum* L.) improved the amount of bioavailable Zn and Se in wheat grain (House and Welch, 1989). Increasing Zn levels via Zn fertilisation has also been shown for navy beans (*Phaseolus vulgaris* L.) as well as other crops (Moraghan, 1994). For Fe, providing more to plants than required to sustain growth does little to further increase the Fe in edible seeds and grains (for example, see Welch and Van Campen, 1975). Interestingly, the micronutrient I, supplied in irrigation water, can greatly increase the levels of I in edible portions of food crops alleviating the debilitating disease, cretinism, as well as other I-deficiency disorders in populations dependent on irrigated food crops grown on low-I soils (Cao et al., 1994). In Finland, Se added to fertilisers and applied to soils increased the Se status of the entire Finnish population (Mäkelä et al., 1993).

The accumulation of micronutrient elements in seeds and grains is controlled by a number of processes including root-cell uptake, root–shoot transfer, and the ability of leaf tissues to load these nutrients into the vascular phloem elements which are ultimately responsible for delivering these nutrients to developing seeds and grains via the phloem sap (Welch, 1986, 1999). Phloem loading and unloading of these nutrients are tightly controlled by poorly understood homeostatic mechanisms in the plant and further research should be carried out to understand these processes if we are to significantly increase certain micronutrient elements, such as Fe, in staple seeds and grains (Welch, 1995).

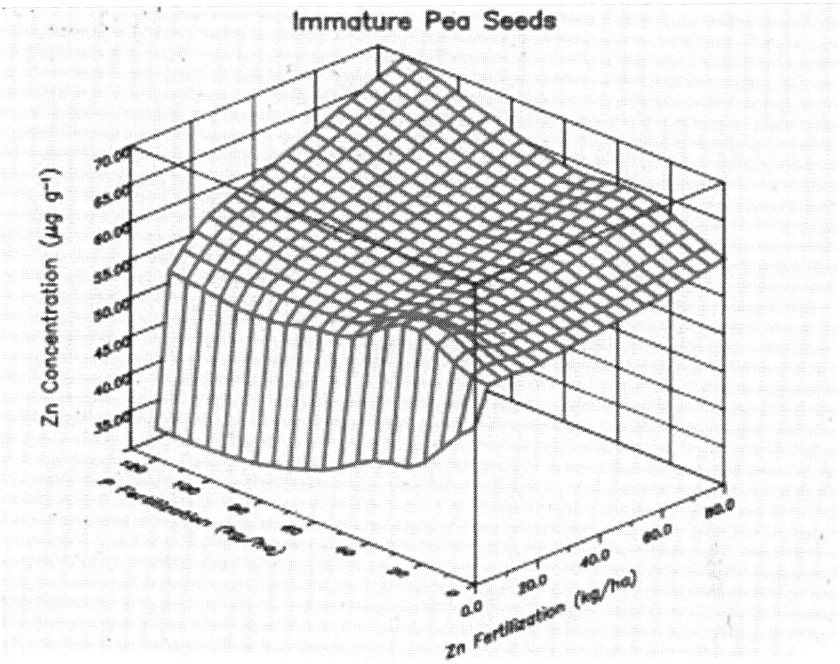

Fig. 12.4 Effects of P and Zn fertilisers on the Zn concentration in immature pea seeds (Peck et al., 1980)

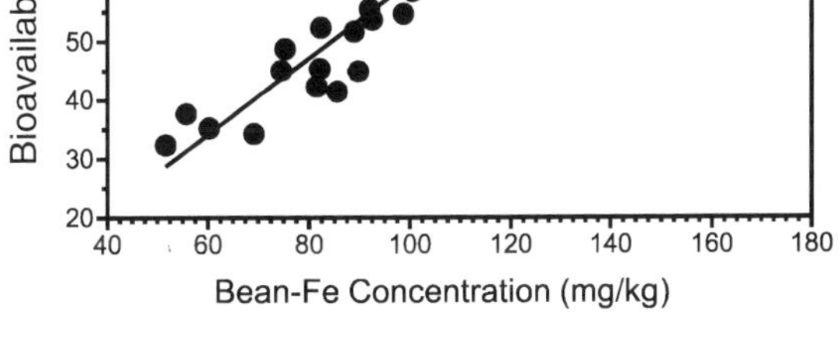

Fig. 12.5 Effect of increasing Fe supply to 24 lines of common bean on the bioavailable levels of Fe in the mature seeds fed to rats (Welch et al., 2000)

Soil amendments are frequently used by farmers to adjust soil pH and to enhance the plant growth properties of soils. Using lime ($CaCO_3$) raises soil pH, permitting acid-intolerant legume species to grow in soils that would otherwise be too acidic for their growth. It is also used to supply Ca to plants. However, adding lime depresses the uptake of Zn, Cu, Fe, and Co, and increases the uptake of Se and Mo by plants. A high soil-pH favors the oxidation of reduced forms of Se such as Se^{2-} and selenite to the more soluble and plant-available selenate anion. Gypsum ($CaSO_4$) and elemental S are used to decrease the pH of alkaline soils as well as to provide S for plant uptake and to ameliorate high-Na alkali soils. Using gypsum on alkaline soils could increase plant-available Fe, Mn, Zn, Cu, and Co by decreasing the soil pH (Sander et al., 1987; Allaway, 1986).

The use of farm yard manure (FYM) and other forms of organic matter can also change plant-available micronutrients by changing both the physical and biological characteristics of the soil (Allaway, 1975, 1986). In many circumstances, these changes improve soil physical structure and water-holding capacity resulting in more extensive root development and enhanced soil microflora and fauna activity, all of which can affect available micronutrient levels in soil to plants (Stevenson, 1991, 1994). However, very few controlled experiments have been done to determine which types of organic matter input practices significantly enhance or depress the levels of trace elements in edible portions of major food crops. More research should be carried out to understand the impact of various types of organic matter on crop nutritional quality with respect to trace elements.

12.3.1.2 Crop Variety Selection

Using micronutrient-dense staple food-crop varieties is one approach that could be exploited to increase the micronutrient output of farms (Welch and Graham, 2002). Although there is substantial evidence in the literature that plant traits for micronutrient efficiency and high micronutrient content of edible parts do exist for various plant species (Gerloff and Gabelman, 1983; Graham, 1984), until recently there has been no systematic survey of staple plant food genomes for these types of traits. However, currently, there is such a global effort underway for surveying the world genomes of rice (*Oryza sativa* L.), bread wheat (*Triticum aestivum* L. and durum wheat (*T. durum* Desf.), maize (*Zea mays* L.), beans (*Phaseolus vulgaris* L.), cassava (*Manihot esculenta* Crantz) and sweet potato (*Ipomoea batatas*, L.) for high micronutrient density traits (Graham et al., 2001). This project is currently known as "HarvestPlus". It is administered by two of the Consultative Groups on International Agricultural Research (CGIAR), Future Harvest Centers, the International Food Policy Research Institute (IFPRI) in Washington, DC and the Centro Internacional de Agricultura Tropical (CIAT) in Cali, Columbia. The program also includes other CGIAR Centers including, Centro Internacional de Mejoramiento de Maiz y Trigo

(CIMMYT) in Mexico, the International Rice Research Institute (IRRI) in the Philippines, the International Institute for Tropical Agriculture (IITA) in Nigeria and the Centro Internacional de la Papa in Lima, Peru. Other CGIAR Centers in Syria and India are also involved (covering other staple food crops), as well as some collaborating institutions (see web site at http://www.harvestplus.org/ for more detailed information).

HarvestPlus, directed at surveying the staple plant food genomes for micronutrient efficiencies and accumulation ability, is based on the premise that increasing the bioavailable concentrations of micronutrients in staple plant foods through traditional plant breeding techniques would be the most efficient and cost-effective means to target micronutrients to people most at risk of developing micronutrient malnutrition, i.e., poor women, infants and children. Historically, most interventions that employ micronutrient supplements or micronutrient food fortification programs have proved not to be sustainable and usually do not effectively reach all of the most at-risk people. Furthermore, these types of interventions are relatively expensive and require sophisticated infrastructures for their creation, management and maintenance to assure compliance (Yip, 1997) The cost of breeding plants with traits that result in significant accumulations of micronutrients in edible portions of staple foods would be a one time cost. Once achieved, these traits can be passed on in breeding programs for future varietal generations and transferred globally to all nations with relatively little additional effort or expense. Thus, this approach to improved micronutrient nutrition is sustainable and cost effective (Graham et al., 2001).

Currently, the CGIAR-sponsored HarvestPlus focuses on two trace elements, Fe, and Zn in staple plant foods, because these plant foods feed most of the world's poor and because deficiencies of these trace elements are known to affect vast numbers of people in developing countries. This plant breeding approach could be expanded to include other nutritionally limiting trace elements such as I, Se, Cu, B, Co and Mo, given the resources.

Recent results of the current effort to identify micronutrient-dense genotypes are encouraging (Welch et al., 2000). For example, Fig. 12.5 presents the results of an experiment in which 24 genotypes of beans (*Phaseolus vulgaris* L.), selected by CIAT for their variability in seed-Fe concentrations, were grown in radio-labeled (i.e., ^{59}Fe) nutrient solutions and subsequently fed to Fe-depleted rats to determine the bioavailability of Fe in the beans. The mature beans harvested varied in Fe concentrations from about 50 to 160 mg kg^{-1} dry weight depending on the genotype. Thus, there was about a threefold range in Fe concentration in the genotypes studied, which demonstrates that beans can be substantially enriched in seed-Fe via genetic selection. There was a positive relationship between the Fe concentration in the beans and Fe bioavailability to rats fed the beans. These data support the contention that increasing the concentration of Fe in bean seeds could be of value in supplying more bioavailable Fe to humans even though beans can contain high levels of certain antinutrients such as phytic acid and tannins (polyphenols) that are known to reduce dietary Fe bioavailability under certain circumstances (see discussion on antinutrients below) (Fairweather-Tait and Hurrell, 1996; Lopez and Martos, 2004; Benito and Miller, 1998).

There is also a large variation in the bioavailable levels of Zn in different populations of wheat. Figure 12.6 shows the data from an experiment in which the bioavailable levels of Zn were tested in 28 lines of wheat grain intrinsically radio-labeled with ^{65}Zn and fed to rats. Clearly, there is a large variation in the bioavailable levels of Zn in different wheat genotypes. Different wheat species and genotypes within this species have been reported to vary greatly in grain-Zn concentrations (i.e., 27–143 mg kg^{-1} Zn) and bioavailable Zn levels (Welch et al., 2005; Genc et al., 2005; Lopez et al., 2003; Cakmak et al., 2001; Balint et al., 2001; Rengel and Romheld, 2000; Monasterio and Graham, 2000).

Thus, it seems feasible for wheat breeders to select for high-Zn and high-Fe density traits in breeding programs. However, further research is needed to determine if the edible portions of Fe and Zn-enriched wheat grain still retain higher levels of Zn after milling and processing, and if enriched levels of Zn in wheat grain are bioavailable to target human populations.

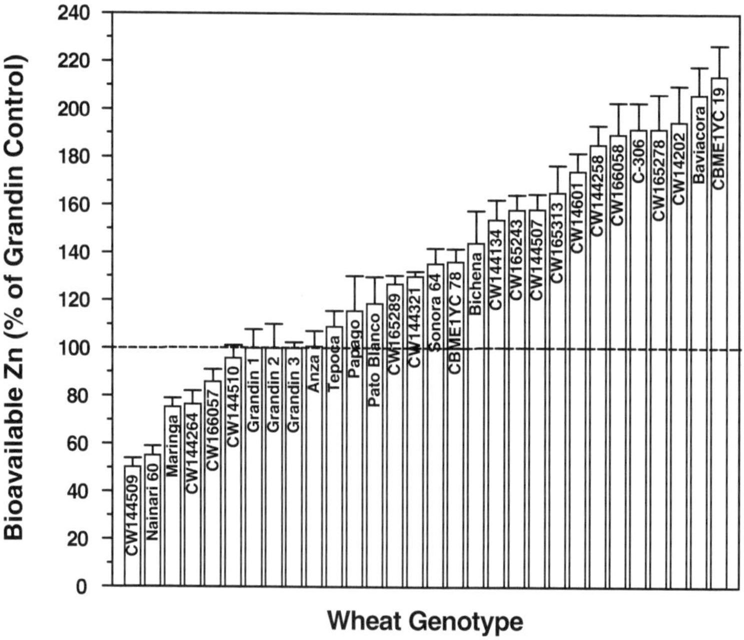

Fig. 12.6 Variation in bioavailable levels of Zn in whole grain fed to rats from 28 wheat genotypes. Data are expressed as a percentage of the Grandin genotype. Error bars represent +SEM (Welch et al., 2005)

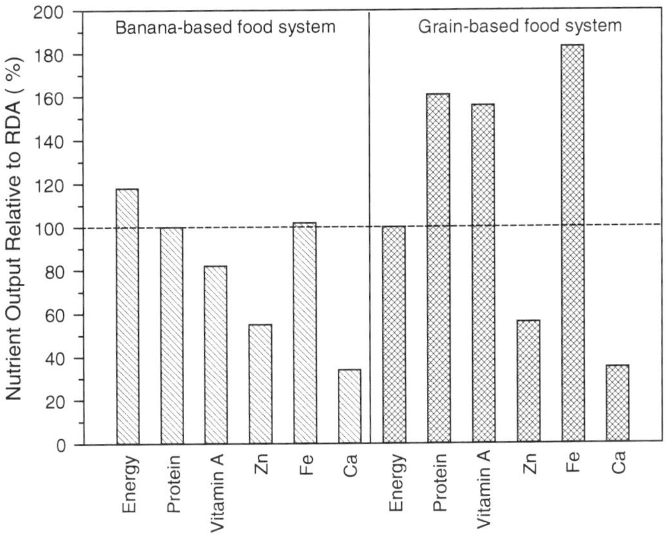

Fig. 12.7 Nutrient output of two Ugandan farming systems (McIntyre et al., 2001). Dashed line represents 100% of required level for each nutrient

Further reports on the genetic variation in some concentrations of trace elements in staple food crops are available (Bãnziger and Long, 2000; Cakmak, 2002; Graham et al., 1998, 1999; Kalayci et al., 1999).

12.3.1.3 Crop Management

Using certain legume crops in rotation with cereal crops can result in substantial increases in the concentration of Zn in cereal grain in areas where soil-Zn is currently limiting wheat production (refer to Holloway, 1997). The selection of crops to avoid micronutrient deficiencies in animals has long been a practice (Mertz, 1987; Underwood, 1971). For instance, in large areas of the USA, soils contain too little Co to meet grazing animal requirements when the livestock are dependent on grasses for feed. However, mixing legumes (that accumulate significantly more Co than grasses) with grasses in pastures is an effective way to supply adequate levels of Co to the grazing animals (Allaway, 1986). Such practices as using pulses in cereal rotations could contribute substantially to increasing the trace element output of farming systems in developing countries. This is not only because it would increase the amount of certain trace elements in cereal grains following pulses, but also it would increase the dietary supply of the more micronutrient-rich pulses for local, regional and national markets lowering the cost to consumers and potentially increasing the consumption of these important micronutrient sources to people at risk of developing micronutrient deficiencies.

12.3.1.4 Increasing the Use of Indigenous and Traditional Food Crops with High Nutritive Value

Within many developing nations, certain indigenous food crops are being displaced and lost as important nutritional components of traditional diets. For instance, in Africa during the past few centuries traditional grains, such as African rice (*Oryza glaberrima* Steudel), Fonio millet (*Digitaria exilis* Stapf and *D. iburua* Stapf), Tef (annual bunchgrass, *Eragrostis tef* Trotter), pearl millet (*Pennisetum glaucum R. Br.*) and sorghum (*Sorghum bicolor* Moench), have been superseded by high-yielding cereals (e.g., rice and wheat) introduced and promoted by agricultural experts from developed nations. The production of many of these traditional crops has decreased even further because of importation of, and subsidies paid for millions of tonnes of wheat, rice and maize that are sold at lower prices. Many traditional crops are much richer sources of micronutrients than the introduced cereal crops that are displacing them (National Research Council, 1996). Increasing the supply of fruits and vegetables to people in many nations would also help reduce the numbers of people afflicted with trace element deficiencies (Mwanga and Cloete, 2003).

12.3.1.5 Designing Cropping Systems to Meet Human Needs

Cropping systems can be designed to maximize nutrient output to meet human needs. An example of such an approach was published by McIntyre et al. (2001). They modeled both a maize-based and a banana-based cropping system in Uganda to provide sufficient quantity to meet household nutritional requirements. Figure 12.7 depicts some of their results. Clearly, the banana-based system does not provide enough vitamin A, Zn and Ca, and the grain-based system does not provide enough Zn and Ca to meet human needs. The Fe values were calculated using the established US National Research Council's RDA value of 15 mg Fe day^{-1}. If one uses the recommended nutrient intake (RNI) value for Fe developed by FAO/WHO of 59 mg Fe day^{-1} to calculate Fe needs, neither system would supply enough Fe to meet nutritional requirements of the people dependent on these cropping systems. These authors explored how these systems could be modified, using locally available resources, to meet human nutritional requirements. They concluded that adequate nutrition, given the same resource base, would require the incorporation of several common but underutilized species into the cropping systems.

Designing cropping systems for maximum nutrient output to improve nutrition and health should become an integral part of agriculture's goals and government policies. Additionally, ways must be found to increase diet diversity among food-insecure people. This would substantially reduce the risk of trace element deficiencies to the most at-risk people. Furthermore, any increase in the production of more micronutrient-rich foods (micronutrient-dense food crops, livestock, dairy or fish) could contribute greatly to finding sustainable solutions to micronutrient malnutrition. Given these axioms, selecting cropping systems not only for their production potential, but also for their ability to supply needed dietary sources of bioavailable trace elements should become a goal of all nations if we are to meet the laudable objectives of better health and prosperity for all.

12.4 Molecular Alterations of Plant Genes to Improve Trace Element Levels

Modern molecular biological techniques can be used to genetically alter food crops to increase their nutritional and health-promoting qualities (Becker and Frei, 2004; Chassy and Mackey, 2003; McCullum et al., 2003; Mackey, 2003; King, 2002; Grusak, 2002, Grusak and DellaPenna, 1999; Schachtman and Barker, 1999; Thompson, 2002). However, this requires detailed knowledge of various physiological and biochemical processes in plants (Grusak and DellaPenna, 1999). Several homoeostatic plant processes must be altered to allow for increased accumulation of nutrients in edible plant products. For trace elements, these processes include increased uptake, increased translocation from roots to shoots, increased remobilization from shoots to reproductive organs, and increased deposition in edible portions of food crops. All of these potential genetic modifications must be done without negatively affecting crop yields, crop quality, food safety, or consumer acceptability. Finally, the trace elements must be in forms that are bioavailable to the people that eat the plant foods in meals that contain numerous other interacting dietary components (Graham et al., 2001; Welch and Graham, 1999). Modern molecular biology technologies, such as the identification of quantitative trait loci (QTL) and the use of marker-assisted selection in combination with traditional breeding strategies, have the potential to greatly increase progress in breeding major staples for improved micronutrient density (Schachtman and Barker, 1999).

12.4.1 Increasing Efficiency of Trace Element Uptake

The mechanisms by which plants accumulate trace elements are under genetic regulation that is influenced by environmental factors. Unfortunately, plant breeders normally have not taken advantage of trace element efficiency traits to enhance the ability of major food crops to absorb essential trace elements from micronutrient-poor soils. Commonly, breeders have used their most productive soils to breed high-yielding, disease- and stress-resistant crops. These fertile soils contain ample available sources of essential trace elements for crops. Because breeders normally use highly fertile soils for their selections, they may have inadvertently lost trace-element efficiency traits during their genetic selection for high-yielding traits because there were no selection pressures to preserve such traits in the breeding process. However, there is ample evidence to show that these traits do exist in plant genomes and that they can be selected for in breeding programs. Graham and colleagues have published extensively on this subject for various micronutrients (e.g., Zn, Mn and Cu) and for several cereal crops including wheat, oats and barley (Graham, 1984, 1988a, b; Graham and Welch, 1996; Rengel and Graham, 1995a, b). Selecting for the ability to accumulate more essential trace elements from nutrient-poor soils is the first step in breeding for trace element dense staple food crops.

12.4.2 Increasing Translocation, Re-mobilisation and Deposition of Trace Elements

The second step in increasing the density of micronutrients in staple foods involves altering the genes that control the translocation of root-accumulated micronutrient elements to shoots. Here also, there is sufficient evidence to suggest that this can be done by genetic selection, but more research is required to more fully delineate the processes and genes involved (Welch, 1986, 1995).

Once higher concentrations of essential trace elements are accumulated in plant shoots, they must be re-translocated out of leaves to reproductive organs before they can be deposited in developing seeds and grains. Re-translocation requires the loading of trace elements from source tissues into vascular phloem elements and the phloem sap, long distance transport within the phloem sap, and unloading of trace elements out of the phloem sap and into sink sites within reproductive organs. The mechanisms that control these processes in plants are not known with any certainty. Further research is needed to determine what these mechanisms are and what genes are responsible for their construction and their regulation.

Knowing how essential trace elements are stored and in what forms they are occurring in edible seeds and grains are also important components (see discussion below) of increasing the bioavailable content of trace elements in edible plant parts. Here also, very little is known about this aspect of these nutrients in plants and much more research should be directed at increasing our knowledge in this area (Welch, 1986, 1999).

12.5 Improving Bioavailability of Essential Trace Elements in Food Crops

Increasing the concentrations of essential trace elements in edible plant foods is only the first step in making these foods richer sources of these nutrients for humans. This is because not all of the trace elements in plant foods are bioavailable to humans that eat these foods. Plant foods can contain substances (i.e., antinutrients) that interfere with the absorption or utilization of these nutrients in humans (Welch and Graham, 1999).

Determining the bioavailability of trace elements in plant foods to humans is pervaded with numerous complexities. A myriad of factors interact to ultimately determine the bioavailability of a particular micronutrient to an individual eating a mixed diet within a given environment (Fairweather-Tait and Hurrell, 1996; House, 1999; Van Campen and Glahn, 1999). Because of this complexity, the data obtained using various bioavailability model systems is always ambiguous. Only data obtained on reducing the prevalence of trace element deficiencies among those afflicted using feeding trials in test populations under free living conditions can delineate the actual efficacy of using trace element-enriched varieties of plant foods as an intervention tool. However, it is impractical to test in this way the bioavailability of selected micronutrients in numerous genotypes of staple plant foods that can be generated in plant breeding programs (Graham et al., 2001; Graham and Welch, 1996).

12.5.1 Inhibitor and Promoter Substances

Plant foods (especially staple seeds and grains) contain various antinutrients (see Table 12.2) in differing amounts depending on both genetic and environmental factors that can reduce the bioavailability of dietary non-heme Fe, Zn and other trace elements to humans (Welch and House, 1984). Dietary substances that promote the bioavailability of trace elements in the presence of antinutrients are also known whose levels are also influenced by both genetic and environmental factors (see Table 12.3). Current plant molecular biological and genetic modification approaches now make it possible to reduce or eliminate antinutrients from staple plant foods, or to significantly increase the levels of promoter substances in these foods (Forssard et al., 2000; Becker and Frei, 2004; King, 2002; Forssard et al., 2000; Theil et al., 1997; Welch, 2002). Given these options (i.e., to decrease antinutrients or increase promoters in staple plant foods), which is the wisest path to pursue?

Table 12.2 Examples of antinutrients in plant foods that reduce the bioavailability of essential trace elements and examples of major dietary sources (Modified from Graham et al., 2001.)

Antinutrients	Essential trace elements inhibited	Major dietary sources
Phytic acid or phytin	Fe, Zn, Cu, Ni	Whole legume seeds and cereal grains
Fiber (cellulose, hemicellulose, lignin, cutin, suberin, etc.)	Fe, Zn, Cu	Whole cereal grain products (e.g., wheat, rice, maize, oat, barley, rye)
Certain tannins and other polyphenolics	Fe	Tea, coffee, beans, sorghum
Hemagglutinins (e.g., lectins)	Fe	Most legumes and wheat
Goitrogens	I	*Brassicas* and *Alliums*
Heavy metals (Cd, Hg, Pb, etc.)	Fe, Zn	Contaminated leafy vegetables and roots

Table 12.3 Examples of substances in foods reported to promote Fe and Zn bioavailability and examples of major dietary sources (Modified from Graham et al., 2001)

Substance	Trace element	Major dietary sources
Certain organic acids (e.g., ascorbic acid, fumarate, malate, citrate)	Fe and/or Zn	Fresh fruits and vegetables
Hemoglobin	Fe	Animal meats
Certain amino acids (e.g., methionine, cysteine, histidine)	Fe and/or Zn	Animal meats
Long-chain fatty acids (e.g., palmitate)	Zn	Human breast milk
Se	I	Sea foods, tropical nuts
β-carotene	Fe	Green and orange vegetables
Inulin and other non-digestible carbohydrates (prebiotics)	Fe, Zn	Chicory, garlic, onion, wheat, Jerusalem artichoke

Plant breeders could breed for genotypes that contain lower concentrations of antinutrients, or molecular biologists could alter plant genes in ways that reduce or even eliminate antinutrients from plant food meals. However, doing so is not without risk and should be done with caution because many antinutrients are major plant metabolites that may play important roles in plant metabolism, in plant abiotic stress resistance and in plant resistance to crop pests or pathogens. Additionally, some of the antinutrients, such as phytate and polyphenols, may play important beneficial roles in human diets by acting as anticarcinogens or by promoting health in other ways such as in decreasing the risk of heart disease or diabetes (Saied and Shamsuddin, 1998; Shamsuddin, 1999; Zhou and Erdman, 1995). Thus, plant breeders and molecular biologists should be aware of the possible negative consequences of changing antinutrients in major plant foods before they attempt to alter food crops in this fashion (Graham and Welch, 1996; Welch, 1996).

Other substances (see Graham et al., 2001 and Welch, 2002 for detailed discussion of this topic) can promote the bioavailability of trace elements in plant foods to humans even in the presence of antinutrients from those foods. Many of these compounds are normal plant metabolites and only small changes in their concentration may have significant effects on the bioavailability of micronutrients. Therefore, it is highly recommended that plant breeders and molecular biologists closely scrutinize the strategy of increasing promoter substances in food crops when attempting to improve food crops as sources of micronutrients for people.

12.6 Summary

There are numerous ways in which agriculture can contribute to improving human nutrition and health. For example, increasing the output of bioavailable, essential trace elements in staple food crops through adopting agricultural practices that are designed to meet human needs can be accomplished from "off the shelf" technologies.

These include choice of cropping systems, agronomic practices, variety selection, as well as the use of modern molecular technologies. Clearly, closely linking agriculture to human health must be accomplished if we are to find sustainable solutions to trace element deficiencies and associated diet-related chronic diseases afflicting the lives and health of massive numbers of people, and the development potential for numerous nations globally.

References

Allaway, W. H. (1975). The effects of soils and fertilisers on human and animal nutrition. USDA-ARS Agriculture Information Bulletin No. 378. Washington, DC, US Government Printing Office. Ref Type: Pamphlet.

Allaway, W. H. (1986). Soil-plant-animal and human interrelationships in trace element nutrition. In Mertz, W. (Ed.), Trace Elements in Human and Animal Nutrition (5th ed., pp. 465–488). Academic, Orlando, San Diego, CA, New York, Austin, London, Montreal, Sydney, Tokyo, Toronto.

Balint, A. F., Kovacs, G., Erdei, L., Sutka, J. (2001). Comparison of the Cu, Zn, Fe, Ca and Mg contents of the grains of wild, ancient and cultivated wheat species. Cereal Research Communications, 29: 375–382.

Bãnziger, M., Long, J. (2000). The potential for increasing the iron and zinc density of maize through plant-breeding. Food and Nutrition Bulletin, 21: 397–400.

Becker, K., Frei, M. (2004). Improving the nutrient availability in rice – biotechnology or biodiversity? Agriculture + Rural Development, 11: 64–65.

Benito, P., Miller, D. (1998). Iron absorption and bioavailability: An updated review. Nutrition Research, 18: 581–603.

Bouis, H. E. (1999). Economics of enhanced micronutrient density in food staples. Field Crops Research, 60: 165–173.

Broek, N. V. D. (2003). Anaemia and micronutrient deficiencies. British Medical Bulletin, 67: 149–160

Cakmak, I. (2002). Plant nutrition research: Priorities to meet human needs for food in sustainable ways. Plant and Soil, 247: 3–24.

Cakmak, O., Ozturk, L., Karanlik, S., Ozkan, H., Kaya, Z., Cakmak, I. (2001). Tolerance of 65 durum wheat genotypes to zinc deficiency in a calcareous soil. Journal of Plant Nutrition, 24: 1831–1847.

Cao, X. Y., Jiang, X. M., Kareem, A., Dou, Z. H., Rakeman, M. R., Zhang, M. L. (1994). Iodination of irrigation water as a method of supplying iodine to a severely iodine-deficient population in Xinjiang, China. Lancet, 344: 107–110.

Chassy, B. M., Mackey, M. (Eds.) (2003). The future of food and nutrition with biotechnology. Journal of the American College of Nutrition, 21 (Supplement), 157S–221S.

Clugston, G. A., Smith, T. E. (2002). Global nutrition problems and novel foods. Asia Pacific Journal of Clinical Nutrition, 11: S100–S111.

Combs, G. F., Jr., Welch, R. M. (1998). Creating Healthful Food Systems: Linking Agriculture to Human Needs. Cornell International Institute for Food, Agriculture and Development. Ithaca, NY.

Copenhagen Consensus (2004). Copenhagen Consensus 2004 Today's challenge – Tomorrow's opportunity. Available at www.copenhagenconsensus.com and http://www.copenhagenconsensus.com/Default.aspx?ID=158

Darnton-Hill, I., Webb, P., Harvey, P. W., Hunt, J. M., Dalmiya, N., Chopra, M. (2005). Micronutrient deficiencies and gender: Social and economic costs. American Journal of Clinical Nutrition, 81: 1198S–1205S.

Fairweather-Tait, S. J., Hurrell, R. F. (1996). Bioavailability of minerals and trace elements. Nutrition Research Reviews, 9: 295–324.

Fordyce, F. M., Zhang, G. D., Green, K., Liu, X. P. (2000). Soil, grain and water chemistry in relation to human selenium-responsive diseases in Enshi District, China. Applied Geochemistry, 15: 117–132.

Forssard, E., Bucher, M., Mächler, F., Mozafar, A., Hurrell, R. (2000). Review: Potential for increasing the content and bioavailability of Fe, Zn and Ca in plants for human nutrition. Journal of the Science of Food and Agriculture, 80: 861–879.

Fresco, L. O. (2000). Scientific and ethical challenges in agriculture to meet human needs. Food, Nutrition and Agriculture, 27: 4–13.

Garcia-Casal, M. N., Osorio, C., Landaeta, M., Leets, I., Matus, P., Fazzino, F. (2005). High prevalence of folic acid and vitamin B12 deficiencies in infants, children, adolescents and pregnant women in Venezuela. European Journal of Clinical Nutrition, 59: 1064–1070.

Ge, K., Yang, G. (1993). The epidemiology of selenium deficiency in the etiological study of endemic diseases in China. American Journal of Clinical Nutrition, 57 (Suppl.): 259S–263S.

Genc, Y., Humphries, J. M., Lyons, G. H., Graham, R. D. (2005). Exploiting genotypic variation in plant nutrient accumulation to alleviate micronutrient deficiency in populations. Journal of Trace Elements in Medicine and Biology, 18: 319–324.

Gerloff, G. C., Gabelman, W. H. (1983). Genetic basis of inorganic plant nutrition. In Lauchli, A., Bieleski, R. L. (Eds.), Inorganic Plant Nutrition (1st ed., pp. 453–480). Springer, Berlin, Heidelberg, New York, Tokyo.

Gibson, R. S. (2003). Concurrent micronutrient deficiencies in developing countries: Problems and solutions. Proceedings of the Nutrition Society of New Zealand, 28: 21–34.

Graham, R. D. (1984). Breeding for nutritional characteristics in cereals. Advances in Plant Nutrition, 1: 57–102.

Graham, R. D. (1988a). Development of wheats with enhanced nutrient efficiency: Progress and potential. In Klatt, A. R. (Ed.), Wheat Production Constraints in Tropical Environments. Proceedings of the International Conference, Maj, Thailand, 19–23 January, 1987 (pp. 305–320). CIMMYT, Mexico DF, Mexico.

Graham, R. D. (1988b). Genotypic differences in tolerance to manganese deficiency. In Graham, R. D., Hannam, R. J., Uren, N. C. (Eds.), Manganese in Soils and Plants (pp. 216–276). Kluwer, Dordrecht, The Netherlands.

Graham, R. D., Welch, R. M. (1996). Breeding for staple-food crops with high micronutrient density (Rep. No. 1). International Food Policy Research Institute., Washington, DC.

Graham, R. D., Senadhira, D., Beebe, S. E., Iglesias, C. (1998). A strategy for breeding staple-food crops with high micronutrient density. Soil Science and Plant Nutrition, 43: 1153–1157.

Graham, R. D., Senadhira, D., Beebe, S., Iglesias, C., Monasterio, I. (1999). Breeding for micronutrient density in edible portions of staple food crops: Conventional approaches. Field Crops Research, 60: 57–80.

Graham, R. D., Welch, R. M., Bouis, H. E. (2001). Addressing micronutrient malnutrition through enhancing the nutritional quality of staple foods: Principles, perspectives and knowledge gaps. Advances in Agronomy, 70: 77–142.

Grunes, D. L., Allaway, W. H. (1985). Nutritional quality of plants in relation to fertiliser use. In Engelstad, O. P. (Ed.), Fertiliser Technology and Use (3rd ed., pp. 589–619). Soil Science Society of America, Madison, WI.

Grusak, M. A. (2002). Enhancing mineral content in plant food products. Journal of the American College of Nutrition, 21: 178S–183S.

Grusak, M. A., DellaPenna, D. (1999). Improving the nutrient composition of plants to enhance human nutrition and health. Annual Review of Plant Physiology, 50: 133–161.

Hotz, C., Brown, K. H. (2004). International Zinc Nutrition Consultative Group (IZiNCG). Technical Document #1. Assessment of the risk of zinc deficiency in populations and options for its control. Food and Nutrition Bulletin, 25: S91–S203.

House, W. A. (1999). Trace element bioavailability as exemplified by iron and zinc. Field Crops Research, 60: 115–141.

House, W. A., Welch, R. M. (1989). Bioavailability of and interactions between zinc and selenium in rats fed wheat grain intrinsically labeled with ^{65}Zn and ^{75}Se. Journal of Nutrition, 119: 916–921.

Holloway, R. E. (1997) Zinc as a subsoil nutrient for cereals. Ph.D. thesis, University of Adelaide, Adelaide, South Australia.

Kalayci, M., Torun, B., Eker, S., Aydin, M., Ozturk, L., Cakmak, I. (1999). Grain yield, zinc efficiency and zinc concentration of wheat cultivars grown in a zinc-deficient calcareous soil in field and greenhouse. Field Crops Research, 63: 87–98.

Kennedy, G., Nantel, G., Shetty, P. (2003). The scourge of "hidden hunger": Global dimensions of micronutrient deficiencies. Food, Nutrition and Agriculture, 32: 8–16.

King, J. C. (2002). Biotechnology: A solution for improving nutrient bioavailability. International Journal for Vitamin and Nutrition Research, 72: 7–12.

Lopez, H. W., Krespine, V., Lemaire, A., Coudray, C., Feillet-Coudray, C., Messager, A. (2003). Wheat variety has a major influence on mineral bioavailability; Studies in rats. Journal of Cereal Science, 37: 257–266.

Lopez, M. A., Martos, F. C. (2004). Iron availability: An updated review. International Journal of Food Science and Nutrition, 55: 597–606.

Mackey, M. A. (2003). The developing world benefits from plant biotechnology. Journal of Nutrition Education and Behavior, 35: 210–214.

Mäkelä, A.-L., Näntö, V., Mäkelä, P., Wang, W. (1993). The effect of nationwide selenium enrichment of fertilisers on selenium status of healthy Finnish medical students living in south western Finland. Biological Trace Element Research, 36: 151–157.

Mason, J. B., Lotfi, M., Dalmiya, N., Sethuraman, K., Deitchler, M., Geibel, S. (2001). The micronutrient report: Current progress in the control of vitamin A, iodine, and iron deficiencies. Micronutrient Initiative/International Development Research Center, Ottowa.

McCullum, C., Benbrook, C., Knowles, L., Roberts, S., Schryver, T. (2003). Application of modern biotechnology to food and agriculture: Food systems perspective. Journal of Nutrition Education and Behavior, 35: 319–332.

McInerney, J. (2002). The production of food: From quantity to quality. Proceedings of the Nutrition Society, 61: 273–279.

McIntyre, B. D., Bouldin, D. R., Urey, G. H., Kizito, F. (2001). Modeling cropping strategies to improve human nutrition in Uganda. Agricultural Systems, 67: 105–120.

Mertz, W. (1987) Trace Elements in Human and Animal Nutrition. Vol. 1 (5th ed.). Academic, San Diego, CA, New York.

Monasterio, I., Graham, R. D. (2000). Breeding for trace minerals in wheat. Food and Nutrition Bulletin, 21: 392–396.

Moraghan, J. T. (1994). Accumulation of zinc, phosphorus, and magnesium by navy bean seed. Journal of Plant Nutrition, 17: 1111–1125.

Müller, O., Krawinkel, M. (2005). Malnutrition and health in developing countries. Canadian Medical Association Journal, 173: 279–286.

Mwanga, R. O. M., Cloete, M. (2003). The role of horticulture: Issues, opportunities and constraints. Horticultural Science in Emerging Economies: Issues and Constraints, 45–51.

National Research Council (1996). Lost Crops of Africa, Grains, Vol. 1. National Academy, Washington, DC.

Nielsen, F. H. (1997). Beyond copper, iodine, selenium and zinc: Other elements that will be found important in human nutrition by the year 2000. In Fischer, P.W. F., L'Abbé, M. R., Cockell, K. A., Gibson, R. S. (Eds.), Trace Elements in Man and Animals – 9. Proceedings of the Ninth International Symposium on Trace Elements in Man and Animals NRC Research Press, Ottawa.

Peck, N., Grunes, D. L., Welch, R. M., MacConald, G. E. (1980). Nutritional quality of vegetable crops as affected by phosphorus and zinc fertilisers. Agronomy Journal, 72: 528–534.

Rengel, Z., Graham, R. D. (1995a). Wheat genotypes differ in Zn efficiency when grown in chelate- buffered nutrient solution. Plant and Soil, 176: 307–316.

Rengel, Z., Graham, R. D. (1995b). Wheat genotypes differ in Zn efficiency when grown in chelate-buffered nutrient solution. 2. Nutrient uptake. Plant and Soil, 176: 317–324.

Rengel, Z., Romheld, V. (2000). Differential tolerance to Fe and Zn deficiencies in wheat germplasm. Euphytica, 113: 219–225.

Rengel, Z., Batten, G. D., Crowley, D. E. (1999). Agronomic approaches for improving the micronutrient density of edible portions of field crops. Field Crops Research, 60: 27–40.

Rouse, T. I., Davis, D. P. (2004). Exploring a vision: Integrating knowledge for food and health. A workshop summary. Board on Agriculture and Natural Resources, Division on Earth and Life Studies, National Research Council of the National Academies of Sciences, The National Academies Press, Washington, DC.

Saied, H. T., Shamsuddin, A. M. (1998). Up-regulation of the tumor suppressor gene p53 and WAF1 gene expression by IP6 in HT-29 human colon carcinoma cell line. Anticancer Research, 18: 1479–1484.

Salunkhe, D. K., Desai, B. B. (1988). Effects of agricultural practices, handling, processing, and storage on vegetables. In Karma, E., Harris, R. S. (Eds.), Nutritional Evaluation of Food Processing (3rd ed., pp. 23–71). Avi Book, Van Nostrand Reinhold, New York.

Sander, D. H., Allaway, W. H., Olson, R. A. (1987). Modification of nutritional quality by environment and production practices. In Olson, R. A., Frey, K. J. (Eds.), Nutritional Quality of Cereal Grains and Agronomic Improvements. Agronomy Monograph no. 28 (pp. 45–82). American Society of Agronomy, Madison, WI.

Sanghvi, T. G. (1996). Economic Rationale for Investing in Micronutrient Programs. A Policy Brief Based on New Analyses. Office of Nutrition, Bureau for Research and Development, United States Agency for International Development, Washington, DC.

Schachtman, D. P., Barker, S. J. (1999). Molecular approaches for increasing the micronutrient density in edible portions of food crops. Field Crops Research, 60: 81–92.

Shamsuddin, A. M. (1999). Metabolism and cellular functions of IP6: A review. Anticancer Research, 19: 3733–3736.

Sobal, J. (1999). Food system globalization, eating transformations, and nutrition transitions. In Grew, R. (Ed.), Food in Global History (pp. 171–193). Westview, Boulder, CO.

Stabler, S. P., Allen, R. H. (2004). Vitamin B12 deficiency as a worldwide problem. Annual Review of Nutrition, 24: 299–326.

Stevenson, F. J. (1991). Organic matter-micronutrient reactions in soil. In Mortvedt, J. J., Cox, F. R., Shuman, L. M., Welch, R. M. (Eds.), Micronutrients in Agriculture (2nd ed., pp. 145–186). Soil Science Society of America, Madison, WI.

Stevenson, F. J. (1994). Humus chemistry: Genesis, Composition, Reactions (2nd ed.). Wiley, New York.

Subbulakshmi, G., Naik, M. (1999). Food fortification in developing countries – Current status and strategies. Journal of Food Science and Technology – Mysore, 36: 371–395.

Theil, E. C., Burton, J. W., Beard, J. L. (1997). A sustainable solution for dietary iron deficiency through plant biotechnology and breeding to increase seed ferritin control. European Journal Biochemistry, 51: S28–S31.

Thompson, J. A. (2002). Research needs to improve agricultural productivity and food quality, with emphasis on biotechnology. Journal of Nutrition, 132: 3441S–3442S.

Tilman, D., Cassman, K. G., Matson, P. A., Naylor, R., Polasky, S. (2002). Agricultural sustainability and intensive production practices. Nature (London), 418: 671–677.

Tontisirin, K., Nantel, G., Bhattacharjee, L. (2002). Food-based strategies to meet the challenges of micronutrient malnutrition in the developing world. Proceedings of the Nutrition Society, 61: 243–250.

Underwood, B. A. (2003). Scientific research: Essential, but is it enough to combat world food insecurities? 11th International Symposium on Trace Elements in Man and Animals. June 2–5, 2002, Berkeley, CA. Journal of Nutrition, 133: 1434S–1437S.

Underwood, B. A., Smitasiri, S. (1999). Micronutrient malnutrition: Policies and programs for control and their implications. Annual Review of Nutrition, 19: 303–324.

Underwood, E. J. (1971). Trace Elements in Human and Animal Nutrition (3rd ed.). Academic, New York, London.

Van Campen, D. R., Glahn, R. P. (1999). Micronutrient bioavailability techniques: Accuracy, problems and limitations. Field Crops Research, 60: 93–113.
Welch, R. M. (1986). Effects of nutrient deficiencies on seed production and quality. Advances in Plant Nutrition, 2: 205–247.
Welch, R. M. (1995). Micronutrient nutrition of plants. Critical Reviews in Plant Science, 14: 49–82.
Welch, R. M. (1999). Importance of seed mineral nutrient reserves in crop growth and development. In Rengel, Z. (Ed.), Mineral Nutrition of Crops. Fundamental Mechanisms and Implication*s* (pp. 205–226). Food Products Press, New York.
Welch, R. M. (2001). Micronutrients, agriculture and nutrition; linkages for improved health and well being. In Singh, K., Mori, S., Welch, R. M. (Eds.), Perspectives on the Micronutrient Nutrition of Crops (pp. 247–289). Scientific Publishers (India), Jodhpur, India.
Welch, R. M. (2002). Breeding strategies for biofortified staple plant foods to reduce micronutrient malnutrition globally. Journal of Nutrition, 132: 495S–499S.
Welch, R. M., Graham, R. D. (1999). A new paradigm for world agriculture: Meeting human needs – Productive, sustainable, nutritious. Field Crops Research, 60: 1–10.
Welch, R. M., Graham, R. D. (2002). Breeding crops for enhanced micronutrient content. Plant and Soil, 245: 205–214.
Welch, R. M., House, W. A. (1984). Factors affecting the bioavailability of mineral nutrients in plant foods. In Welch, R. M., Gabelman, W. H. (Eds.), Crops as Sources of Nutrients for Humans (pp. 37–54). American Society of Agronomy, Madison, WI.
Welch, R. M., Van Campen, D. R. (1975). Iron availability to rats from soybeans. Journal of Nutrition, 105: 253–256.
Welch, R. M., House, W. A., Allaway, W. H. (1974). Availability of zinc from pea seeds to rats. Journal of Nutrition, 104: 733–740.
Welch, R. M., Combs, G. F., Jr., Duxbury, J. M. (1997). Toward a "Greener" revolution. Issues in Science and Technology, 14: 50–58.
Welch, R. M., House, W. A., Beebe, S., Cheng, Z. (2000). Genetic selection for enhanced bioavailable levels of iron in bean (*Phaseolus vulgaris* L.) seeds. Journal of Agricultural and Food Chemistry, 48: 3576–3580.
Welch, R. M., House, W. A., Ortiz-Monasterio, I., Cheng, Z. (2005). Potential for improving bioavailable zinc in wheat grain (*Triticum* Species) through plant breeding. Journal of Agricultural and Food Chemistry, 53: 2176–2180.
World Health Organization (1999). Malnutrition Worldwide. Available at http://www.who.int/nut/malnutrition_worldwide.htm
World Health Organization (2002). The World Health Report 2002. Reducing Risks, Promoting Healthy Life World Health Organization, Geneva.
World Health Organization (2003). Joint WHO/FAO Expert Consultation on Diet nutrition and the prevention of chronic diseases (2002: Geneva, Switzerland) (World Health Technical Report Series, No. 916). World Health Organization, Geneva.
World Health Organization (2004). Global strategy on diet, physical activity and health. Fifty-seventh World Health Assembly Agenda item 12.6, 22 May 2004 (Rep. No. WHA57.17). World Health Organization, Geneva.
Yip, R. (1997). The challenge of improving iron nutrition: Limitations and potentials of major intervention approaches. European Journal of Clinical Nutrition, 51: S16–S24.
Zhou, J. R., Erdman, J. W., Jr. (1995). Phytic acid in health and disease. Critical Reviews in Food Science and Nutrition, 35: 495–508.

Appendix 1

Table A1.1 Botanic names of crop plants

Common name	Botanic name
Alfalfa (Lucerne)	*Medicago sativa* L.
Almond	*Prunus dulcis* (Mill.) D.A. Webb or *Amygdalus communis*
Apple	*Mallus sylvestris* Mill.
Avocado	*Persea americana* Mill.
Banana	*Musa x paradisiaca* L.
Barley	*Hordeum vulgare* L.
Bean, broad	*Vicia faba* L.
Bean, French	*Phaseolus aconitifoliu* Jacq
Bean, green	*Phaseolus vulgaris* L.
Bean, mung	*Vigna radiata* (L.) Wilczek var *radiata*
Black gram	*Phaseolus mungo* L.
Cabbage	*Brassica oleracae* var capitata (L.)
Carrot	*Daucus carota* L.
Cassava (Manioc)	*Manihot esculenta* Crantz.
Cauliflower	*Brassica oleracae* var botrytis L.
Cherry, sweet	*Prunus avium* L.
Chickpea	*Cicer arietinum* L.
Citrus, lemon	*Citrus limon* L.
Citrus, orange (sweet)	*Citrus sinensis* L.
Clover, berseem	*Trifolium alexandrinum* L.
Clover, subterranean	*Trifolium subterraneum* L.
Coconut	*Cocos nucifera* L.
Coffee	*Coffea Arabica* L.
Cotton	*Gossypium hirsutum* L
Cowpea	*Vigna sativa* Lens (or *Vigna unguiculata* L.)
Dhaincha (dunchi fiber)	*Sesbania bispinosa* Jacq.
Flax/linseed	*Linum usitatissimum* L.
French bean	*Phaseolus aconitifolius* Jacq.
Grape, muscadine	*Vitis rotundifolia* Michx.
Grape, wine	*Vitis vinifera* L.
Green gram	*Phaseolus aureus* Roxb.
Groundnut/peanut	*Arachis hypogea* L.
Guava	*Psidium* spp.
Kiwi fruit (Chinese gooseberry)	*Actinidia chinensis* Planch

(continued)

Table A1.1 (continued)

Common name	Botanic name
Lentil	*Lens esculenta* Moench (or *Lens culinaris* Medik)
Lettuce	*Lactuca sativa* L.
Lupin, sweet	*Lupinus angustifolius* L.
Maize (corn)	*Zea mays* L.
Mango	*Mangifera indica* L.
Medic, black	*Medicago lupulina* L.
Millet, finger	*Eleusine coracana* L.
Millet, fonio	*Digitaria exilis* Stapf and *D. iburua*Stapf
Millet, pearl	*Pennisetum americanum* L. Leeke, or *Pennisetum typhodium* Stapf., or *Pennisetum glaucum* L.
Mustard, black	*Brassica nigra* L.
Mustard, white	*Sinapis alba* L.
Mustard, Indian	*Brassica juncea* Coss
Oats	*Avena sativa* L.
Oilseed rape (Canola)	*Brassica napus* L.
Okra	*Abelmoschus esculentis* L Moench
Olive	*Olea europaea* L.
Onion	*Allium cepa* L
Pea	*Pisum sativum* L.
Peach	*Prunus dulcis* L.
Pear	*Pyrus communis* L.
Pecan	*Carya illinoensis* (Wagenh.) K. Koch
Pepper, black	*Piper nigrum* L.
Pepper, Indian long	*Piper longum* L.
Pepper, tabasca	*Capsicum frutescens* L.
Pigeon pea	*Cajanus cajan* L. (Huth.)
Pineapple	*Ananas comosus* L.
Pistachio	*Pistacia vera* L.
Potato	*Solanum tuberosum* L.
Raya	*Brassica campestris* L.
Rice	*Oryza sativa* L.
Rice, African	*Oryza glaberrima* Steudel
Rubber	*Hevea brasiliensis*
Rye	*Secale cereale* L.
Sesame	*Sesamum indicum* L.
Siratro	*Phaseolus atropurpureus* DC.
Sorghum	*Sorghum bicolour* Moench
Soya bean	*Glycine max* L.
Sugar beet	*Beta vulgaris* L.
Sugar cane	*Saccharum officinarum* L.
Sunflower	*Helianthus annuus* L.
Sweet potato	*Ipomoea batatas*, L.
Tea	*Camellia sinensis* L. Ktze
Teff	*Eragrostis tef* (Zuccagni) Trotter
Tobacco	*Nicotiana tabacum* L.
Tomato	*Lycopersicum esculentum* Mill.

(continued)

Table A1.1 (continued)

Common name	Botanic name
Triticale	x *Triticosecale* (*Triticum* x *Secale*)
Turnip	*Brassica rapa* L.
Walnut	*Juglans regia* L. (or *Juglans nigra*)
Wheat – bread	*Triticum aestivum* L.
Wheat – durum	*Triticum durum* Desf. (or *T. turgidum* var durum)
Yam, composite	*Discorea composita* Hemsl.

Appendix 2

Comparison of the Major Soil Groups in the FAO-UNESCO/World Reference Base for Soil Resources and USDA Soil Taxonomy Classifications (Derived from Kabata-Pendias, 2001 and ISSS, 1998)

FAO-UNESCO/ WRB Unit	World Area (Mha)	USDA Soil Taxonomy Equivalent	Other Names/ Descriptions
Fluvisols	350	Fluvents	Alluvial soils
Gleysols	720	Aquic –suborders	Hydromorphic soils with characteristic pale gleyic colours due to Fe (II)
Regosols	258	Entisols Orthents, Psamments	Skeletal soils, deep soils on unconsolidated parent materials
Arenosols	900	Psamments (part)	Sandy soils, red and yellow sand soils
Leptosols (WRB)	1,655	Entisols, lithic subgroups Rendolls	Lithosols, Rankers, Rendzinas, shallow soils over hard rock or material with high $CaCO_3$
Rendzinas (FAO-UNESCO)	-	Rendolls	Calcic Leptosols (in WRB)
Andosols	110	Andepts	Soils on volcanic ash
Vertisols	335	Vertisols	Shrink-swell, deep cracking clay soils
Solonchaks	<340	Salorthids	Salt-affected soils, high salt content
Solonetz	135	Natric Great Group	Distinct Natric horiz',
Yermosols (FAO-UNESCO)	-	Typic Aridisol	Aridisols, Desert soils
Xerosols (FAO-UNESCO)	-	Mollic Aridisol	Grey soils
Gypsisols (WRB)	90	Aridisols, Gypsiorthids	Yermosols, Xerosols, with gypsum horiz'
Kastanozems	465	Ustolls, Borolls	Chestnut soils of dry (short grass) steppes
Chernozems	230	Borolls (part)	Soils of tall grass steppes, calc' black soils
Phaeozems	190	Udolls (part), Aquoll,	More leached than Kastan' & Chernozems
Greyzems (FAO-UNESCO)	-	Borolls (part)	Dark surface, bleached E horiz' and textural B horizons, cool temp' forest soils
Cambisols	1,500	Inceptisols (part) Dystro- Eutrochrepts	Brown forest soils, brown earths

(continued)

Appendix 2 (continued)

FAO-UNESCO/ WRB Unit	World Area (Mha)	USDA Soil Taxonomy Equivalent	Other Names/ Descriptions
Luvisols	650	Alfisols (part)	Sols lessivés, leached brown soils
Albeluvisol (WRB)	320	Glossic', Alfisol Great Group	Acid soils with bleached horiz' above clay- rich subsoil (Podzoluvisol: FAO-UNESCO)
Podzols	485	Spodosols (part)	Highly altered profile due to severe leaching
Umbrisols (WRB)	100	Umbrepts, Humitotropepts	Brown podzolic, Humic ochric brown soils
Planosols	130	Albaqualfs, Albaquults	Mineral hydromorphic soils
Allisols (WRB)	100	Aquults, Humults, Udults	Red yellow soils with high clay contents, distinguished by argic B horiz' in trop', sub-trop' and Mediterranean regions
Lixisols	435	Latosols, oxic subgroups of Alfisols	Red yellow podzolic, red & yellow earths, argic horiz' with low activity clay, seasonally dry trop', sub-trop' & temp' regions
Acrisols	1000	Ultisols (part)	Strongly leached with clay horiz'
Nitosols	200	Some Ultisols, some Alfisols	Trop' soils (transition between Acrisols and Ferralsols) especially in E. Africa
Ferralsols	750	Oxisols	Deep trop' soils, Lateritic soils, Latosols
Plinthosols (WRB)	60	Plinthaquox	Plinthic Ferralsols, groundwater laterite soils, have 'plinthite' Fe-rich horiz'
Histosols	275	Histosols	Organic soils, peat soils
Anthrosols	0.5	Anthr' suborder, in Great/subgroups	Soils conditioned by human activity, Hortisols

Abbreviations: horiz' = horizon; calc' = calcareous; temp' = temperate; trop' = tropical; sub-trop'= sub-tropical; Kastan' = Kastanozems; WRB= Soil Group featured in the World Reference Base for Soil Resources (1998) not used in the original FAO-Unesco classification for the Soil Map of the World (1974); FAO-UNESCO = Soil Group used in the FAO-UNESCO soil classification (1974), but not used in the World Reference Base for Soil Resources (included in this list because still widely used).

References and Further Reading

Brady, N.C.,Weil, R.R. (2002) The Nature and Properties of Soils. (13th edition) Prentice Hall, Upper Saddle River, NJ. pp 79-120.

ISSS Working Group RB. (1998) World Reference Base for Soil Resources: Introduction. (Deckers, J.A., Nachtergaele, F.O., Spaargaren, O.C. (Eds.) International Soil Reference and Information Centre (ISRIC) and Food and Agriculture Organization of the United Nations (FAO), Acco Publishers, Leuven.

Kabata-Pendias, A. (2001) Trace Elements in Soils and Plants (3rd edition). CRC Press, Boca Raton Fl.pp 28-29.

Soil Survey Staff (1999) Soil Taxonomy: A basic system of soil classification for making and interpreting soil surveys (2nd ed.). Agricultural Handbook 436, Natural Resources conservation Service, USDA, Washington DC. USA, pp. 869.

Appendix 3

Table A3.1 Average national yields for maize, rice and wheat in 2003 for selected countries and the world (t ha^{-1})

	Maize	Rice	Wheat
Argentina	6.47	5.40	2.06
Australia	8.23	10.29	2.00
Brazil	3.7	3.24	2.37
China	4.85	6.07	3.91
Egypt	7.71	9.43	6.15
France	7.14	5.61	6.23
India	2.11	2.62	3.00
Indonesia	3.25	4.54	–
Iran	8.57	5.89	1.98
Mexico	2.53	3.79	4.79
Myanmar	2.50	3.71	1.11
Pakistan	1.46	3.05	2.38
Russian Federation	3.26	2.97	3.17
South Africa	2.90	–	1.78
Syria	4.00	–	2.74
Tanzania	1.54	1.96	1.18
Turkey	4.87	5.31	2.02
USA	8.92	7.45	2.97
Vietnam	3.22	4.63	–
World	3.41 (0.54–23.6)[a]	3.37 (0.7–10.29)	2.75 (0.35–7.83)

[a]Lowest and highest average national yields

FAO (2003) FAOSTAT Production Statistics, Food and Agriculture Organisation, Rome

Glossary

adsorption (desorption)	Retention of ions on solid surfaces in the soil by a combination of mechanisms: ion exchange, specific adsorption, precipitation and organic complexation (also referred to as 'sorption') – desorption is the release of ions from the various retention mechanisms.
aeolian deposits	Particles of fine sand and silt transported by the wind and deposited at distance from the source (also called loess).
Albeluvisols	Formerly called Podzoluvisols, soil with a brown or bleached topsoil and an argic subsurface horizon with an irregular boundary – World Reference Base classification (Glossoboralfs, Fragiboralfs, Fragiaqualfs and Ferrudalfs in USDA Soil Taxonomy).
Alfisols	Moderately leached soils, more highly weathered than Mollisols (Chernozems) but less than Spodosols (Podzols) – occur widely from cool to hot humid areas and also in semiarid tropics and Mediterranean climates (USDA Soil Taxonomy – see Appendix 2)
aqua regia	Mixture of concentrated nitric and hydrochloric acids, used for the digestion of soil samples to provide "pseudo-total" concentrations of elements (cf., true "total" analysis with hot HNO_3, $HClO_4$ and HF acids).
arable land	Land which is cultivated for crops.
Arenosols	Sandy soils with >65% sand-sized (>2 mm) particles. These soils are very freely drained, have low concentrations of most elements and are highly prone to causing deficiencies of micronutrients in crops – FAO-UNESCO/World Reference Base classification. (Psamments and Psammaquents in USDA Soil Taxonomy).

argic horizon	Horizon with an accumulation of clay (>3% in sandy soils and >8% in clayey soils) compared to overlying horizons.
Aridisols	Dry soils of arid/semiarid regions, usually with ochric surface horizon (USDA Soil Taxonomy – see Appendix 2).
auxin	A plant growth regulating substance (e.g., indole acetic acid – IAA).
biogas slurry	Organic residue from methane generation using manures as a carbon source.
biosolids	Alternative name for sewage sludge (solid residues from waste water treatment).
broad acre crops	Term used mainly in Australia for relatively low-intensity field crops such as cereals, oilseeds and pulses. The next more intensive group of crops ("semi-intensive crops") are cotton, rice, sugar cane and potatoes.
C3 plants	Plants with a basic photosynthesis mechanism which fixes CO_2 in only one stage. Examples: wheat, rice and soya beans.
C4 plants	Plants which fix CO_2 in two stages can raise its concentration in their leaves above ambient levels. Examples: maize, sugarcane and sorghum.
Calcareous soils	Soils with a high content (>15%) of free $CaCO_3$ either developed on limestones, or that have become calcified by the deposition of $CaCO_3$ in pores and voids as a result of the evaporation of soil solution in arid environments. These soils generally have neutral or alkaline pH values and trace elements tend to be held in a sparingly available form. Classed as Calcisols in the FAO-UNESCO/World Reference Base classification.
calcifuge (plants)	Plants which are unable to grow on calcium-rich soils (adapted to more acid soils).
Calcisols	Soils with a high content (>15%) of free $CaCO_3$ occur widely in Mediterranean climates and semiarid tropics (see calcareous soils) – FAO-UNESCO/World Reference Base classification. USDA Soil Taxonomy equivalents: Calciorthids, and pale-coloured groups of Alfisols and Aridisols.
calcite	Crystalline calcium carbonate ($CaCO_3$); a mineral.
calcrete	$CaCO_3$ deposit in soil as concretions, nodules or layers.

Cation Exchange Capacity (CEC)	The ability of a soil or soil constituent (e.g., clay mineral or humus) to adsorb cations on permanent, or pH-dependent negatively charged sites on surfaces. Cations of different elements can replace each other as counter-ions to the negative charges.
chlorosis	Pale green, yellow or white patches/stripes on leaves due to a lack of chlorophyll.
clay mineral	A sheet-like alumino-silicate mineral with a relatively large surface area and variable ability to adsorb cations (e.g., kaolinites, illites, smectites and vermiculite).
coir	Course fibre extracted from the fibrous outer shell of the coconut.
cultivar (CV)	Cultivated variety of a crop.
dicotyledon	Members of the flowering plants (Angiosperms) which have two cotyledons in the seed crops (dicots) and have forked veins in their leaves (cf. monocotyledon crops '*Gramineae*', the grasses, including cereals) which have parallel veins in their leaves.
efficient (micronutrient-efficient)	Relative ability for a cultivar to grow in a medium with medium to low/marginal supplies of (genotypes) available micronutrients (e.g., Cu, Mn, Zn, etc). cf., -inefficient genotypes.
essential trace elements	They are essential because (a) a deficiency of the element makes it impossible for the plant to complete its life cycle; (b) the deficiency is specific for the element in question, and (c) the element is directly involved in the nutrition of the plant; e.g., a constituent of an essential metabolite or required for the action of an enzyme system.
FAO-UNESCO Classification	The soil classification system developed for the joint project by the UN Food and Agriculture Organization and UNESCO to produce a Soil Map of the World (1981).
FAO-UNESCO Soil Map the World	Soil map (1:5,000,000) published in 1981 which was the first map to cover the soils of the whole world – updated in 1998 to become the World Reference Base of Soil Resources.
ferritin	A globular protein complex consisting of 24 protein subunits is the main intracellular iron storage protein keeping it in a soluble and non-toxic form (Wikipedia).

fine fractions (of soils)	Clay (<0.0002 mm diameter) and silt (0.02 – 0.002 mm) contents (have higher adsorptive capacities for cations and add to the moisture retention properties of soils.
gleyed soils	Soils with reducing conditions caused by permanent or intermittent waterlogging at (Gleysols) shallow depth (hydromorphic soil) characterized by pale colours and low concentrations of iron oxides (World Reference Base classification).
Glyphosate	Broad spectrum herbicide – the most widely used herbicide in the world (constituent of RoundUp) (Monsanto Corp. USA).
Grain Harvest Index (GHI)	Ratio of grain to above-ground dry matter (also called harvest index – HI).
Green Revolution	The transformation of agriculture around the world, especially in developing countries, through the introduction of new varieties of wheat and rice from the High Yielding Varieties Program (1940s–1970s). It also promoted the use of irrigation, fertilisers, pesticides and mechanisation and brought about vast increases in food crop production (term "Green Revolution" first used in 1968).
haem	a prosthetic group that consists of an iron atom contained in the center of a large heterocyclic organic ring called a porphyrin (Wikipedia).
HarvestPlus	A global alliance of institutions and scientists seeking to improve human nutrition by breeding new varieties of staple food crops consumed by the poor that have higher levels of micronutrients, through a process called biofortification (www.harvestplus.org).
Histosols	A unique group of soils composed predominantly of organic matter (derived from peat); includes both peat and muck soils.
hookworm	A parasitic nematode worm that lives in the small intestine of its host and is a major cause of maternal and child morbidity in tropical and subtropical developing countries, where its most common effect on health is to cause anaemia and loss of iron (Wikipedia).
humose	Humus-rich (soil).
humus	Fraction of the soil organic matter produced by the action of soil micro-organisms. It comprises a series of moderately high molecular weight compounds which have a high adsorptive capacity for ions by both ion exchange and complexation.

hydromorphic	Soil whose properties are dominated by it being regularly saturated with water (a gley soil, see Gleysol).
hypertrophy	Excessive development and enlargement of cells and tissue (e.g., of the tapetum in anthers of copper-deficient cereals).
Inceptisols	Embryonic soils with few diagnostic features (USDA Soil Taxonomy – see Appendix 2).
interveinal	Between the veins on plant leaves (often applied to chlorosis symptoms).
IRRI	International Rice Research Institute (at Los Banos, Philippines).
Kastanozems	Soils of the drier warmer areas of the steppe, which are neutral to slightly alkaline in reaction with an organic matter content of 2–4% (FAO-UNESCO/World Reference Base classification). They are equivalent to Ustolls and Borols in the USDA Soil Taxonomy.
kharif	Monsoon season crops (India).
kraal	Pen (enclosure) for holding livestock in southern Africa (Afrikaans word).
kriging	Use of a geostatistical technique to interpolate the value of a random field at an unobserved location from observations of its value at nearby locations (Wikipedia).
Leptosols	Soils with various types of surface horizons, limited in depth (<25 cm) by continuous hard rock or highly calcareous material (FAO-UNESCO/World Reference Base classification). Equivalent groups in USDA Soil Taxonomy are Entisols, Lithic subgroups and Rendolls.
lignification	Process involving the strengthening of plant cell walls due to the formation of lignin (a polyphenolic compound).
Luvisols	Soil in which the surface horizon has been depleted in clay; this is underlain by a subsurface horizon which is enriched in clay ('argic' horizon) – FAO-UNESCO/ World Reference Base classification. The equivalent USDA Soil Taxonomy group is Alfisols.
Malaria	Vector-borne infectious disease caused by protozoan parasites of the genus *Plasmodium*. It is widespread in tropical and subtropical regions.
micronutrient	Essential trace element required in plant tissue in small but critical concentrations (between 5 and 100 mg kg^{-1}). The elements: Cl, B, Cu, Fe, Mn, Mo, Ni and Zn are required by plants).

mineralisation (in soils)	Decomposition of organic debris in soils by the action of soil fauna and microorganisms.
Mollisols	Dark soils of natural grasslands, high base status soils with Mollic epipedons – USDA Soil Taxonomy (see Appendix 2).
Monocotyledon plants	Sub-class of the Angiosperms having a single cotyledon in the seed, most are herbaceous, with adventitious roots systems and parallel veined leaves – the *Gramineae* (grasses and cereals) are the most important representatives (cf. dicotyledons).
muck soils	Organic soils in which the organic parent material has decomposed so that plant fragments are not readily visible (unlike peat soils) (Histosols).
mycorrhizae	Fungi which colonize the outer layers of plant roots and whose external mycelium effectively increases the effective absorptive volume of the roots.
necrosis	Abnormal death of part of leaf or other plant tissue (i.e., necrotic spots).
organic matter	Organic fraction of soil, comprising humus and partially/un-degraded plant and animal material (litter) – major store of carbon, and plant macro- and micronutrients.
paddy (rice)	Rice production under flooded conditions, also called lowland rice.
paddy field	Area of land with surface water (flooded) used for growing rice.
parent material	Weathered rock (or exposed peat) on which a soil develops.
peat	Partially decomposed plant material which forms the parent material of organic-rich soils (Histosols); the organic parent material in peat soils is less decomposed than in muck soils and fragments of the plants from which the peat was derived can be identified.
phytate	Phosphorus-containing compound (inositol hexaphosphate) found in cereal grains which can bind zinc and reduce its availability to humans (an 'antinutrient').
Phytosiderophores	Organic compounds released by the roots of some plants suffering from a deficiency of iron or certain other micronutrients, which can mobilize iron and elements co-precipitated onto iron oxides and render them available for uptake by the plant.
phytotoxicity	When plants suffer from toxic doses of a chemical.

Podzol	Type of soil found in cool, humid environments on freely drained parent materials usually under coniferous trees or ericaceous vegetation in humid climates. They have a distinct combination of horizons including organic-rich topsoil, a horizon depleted in clay, iron and aluminium and a lower horizon in which iron and aluminium have accumulated –FAO-UNESCO/World Reference Base classification (called Spodosols in the USDA Soil Taxonomy).
Podzoluvisols	Intergrade soil with properties of both Podzols and Luvisols; basically a Luvisol which had both clay and iron and aluminium leached down the profile (more severely leached than normal Luvisols) in FAO-UNESCO classification, but now called Albeluvisols in the World Reference Base classification.
pollen sterility	Failure of viable pollen to develop in the anthers of plants – linked to copper, zinc and boron deficiencies. In self-pollinating plants such as wheat and barley, pollen sterility results in unfilled grain positions in the ears (spikes) and a significantly reduced grain yield.
polyols	Alcohols with several hydroxyl radicals (polyhydric alcohols).
proteoid roots	Proteoid roots have discrete clusters of closely spaced lateral rootlets along their lengths that greatly increase the surface area for nutrient uptake and release exudates that can solubilize phosphorus and some other elements.
rabi	Winter season crops (India).
Rendzina	Shallow calcareous soil over limestone (especially Chalk), often with organic-rich A-horizon (topsoil), FAO-UNESCO classification. Now in Calcisols and Leptosols in World Reference Base classification.
rhizosphere	Thin layer (approximately 2 mm thick) around plant roots which is a zone of intense microbial activity due to root secretions.
rice–wheat cropping system	Growing alternating crops of rice and wheat on same land (important on the Indo-Gangetic Plain and in other parts of Asia).
RoundUp	Brand name of Glyphosate herbicide, manufactured by Monsanto Corporation

saline soils	Soils whose chemical properties are dominated by a high concentration of salts, especially sodium (see also Solonchaks and Solonetz).
sand	Mineral material particles (usually quartz grains) between 0.02 and 2 mm in diameter.
sandy soils	Soils with a loamy sand or sand texture (see Arenosol) – usually >15% sand.
Soil Taxonomy	USDA Soil Classification system (Soil Survey Staff, 1999).
Solonchaks	Salt-affected soil with high concentrations of salts in the topsoil at some time of the year (not so morphologically distinct as the Solonetz profile) – FAO-UNESCO/World Reference Base classification
Solonetz	Salt-affected soil with a humus-rich surface soil and a saline subsoil ('natric' horizon) – FAO-UNESCO/ World Reference Base classification.
Solonised soils	Australian term – soils with a calcareous horizon consisting of 20–50% hard calcrete fragments, carbonate nodules or concretions of carbonate-coated gravel. In the Australian Soil Classification these soils are "Supracalcic Calcarosols". Equivalents: Calcic Luvisols, Calciorthids (in World Reference Base and USDA Soil Taxonomy, respectively).
subclinical	Deficiency of a micronutrient in plants or animals without the deficiency appearance of obvious symptoms (also called hidden or latent deficiency).
superphosphate	Phosphatic fertiliser (21% P_2O_5) usually with significant concentrations of metal (ordinary) impurities (e.g., Zn).
tapetum	Layer of cells in the anthers of cereals which nourish developing pollen grains.
tarai soils	Shallow water table soils found in the foothills of mountain ranges (Aquic hapludoll of the Mollisol Group, and Hapludalfs of the Alfisol Group – USDA Soil Taxonomy classification). Term used in India and adjacent countries.
thalassaemia	Inherited blood disorder – incidence of carriers varies greatly throughout the world, very common in the Middle East, the Indian subcontinent and through out South East Asia and parts of the southern Mediterranean. These regions coincide with areas where malaria occurs and thalassaemia is believed to offer some resistance to this disease.

total (analysis)	Complete solution of the mineral matrix of the soil by digestion in hot concentrated nitric acid (HNO_3), perchloric acid ($HClO_4$) and hydrofluoric acid (HF); cf. "pseudototal" analysis in *aqua regia*.
trace elements	Elements present in the Earth's crust at low concentrations (<1%) and normally present in plant tissues at concentrations of below 100 mg kg^{-1}.
turgor	Pressure of cell sap which keeps leaves rigid and fully exposed to light and able to participate in evapotranspiration.
Vertisols	Deep black, clayey soils (>30% clay) that expand upon wetting and shrink on drying to form prominent cracks to a depth of at least 50 cm. Topsoil falls down these cracks and, in time the profile becomes inverted – in both FAO-UNESCO/World Reference Base and the USDA Soil Taxonomy. Vertosols in the Australian Soil Classification. These soils are also called 'swell-shrink' soils and 'self-mulching soils' (in India and Australia, respectively).
World Reference Base for Soil Resources	A soil classification system, data base and atlas produced by the International Society of Soil Science in 1998 (ISSS Working Group RB, 1998) – it largely supersedes the FAO-UNESCO Soil Resources Map of the World (1981).
Zadoks' Scale	Cereal development (growth stage) scale proposed by J.C. Zadoks (1974). It is a decimal scale divided into: 0–09 germination, 1–19 seedling growth, 20–29 tillering, 30–39 stem elongation, 40–49 booting, 50–59 inflorescence emergence, 60–69 anthesis, 70–77 'milk' development, 80–87 dough development, and 90–99 for ripening.

Colour Plates

Fig. 1.2 View of a field trial with copper on wheat growing on a Rendzina soil in France showing the copper-treated area (taller and pale coloured) in the distance. The copper-deficient crop in the foreground shows a lower density of ear-bearing tillers and a darker colour due to melanism (From B.J. Alloway)

Fig. 1.3 Ears of wheat from the field experiment shown in Fig. 1.2. Normal ears from copper-sufficient plants on the left and partially filled ears showing some melanism from copper-deficient plants on the right (From B.J. Alloway)

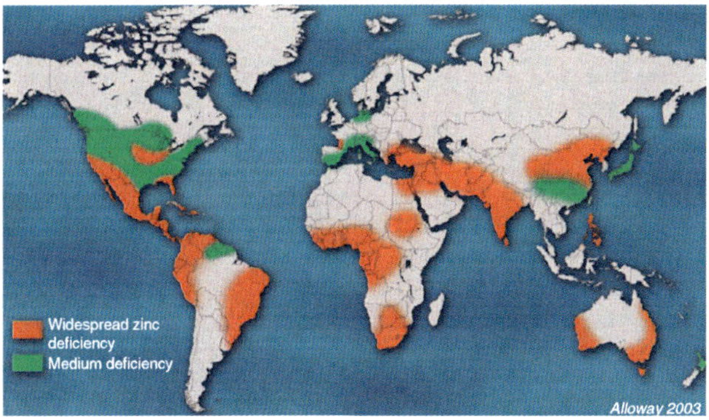

Fig. 2.3 Major areas of reported zinc deficiency in world crops (Alloway, 2004)

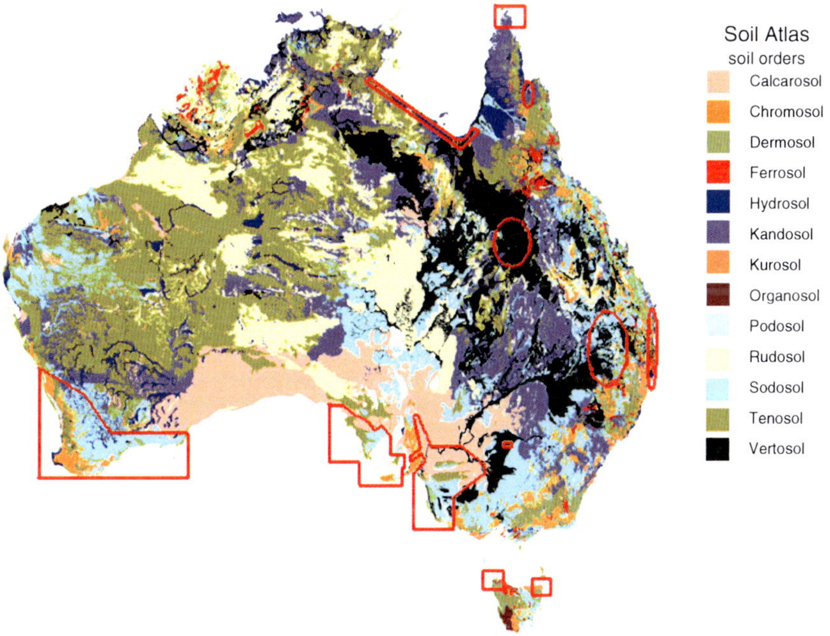

Fig. 3.1 Soil Atlas of Australia showing the major soil orders described in the Australian Soil Classification System (Isbell, 1996). The areas outlined in red indicate areas of potential Zn deficiency[1] (Commonwealth of Australia 2001, National Land and Water Resources Audit 2001; Bureau of Rural Sciences after Commonwealth Scientific and Industrial Research Organisation (1991). Digital Atlas of Australian Soils (ARC/INFO vector format). Available at html: http://www.brs.gov.au/data/datasets)

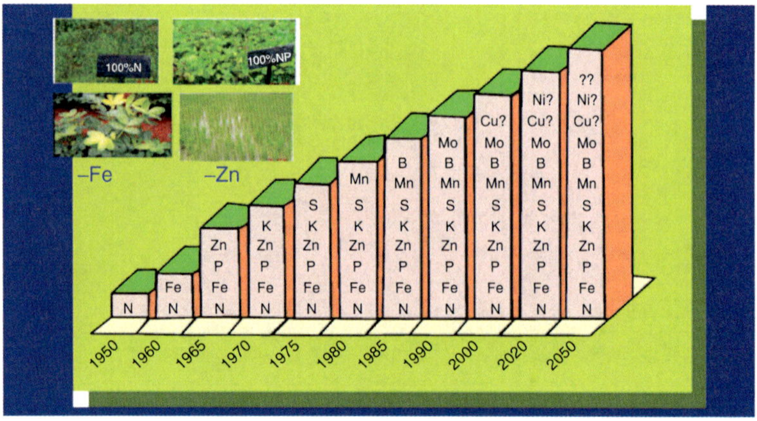

Fig. 4.1 Impact of the green revolution on the emergence of micronutrient deficiencies in crops in India

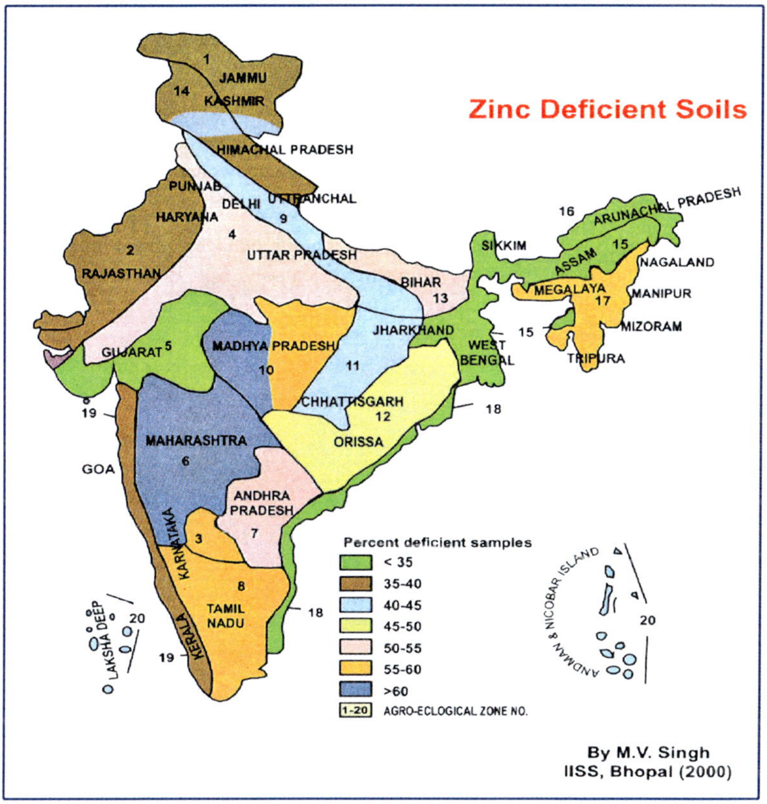

Agro-ecological Zones

1. Western Himalayas (cold arid), 2. Western Plain (hot arid), 3. Deccan Plateau (hot arid), 4. Northern Plain and Central Highlands (hot semi-arid), 5. Central (Malwa) Highlands, Gujarat plains and Kathiawar Peninsula, 6. Deccan Plateau (hot semi-arid), 7. Deccan Plateau (Telangana) and Eastern Ghat (hot semi-arid), 8. Eastern Ghats, Tamil Nadu Uplands and Deccan (Karnataka) Plateau (hot semi-arid), 9. Northern Plan (hot subhumid/dry), 10. Central Highlands (Malwa and Bundelkhand) (hot subhumid/dry), 11. Chattisgarh/Mahanadi Basin, 12. Eastern Plateau (Chhotanagpur) and Eastern Ghats (hot subhumid), 13. Eastern Plain (hot subhumid/moist), 14. Western Himlayas (warm subhumid to humid/perhumid), 15. Assam and Bengal Plain (hot suhumid to humid/perhumid), 16. Eastern Himalayas (warm perhumid), 17. North-Eastern Hills (Purvachal) (warm perhumid), 18. Eastern Coastal Plain (hot subhumid to semi-arid), 19. Western Ghats and Coastal Plain (hot humid/perhumid), 20. Islands of Andaman-Nicobar and Lakshadweep (hot humid/perhumid).

Fig. 4.2 Percentage of zinc-deficient soils in the different agro-ecological zones of India (Singh, 2001b)

Fig. 5.8 Zinc deficiency in maize grown in the north of China (*left*) and in lowland rice grown on paddy soil in the south of China (*right*) (F.S. Zhang and S.H. Lv, 1999, personal communication)

Fig. 5.9 Special practices to ameliorate iron deficiency in fruit trees. *Left*: Bag fertilisation; *Right*: Root suction (S.H. Lv and J.J. Xue, 2001, personal communication)

Fig. 5.10 Fe chlorosis of peanut in a monocropping system on a calcareous soil (*left*) and its improvement by intercropping with maize (*right*) (From Zuo et al., 2000)

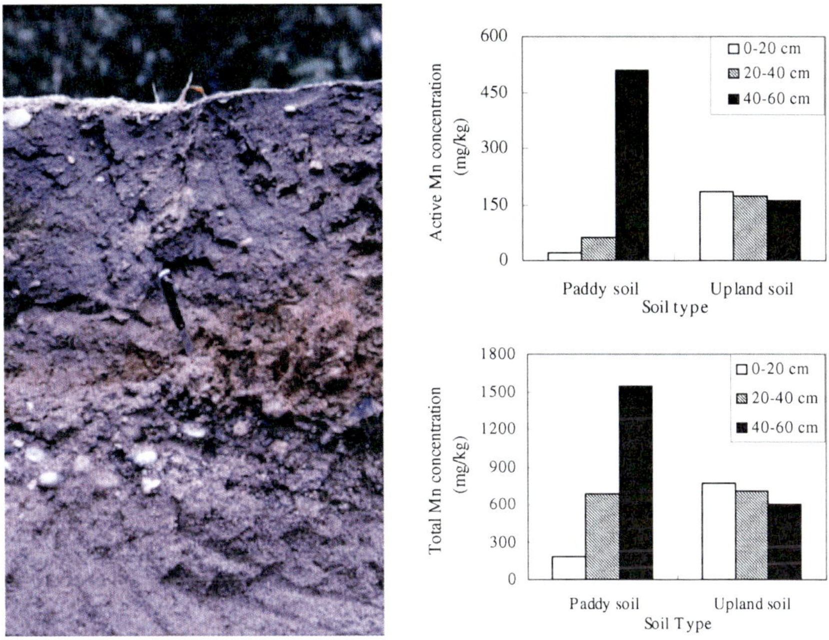

Fig. 5.11 Spatial distribution of Mn in a paddy soil with rice–wheat rotation compared with an upland soil (Lv et al., 2002)

Fig. 5.12 Molybdenum deficiency in winter wheat cultivated in the field (*left*) and special symptom of the leaf (*right*) (Y.H. Wang, 1997, personal communication)

Fig. 5.13 Comparison of three rice production systems. *Left*: Paddy rice production system, flooding; *Middle*: Ground cover rice production system, no flooding; *Right*: Aerobic rice production system (H.Q. Wang, 2002, personal communication)

Fig. 5.14 Manganese deficiency of rice plants cultivated with plastic (*left*) and wheat straw mulch (*right*) (S.H. Lv, 2001, personal communication)

Fig. 7.3 Growth of wheat plants on a Zn-deficient calcareous soil (DTPA-Zn 0.1 mg kg^{-1}) in Eskisehir Province, Central Anatolia showing details of plants with (+Zn) and without (–Zn) fertilisation with 50 kg $ZnSO_4$ ha^{-1} (H. Braun, CIMMYT)

Fig. 8.3 Severe P-deficiencies in maize masking possible micronutrient deficiencies in a subsistence agriculture field in Mpumalanga province, South Africa (From J.H van der Waals)

Fig. 8.4 The limited geographical extent of a subsistence farmer's maize field in the Mpumalanga province, South Africa. Note the uneven colour through the field due to variable soil conditions and fertiliser application (From J.H van der Waals)

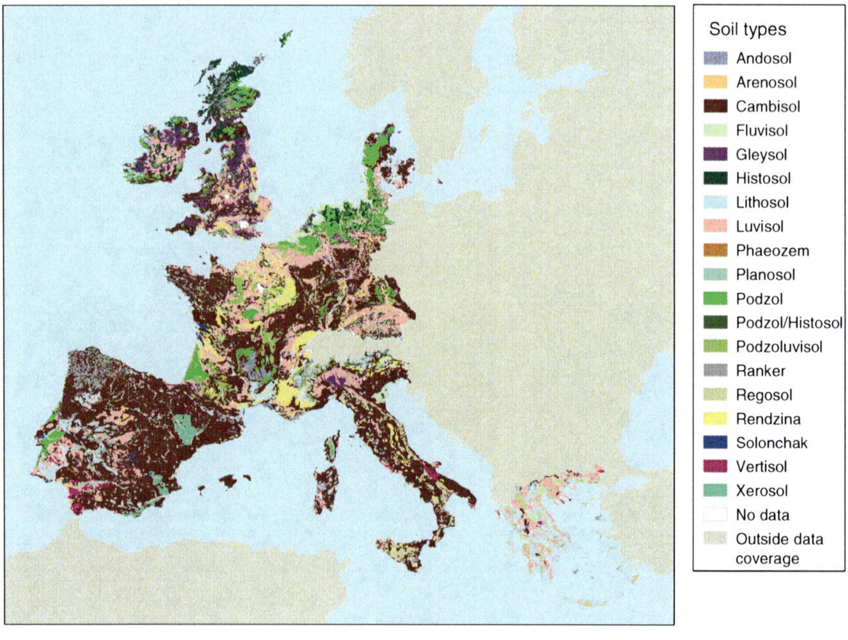

Fig. 9.2 The distribution of soil types in Western Europe (http://dataservice.eea.eu.int/atlas/viewdata/viewpub.asp?id=11)

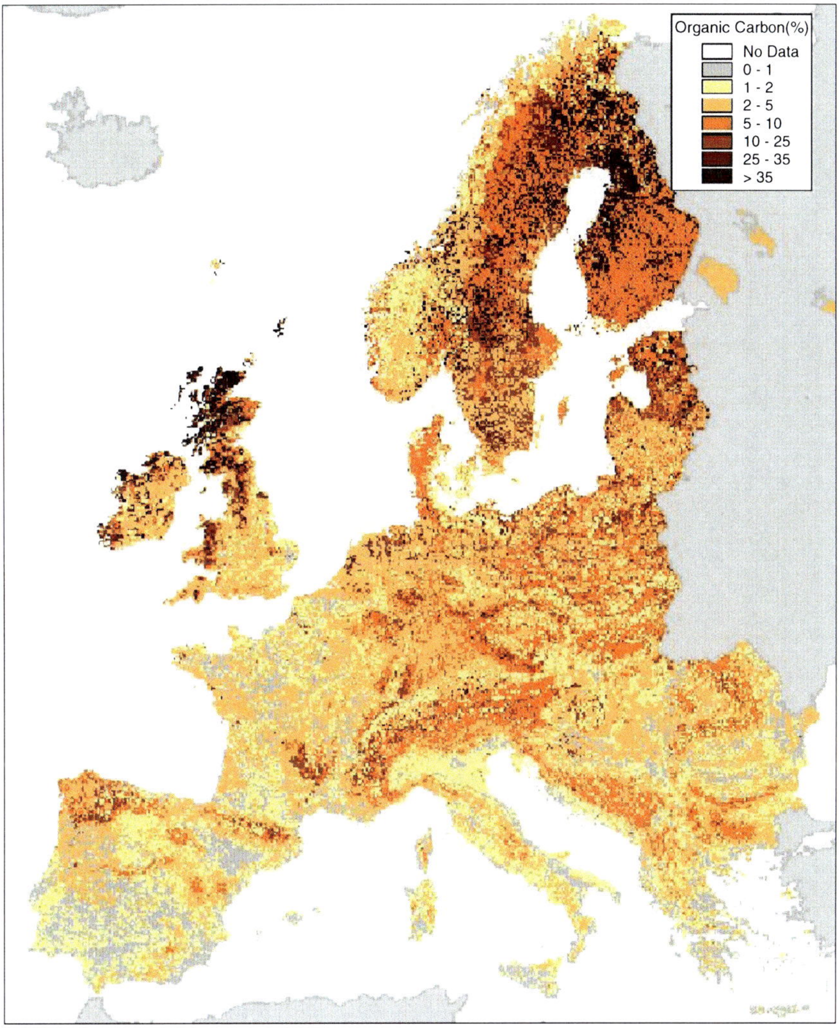

Fig. 9.3 Topsoil organic carbon concentrations in Europe (Jones et al., 2005)

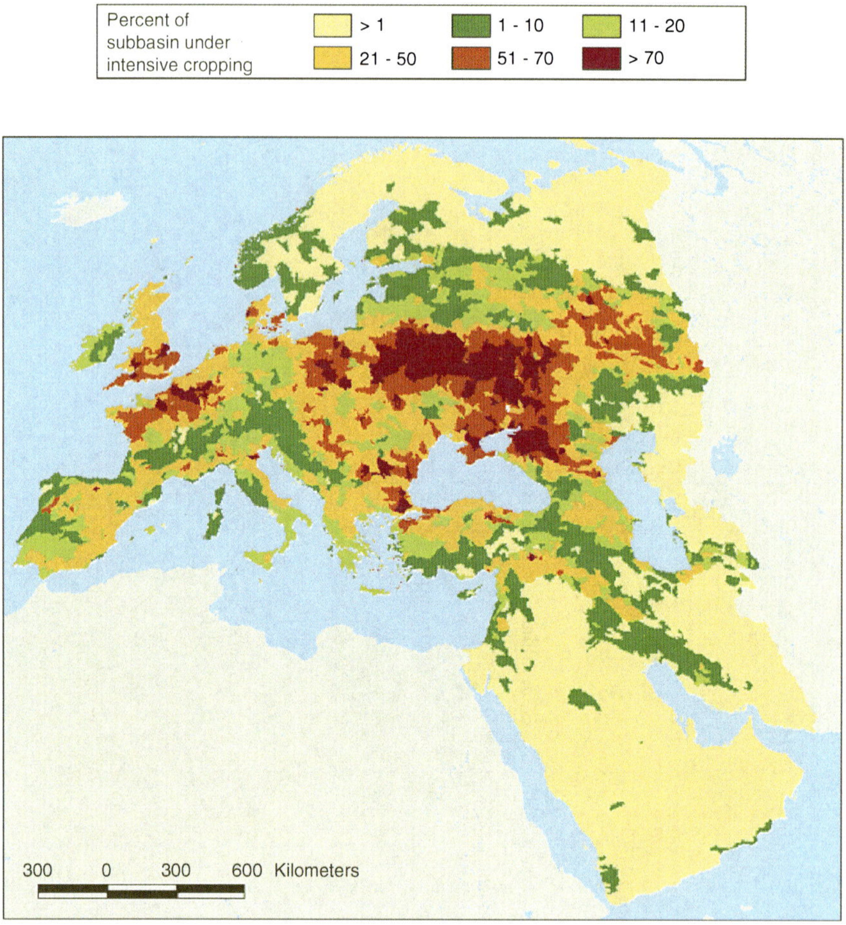

Fig. 9.4 Intensive agricultural land use by river sub-basin in Europe and the Middle East. (World Resources Institute (Page, 2000))

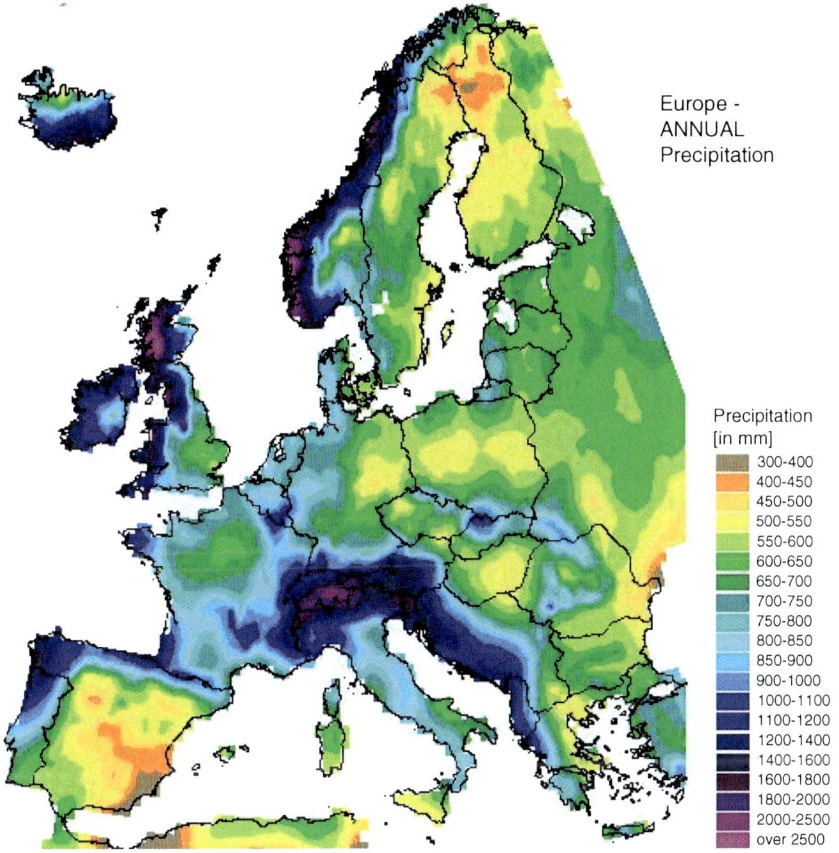

Fig. 9.5 Map of annual precipitation across Europe (http://www.iiasa.ac.at/Research/LUC/GIS/img/eur_prcy.jpg)

11.1a: Pistachio (*Pistacia vera*)

11.1b: Peach (*Prunus dulcis*)

Fig. 11.1 Characteristic symptoms of B toxicity in species in which B is phloem-immobile (a-Pistachio) or phloem mobile (b-Almond). Boron toxicity in phloem immobile species is expressed as marginal leaf scorch on the oldest leaves (1a), in species exhibiting phloem B mobility toxicity is expressed as death of youngest growing tips, necrosis of vascular cambium and deformation of fruits. For a full list of species in each symptom category see Brown et al. (1999)

Fig. 11.3 Branches of Ni sufficient (*left*) and deficient (*right*) pecan (*Carya illinoinensis*). Symptoms include delayed and decreased leaf expansion, poor bud-break, leaf bronzing and chlorosis, rosetting, leaf tip necrosis (Photo courtesy of B. Wood)

Subject Index

Q

R

S

T

Printed by Publishers' Graphics LLC
ICISO140607.23.34.1